Lecture Notes in Artificial Intelligence   1669

Subseries of Lecture Notes in Computer Science
Edited by J. G. Carbonell and J. Siekmann

Lecture Notes in Computer Science

Edited by G. Goos, J. Hartmanis and J. van Leeuwen

T0223270

Springer

Berlin
Heidelberg
New York
Barcelona
Hong Kong
London
Milan
Paris
Singapore
Tokyo

Xiao-Shan Gao   Dongming Wang
Lu Yang (Eds.)

Automated Deduction in Geometry

Second International Workshop, ADG'98
Beijing, China, August 1-3, 1998
Proceedings

Springer

Series Editors

Jaime G. Carbonell, Carnegie Mellon University, Pittsburgh, PA, USA
Jörg Siekmann, University of Saarland, Saarbrücken, Germany

Volume Editors

Xiao-Shan Gao
Institute of Systems Science, Academia Sinica
Beijing 100080, China
E-mail: xgao@mmrc.iss.ac.cn

Dongming Wang
Laboratoire LEIBNIZ, Institut IMAG
46, avenue Félix Viallet, F-38031 Grenoble Cedex, France
E-mail: dongming.wang@imag.fr

Lu Yang
Laboratory for Automated Reasoning and Programming
Chengdu Institute of Computer Applications, Academia Sinica
Chengdu 610041, China
E-mail: cdluyang@public.sc.cninfo.net

Cataloging-in-Publication data applied for

Die Deutsche Bibliothek - CIP-Einheitsaufnahme

Automated deduction in geometry : proceedings / ADG'98, Beijing, China, August 1
- 3, 1998. Xiao-Shan Gao ... (ed.). - Berlin ; Heidelberg ; New York ;
Barcelona ; Hong Kong ; London ; Milan ; Paris ; Singapore ; Tokyo : Springer,
1999
   (Lecture notes in computer science ; Vol. 1669 : Lecture notes in artificial
intelligence)
ISBN 3-540-66672-9

CR Subject Classification (1998): I.2.3, F.4.1, I.3.5, I.5, G.2

ISBN 3-540-66672-9 Springer-Verlag Berlin Heidelberg New York

© Springer-Verlag Berlin Heidelberg 1999
Printed in Germany

Typesetting: Camera-ready by author
SPIN 10705416   06/3142 – 5 4 3 2 1 0   Printed on acid-free paper

Preface

The Second International Workshop on Automated Deduction in Geometry (ADG '98) was held in Beijing, China, August 1-3, 1998. An increase of interest in ADG '98 over the previous workshop ADG '96 is represented by the notable number of more than 40 participants from ten countries and the strong technical program of 25 presentations, of which two one-hour invited talks were given by Professors Wen-tsün Wu and Jing-Zhong Zhang. The workshop provided the participants with a well-focused forum for effective exchange of new ideas and timely report of research progress. Insight surveys, algorithmic developments, and applications in CAGD/CAD and computer vision presented by active researchers, together with geometry software demos, shed light on the features of this second workshop. ADG '98 was hosted by the Mathematics Mechanization Research Center (MMRC) with financial support from the Chinese Academy of Sciences and the French National Center for Scientific Research (CNRS), and was organized by the three co-editors of this proceedings volume.

The papers contained in the volume were selected, under a strict refereeing procedure, from those presented at ADG '98 and submitted afterwards. Most of the 14 accepted papers were carefully revised and some of the revised versions were checked again by external reviewers. We hope that these papers cover some of the most recent and significant research results and developments and reflect the current state-of-the-art of ADG.

We want to thank all those members of MMRC who contributed to the local organization of ADG '98 and the referees for their time and help. We also want to acknowledge some of the previous workshops and conferences and the corresponding publications listed below which have stimulated the continuing activities on ADG.

- International Workshop on Automated Deduction in Geometry (ADG '96) (Toulouse, France, September 27-29, 1996) — LNAI, vol. 1360 (edited by D. Wang), Springer-Verlag, Berlin Heidelberg New York, 1997

- Workshop on Algebraic Approaches to Geometric Reasoning (Castle of Weinberg, Austria, August 17-19, 1992) — Special issue of Annals of Mathematics and Artificial Intelligence, vol. 13, nos. 1-2 (edited by H. Hong, D. Wang and F. Winkler), Baltzer Sci. Publ., Amsterdam, 1995

- International Workshop on Mathematics Mechanization (Beijing, China, July 16-18, 1992) — Proceedings (edited by W.-T. Wu and M.-D. Cheng), Int. Academic Publ., Beijing, 1992

- Workshop on Computer-Aided Geometric Reasoning (INRIA Sophia-Antipolis, June 22-26, 1987) — Proceedings (edited by H. Crapo), INRIA Rocquencourt, France, 1987

- International Workshop on Geometric Reasoning (Oxford, UK, June 30 –
  July 3, 1986) — Special issue of Artificial Intelligence, vol. 37, nos. 1–3
  (edited by D. Kapur and J. L. Mundy), MIT Press, Cambridge, 1988
- Asian Symposia on Computer Mathematics: ASCM '98 (Lanzhou, China,
  August 6–8, 1998) — Proceedings (edited by Z. Li), Lanzhou Univ. Press,
  Lanzhou, 1998; ASCM '96 (Kobe, Japan, August 20–22, 1996) — Proceed-
  ings (edited by H. Kobayashi), Scientists Inc., Tokyo, 1996; ASCM '95 (Bei-
  jing, China, August 18–20, 1995) — Proceedings (edited by H. Shi and
  H. Kobayashi), Scientists Inc., Tokyo, 1995

July 1999   Xiao-Shan Gao
                                                         Dongming Wang
                                                         Lu Yang

Table of Contents

Automatic Geometry Theorem-Proving and Automatic Geometry Problem-Solving

Wen-tsün Wu

Institute of Systems Science, Academia Sinica, Beijing 100080, P. R. China
wtwu@mmrc.iss.ac.cn

1 Principle and Method

Studies in geometry involve both theorem-proving and problem-solving. So far the latter is concerned, we may cite for example geometrical locus-determination and geometrical constructions by ruler-compass which were much studied in ancient Greece. In particular we may mention the regular polygon construction and the three famous difficult problems of angle-trisection, cube-duplication and circle-squaring.

A turning point of geometrical studies occurred in 1637 when R. Descartes published his classic on geometry. The introduction of coordinate system turned then geometrical studies to algebraic ones in replacing pure geometrical reasonings by manipulation of polynomials and solving of polynomial equations. This paved the way to the final solutions of the geometrical-construction problems mentioned above.

Our method of manipulating polynomials and solving arbitrary systems of polynomial equations in the form of Well-Ordering Principle and Zero-Decomposition Theorems permits to treat geometrical studies even in an *automated* way. The germ of such studies as well as their wide-range applications lies in our general method of polynomial equations-solving. To be precise, let the unknowns of the polynomials in question be definite ones forming a finite set X and their coefficients be in a fixed field $\mathbf{K}$ of characteristic 0. By a polynomial set (abbr. *polset*) $PS \subset \mathbf{K}[X]$ we shall mean one consisting of a finite number of non-zero polynomials in $\mathbf{K}[X]$ unless otherwise stated. For such a polset PS the solving of equations $PS = 0$ is then equivalent to the determination of the set Zero(PS) consisting of the totality of solutions of $PS = 0$ or zeros of PS in an arbitrary extension field of $\mathbf{K}$. For any polynomial $G \in \mathbf{K}[X]$, we shall denote the set-difference Zero(PS) \ Zero(G) by Zero(PS/G). We remark in passing that the set Zero(PS) is nothing else but an affine algebraic variety, irreducible or not, in the usual sense. Now the difficulty of solving polynomial equations arises in essence from the fact that the polynomials in question are in general given in confused order. The key point to overcome such difficulties is thus: The given polset should be so manipulated in a way that the resulting re-arranged polynomials will be in some well-ordered form for which the corresponding solutions are almost self-evident and are closely related to those of the original ones to be sought for. In fact, in the case where all polynomials are linear in the unknowns X, the usual elimination method (say the Gaussian one) is just such a method which, after due manipulations, will lead to a set of well-ordered linear forms from which the solutions of the corresponding equations are immediately given.

With this understood we shall introduce a well-ordering procedure of polynomials and polsets in the following way.

1. Ordering of Unknowns. First of all we shall arrange the unknowns X in an arbitrary but fixed order, denoted henceforth by $x_1, x_2, \ldots, x_n$ in the natural order

$$x_1 \prec x_2 \prec \cdots \prec x_n.$$

2. Partial Ordering of Polynomials. For a non-constant polynomial $P \in K[X]$ let x_c be the unknown actually occurring in P with greatest subscript c. Then P may be written in the canonical form below:

$$P = I x_c^d + J_1 x_c^{d-1} + \cdots + J_{d-1} x_c + J_d,$$

in which the coefficients I, J_i are themselves polynomials in $K[x_1, \ldots, x_{c-1}]$ with $I \neq 0$. We call $c \ (> 0)$ the *class* of P, $d \ (\geq 1)$ the *degree* of P, $I \ (\neq 0)$ the *initial* of P, and the formal partial derivative $\partial P / \partial x_c$ the *separant* of P. In case P is a non-zero constant in K, we set the class of P to be 0, and leave the degree etc. undefined. We now introduce a partial ordering for polynomials in $K[X]$ first according to the class, and then to the degree.

3. Partial Ordering of Polsets of Particular Form. Consider now a polset AS of particular form in which the polynomials may be arranged in an ascending order:

$$A_1 \prec A_2 \prec \cdots \prec A_r. \tag{1}$$

We suppose that either $r = 1$ and A_1 is a non-zero constant in K, or $r \geq 1$ with corresponding classes c_i of A_i steadily increasing:

$$0 < c_1 < \cdots < c_r.$$

For technical reasons we shall require that for each pair (i, j) with $0 < j < i$, the degree of x_{c_j} in the initial I_i of A_i should be less than the degree of A_j. Under such conditions the polset $AS = \{A_1, \ldots, A_r\}$ is called an *ascending set* (abbr. *asc-set*). The product of all the initials I_i (resp. all the initials and separants) will be called the *initial-product* (resp. the *initial-separant-product*) of AS. For any polynomial $P \in K[X]$, we shall get by successive divisions with respect to $A_r, \ldots, A_2, A_1$ a unique *remainder* R given in the form

$$R = I_1^{s_1} \cdots I_r^{s_r} P - \sum_i Q_i A_i,$$

with $Q_i \in K[X]$, $s_i \ (\geq 0)$ as small as possible, so that R is *reduced* with respect to AS in the following sense: the degree of x_{c_i} in R is less than that of A_i for each i with $0 < i \leq r$. This remainder R will be denoted as $\mathrm{Remdr}(P/AS)$.

A partial ordering for all asc-sets is now introduced which we shall not enter. A consequence of the ordering is the following critical theorem:

A sequence of asc-sets steadily decreasing in order is necessarily finite.

4. Partial Ordering of Polsets. For any non-empty polset $PS \subset K[X]$ let us consider asc-sets contained in PS which necessarily exist. Any such asc-set of lowest ordering is called a *basic-set* of PS. We may then unambiguously introduce a partial ordering for all polsets according to their basic-sets. Again a critical theorem says that, the ordering of a polset will be lowered by adjoining to it a polynomial reduced with respect to any basic-set of the given polset. Furthermore, as mentioned above, such reduced polynomials may be gotten by forming remainders of polynomials with respect to such basic-sets.

Our method of solving polynomial equations is based on the following observation.

Let AS be an asc-set as in (1). For $r = 1$ and A_1 a non-zero constant in K or $c_1 = 0$ it is clear that $\text{Zero}(AS) = \emptyset$. In the general case $\text{Zero}(AS)$ may also be considered as well-determined in the following sense. Let $U = \{x_j \mid j \neq \text{any } c_i\}$ be considered as an arbitrary parameter set. From $A_i = 0$ each x_{c_i} may then successively be considered as a well-defined algebraic function of x_k with $k < c_i$. For given definite values in K of U, we may solve $A_i = 0$ successively for x_{c_i}. Hence in any case $\text{Zero}(AS)$, and $\text{Zero}(AS/G)$ too for any $G \in K[X]$, may be considered as well-determined.

With the above understood our underlying principle and method of polynomial equations-solving consist in determining an asc-set AS from a given polset PS such that $\text{Zero}(AS)$ is closely related to $\text{Zero}(PS)$, which will give at least a partial determination of an important part of $\text{Zero}(PS)$ and serves to lead to a complete determination of $\text{Zero}(PS)$. By the critical theorems described above such an asc-set may be gotten by successively enlarging the given polset in adjoining remainders with respect to some relevant asc-sets.

The whole procedure of getting such an asc-set from a polset PS may be given in the form of a concise diagram below:

$$
\begin{array}{llllll}
PS = & PS_0 & PS_1 & \cdots & PS_i & \cdots & PS_m \\
& BS_0 & BS_1 & \cdots & BS_i & \cdots & BS_m = CS \\
& RS_0 & RS_1 & \cdots & RS_i & \cdots & RS_m = \emptyset.
\end{array}
\tag{2}
$$

In the scheme (2) all the remainder sets RS_k are non-empty for $k < m$ and we have put

$$
\begin{aligned}
BS_i &= \text{a basic set of } PS_i, \\
RS_i &= \{\text{Remdr}(P/BS_i) \mid P \in PS_i \setminus BS_i\}, \\
PS_i &= PS_0 \cup BS_{i-1} \cup RS_{i-1}.
\end{aligned}
$$

From the critical theorems the basic-sets BS_i will be steadily decreasing in order so that the procedure will end at some stage m with final $RS_m = \emptyset$. The final basic-set $BS_m = CS$ is then the asc-set as required and is called a *characteristic set* (abbr. *char-set*) of the given polset PS. The importance of the notion of char-set may be seen from the following theorem.

4

Well-Ordering Principle. For the char-set CS of PS let IP (or ISP) be the initial-product (or the initial-separant-product) of the asc-set CS. Then

$$\text{Zero}(PS) = \text{Zero}(CS/IP) \bigcup \text{Zero}(PS \cup \{IP\}), \text{ or} \qquad (3)$$

$$\text{Zero}(PS) = \text{Zero}(CS/ISP) \bigcup \text{Zero}(PS \cup \{ISP\}). \qquad (4)$$

By reasoning about $PS \cup \{IP\}$ (or $PS \cup \{ISP\}$) in the same manner as PS further and further we get finally the following theorems.

Zero-Decomposition Theorem. From a given polset PS one may determine in an algorithmic way a set of asc-sets CS_i (or CS'_j) with initial-products IP_i (or initial-separant-products ISP'_j) such that

$$\text{Zero}(PS) = \bigcup_i \text{Zero}(AS_i/IP_i), \text{ or} \qquad (5)$$

$$\text{Zero}(PS) = \bigcup_j \text{Zero}(AS'_j/ISP'_j). \qquad (6)$$

Define an asc-set to be *irreducible* in some natural way; then we may refine the above decomposition theorem to the following form:

Variety-Decomposition Theorem. From a given polset PS one may determine in an algorithmic way a set of irreducible asc-sets IRR_k such that for some so-called *Chow basis* $CB_k \subset K[X]$ which may be determined algorithmically from IRR_k, we have

$$\text{Zero}(PS) = \bigcup_k \text{Var}[IRR_k], \qquad (7)$$

in which $\text{Var}[IRR_k] = \text{Zero}(CB_k) \subset \text{Zero}(IRR_k)$.

These theorems furnish us a general method (to be called the *char-set method*) for polynomial equations-solving. It serves also as the basis of applications to various kinds of problems arising from science and technology, besides those from mathematics itself. For more information we refer to relevant papers due to various authors mainly at MMRC as well as a forthcoming book by the present author. The book is entitled 《 *Mathematics Mechanization: Geometry Theorem-Proving, Geometry Problem-Solving, and Polynomial Equations-Solving* 》 which will be published by Science Press, Beijing.

We have to point out that the idea of introducing ordering of unknowns and polynomials as well as the elimination procedure for solving polynomial equations were both originated in works of mathematicians in ancient China; see Section 3 below. In following the line of thought of our ancestors and applying some techniques of modern mathematics mainly due to J. F. Ritt we arrive finally at the Zero-Decomposition Theorems which settle completely the problem of polynomial equations-solving at least in some theoretical sense. We remark also that the above is not the only way of ordering polynomials and polsets. In fact, in introducing partial orderings in some other manner, working in some way as described in the Riquier-Janet theory on partial differential equations, and proceeding somewhat parallel to the case of the char-set method, we arrive finally at some notion which is equivalent to the well-known *Gröbner basis* of an ideal. For more details we refer again to the writings mentioned above.

2 Applications and Examples

Among the applications which have been made in recent years by myself and my colleagues notably at MMRC we may mention in particular the following:

- Automated Geometrical Constructions.
- Automated Determination of Geometrical Loci.
- Automated Discovering of Unknown Relations.
- Automated Design of Mechanisms.
- Automated CAGD (Computer-Aided Geometric Design).
- Automated Geometry Theorem-Proving and -Discovering.
- Automated Proving and Discovering of Inequalities.
- Automated Global Optimization.

For brevity we shall consider below the application to automated geometry theorem-proving and thus also automated geometry theorem-discovering alone in more details.

The creation of coordinate geometry turned the Euclidean fashion of proving geometry theorems one-by-one by logical reasonings into the Cartesian fashion of proving classes of geometry theorems as a whole by algebraic computations. However, it is not evident at all whether a geometric statement, in whatever geometry under an axiomatic system, Euclidean, non-Euclidean, affine, projective, etc., may be turned into *polynomial* equations or inequations (and eventually also inequalities). It is true that this is indeed the case for most of the ordinary geometries we are well-acquainted with. However, the proof of such a fact is not trivial at all. On the other hand, in the hands of geometers notably in the 19th century, the realm of geometry has been so broadened that we may develop various kinds of geometries directly from number systems without any reference to axiomatic systems, e.g. the Plückerian line geometry and geometries in the sense of F. Klein. Even more, for differential geometries, one starts usually in introducing curves, surfaces, and entities of geometrical interest directly by equations. The present author is even ignorant of any axiomatic systems over which some kind of differential geometry is relied upon.

This being understood we shall restrict ourselves to such geometry theorems for which the hypothesis and conclusion are both expressible by equations and inequations (eventually also inequalities), all in the form of polynomial ones, leaving aside the possible underlying geometries and axiomatic systems. For further simplification, we shall restrict ourselves to theorems of the form $T = \{HYP, CONC\}$, for which the hypothesis and the conclusion are respectively given by the equations $HYP = 0$ and $CONC = 0$ with $HYP \subset \mathbf{K}[X], CONC \in \mathbf{K}[X]$, $\mathbf{K}$ being a field of characteristic 0 and $X = \{x_1, \ldots, x_n\}$ as before.

The basic principle for proving theorems of the above type is as follows.

Each zero X^0 in Zero(HYP) is a geometrical configuration in eventually an extended affine space $\mathbf{K}'^n$, with coordinates X, $\mathbf{K}'$ being an extension field of $\mathbf{K}$, which verifies the hypothesis of the theorem T. The conclusion is true for such a geometrical configuration if $CONC = 0$ at the corresponding zero X^0. Proving a geometrical theorem thus amounts to seeking for some significant part P of Zero(HYP) such that $CONC = 0$ on that part P. The theorems on zero-set structures of Zero(HYP) which give us methods of polynomial equations-solving will serve to offer means of finding such part P and to verify whether $CONC = 0$ on P. In this sense our method of geometry theorem-proving may be considered as a mere application of our general method of polynomial equations-solving.

Each of the theorems for polynomial equations-solving will thus lead to a principle and a method of automated geometry theorem-proving.

Well-Ordering Principle (3) $\Longrightarrow$ Principle-Method I. For a given theorem $T = \{HYP, CONC\}$ let CS be a char-set of the hypothesis polset HYP and IP be the initial-product of CS. Then

$$\mathrm{Remdr}(CONC/CS) = 0 \qquad (8)$$

is a sufficient condition for

$$\mathrm{Zero}(HYP/IP) \subset \mathrm{Zero}(CONC).$$

In other words (8) is a sufficient condition for the theorem T to be true so far

$$IP \neq 0. \qquad (9)$$

Moreover, the condition (8) is also a necessary one for the above to be true in case CS is an irreducible asc-set.

Zero-Decomposition Theorem (5) $\Longrightarrow$ Principle-Method II. For a given theorem $T = \{HYP, CONC\}$ let $\{AS_i\}$ be a set of asc-sets determined from the hypothesis polset HYP such that

$$\mathrm{Zero}(HYP) = \bigcup_i \mathrm{Zero}(AS_i/IP_i),$$

in which each IP_i is the initial-product of the corresponding asc-set AS_i. If

$$\mathrm{Remdr}(CONC/AS_i) = 0, \qquad (10)$$

then the theorem T will be true on that part Zero(AS_i/IP_i) of Zero(HYP) or

$$\mathrm{Zero}(AS_i/IP_i) \subset \mathrm{Zero}(CONC). \qquad (11)$$

Conversely, if AS_i is irreducible and the theorem T is true on the part Zero(AS_i $/IP_i$) or (11) is true, then we have (10).

Variety-Decomposition Theorem (7) $\Longrightarrow$ **Principle-Method III.** For a given theorem $T = \{HYP, CONC\}$ let $\{IRR_k\}$ be a set of irreducible asc-sets determined from the hypothesis polset HYP such that

$$\text{Zero}(HYP) = \bigcup_k \text{Var}[IRR_k].$$

Then the theorem T will be true on that irreducible component $\text{Var}[IRR_k]$ if and only if

$$\text{Remdr}(CONC/IRR_k) = 0. \tag{12}$$

We add some remarks below.

In Principle-Method I the condition (9) is usually inevitable for the theorem $T = \{HYP, CONC\}$ in question to be true. In that case we shall say that the theorem T is *generically* true under the *non-degeneracy conditions* (9).

For theorems of form $T = \{HYP, CONC\}$ Principles I–III give three different methods of geometry theorem-proving. Principle I is the simplest to apply in that geometrically uninteresting or even meaningless degeneracy cases $IP = 0$ have been deprived off at the outset, though somewhat incomplete in that theorems will only be proved in the *generic* sense. The method in Principle III is the most complete in that no non-degeneracy conditions will be involved but it is not so easy to apply. Remark however that it requires only the simple algebraic verification of (12), but not the usual difficult determination of the Chow basis.

We have restricted ourselves to the case of **K** being of characteristic 0. However, so far theorem-proving is concerned, the proving of theorems in *finite* geometries for which the basic field **K** is of prime characteristic is of high interest in itself. The more so since finite geometries are closely related to such technological problems in coding theory and cryptography. We shall nonetheless leave this topic in referring only to some interesting original papers of D. Lin and Z.-J. Liu.

In the above we have also restricted ourselves to theorems involving equations and inequations alone. However, in the case of **K** being the real field **R**, theorems involving inequalities are of high interest in themselves. It is well-known that by A. Seidenberg's technique we can reduce such theorems to those involving equations and inequations alone by introducing new unknowns if necessary. In appearance the above Principle-Methods may still seems to be applicable. However, the actual carrying-out becomes so intricate that only very limited success has ever been achieved. Therefore we shall satisfy ourselves by stating only the interesting theorem below.

8-Triads Theorem. From 8 mutually different points A_i, $i = 1, \ldots, 8$, in circular order with $A_{i+8} = A_i$, there may be formed 8 triads $T_i = A_i A_{i+1} A_{i+3}$, $i = 1, \ldots, 8$. If all of these triads are collinear, then *generically* all the 8 points will lie on the same line.

Our methods permit to give the non-degeneracy conditions automatically and explicitly. These conditions are however so intricate to have any evident geometrical meaning and to be foreseen. In the case of complex affine plane there are thus ∞^8 8-triad configurations such that all the points A_i are different from each other and all the 8 triads $A_i A_{i+1} A_{i+3}$ are collinear but $A_1 A_2 \cdots A_8$ are not on the same line. The situation will however be entirely different if we restrict the configurations to be in the *real* affine plane. In fact S.-C. Chou, X.-S. Gao and others have proved the following beautiful theorem and discovered all the possible degenerate 8-triad configurations in the real affine plane.

Theorem. In the *real* affine plane there exist no 8-triad configurations for which A_i $(i = 1, \ldots, 8)$ are different from each other and all triads $A_i A_{i+1} A_{i+3}$ are collinear while $A_1 A_2 \cdots A_8$ are not on the same line.

Finally we remark that all the above are possible owing to the introduction of coordinate systems so that our methods of polynomial manipulations become available. There are so many beautiful theorems which cannot even be stated without the intervention of coordinate systems. Among these we may cite, for example, the syzygy-theorems of triads of points on a complex planar cubic curve and the 27-line theorem of cubic surface in a complex projective space. Remark that the last one has been considered as one of the gems hidden in the rag-bag of projective-geometry theorems (quite beautiful, however) discovered in the 19th century.

The study of geometrical problems, including in particular problems of geometrical constructions and geometrical locus-determinations, may also in general be reduced to polynomial equations-solving in introducing suitable coordinate systems. Besides those problems in ancient Greek already mentioned at the beginning of the present paper as well as the classical ones like the Appolonius Problem and the Malfatti Problem, we shall mention a few further problems below which have been notably studied by members of MMRC.

Zassenhaus Problem. Given 3 bisectors, whether interior or exterior, of the 3 angles of a triangle, construct the triangle.

It is trivial to construct bisectors of angles of a triangle by ruler-compass alone. The converse, as stated in the above Zassenhaus Problem, is not an easy matter. In fact, X.-S. Gao and D.-K. Wang gave the answer in negative in that such triangles cannot be constructed by ruler-compass alone. The proof consists in showing that the side-squares of the triangle should satisfy some irreducible equations of degree 10 with coefficients in the given bisectors.

Erdos Problem. Given in the plane 10 points A_{ij} $(i, j = 1, \ldots, 5; i < j)$, determine 5 points B_i, $i = 1, \ldots, 5$, such that for each of the above pair (i, j), A_{ij} is on the line $B_i B_j$.

The final answer due to H. Shi and H. Li is that there are generically 6 such quintuples of points B_i which can be explicitly determined from the given A_{ij}.

Pyramid Problem. Given a regular pyramid of square base, determine planes which will cut the pyramid into a regular pentagon.

The problem was solved by D.-K. Wang in giving a necessary and sufficient condition for the pyramid to have such regular-pentagonal cuts. Such planar cuts, in case existent, may be explicitly constructed. Moreover, in that case there will be regular-star cuts too which may also be explicitly constructed.

We remark that, all the problems above are solved by our methods of polynomial equations-solving, after some suitable coordinate system has been chosen. Clearly an immense number of such problems may be treated in the same manner. In fact, various problems arising from partial differential equations, celestial mechanics, theoretical physics, and technologies have been successfully treated in this way.

So far we have not mentioned at all the theorems (4) and (6) involving separants. In fact, the concept of separant plays an important role in studying singularities of algebraic varieties. Moreover, based on theorem (6), the present author has shown that for optimum problems with real object function and real restraint conditions all *polynomial* ones, there is some *finite* set of real values, whose greatest or least value is just the greatest or least value of the original optimum problem, so far it is supposed to exist. This gives in particular a general method of solving problems in non-linear programming and proving a lot of inequalities, algebraic, geometric, trigonometric, etc. For more details we refer to relevant papers and a forthcoming book of the present author.

3   Some Historical Discussions

From the above we see that both geometrical theorem-proving and geometrical problem-solving may be reduced to polynomial equations-solving which paves the way to the grandiose state of wide applications. We owe this much to the classic of Descartes in 1637. In his classic Descartes emphasized geometry problem-solving instead of geometry theorem-proving and showed how such geometry problem-solving may be reduced to polynomial equations-solving. Moreover, in a posthumous work Descartes had even the idea of reducing any kind of problem-solving, not merely geometrical ones, to the solving of polynomial equations in the form of some general scheme. Below we shall reproduce some paragraphs from a classic of G. Polya, viz. 《 *Mathematical Discovery* 》. In that classic is described such a rough outline of Cartesian scheme:

[First, reduce any kind of problem to a mathematical problem.
Second, reduce any kind of a mathematical problem to a problem of algebra.
Third, reduce any problem of algebra to the solution of a single equation.]

The third step may be further clarified as shown in the paragraph below which we again reproduce from the above classic of Polya:

[Descartes advises us to set up as many equations as there are unknowns. Let n stand for the number of unknowns, and $x_1, x_2, \ldots, x_n$ for the unknowns themselves; then we can write the desired system of equations in the form

$$r_1(x_1, x_2, \ldots, x_n) = 0,$$
$$r_2(x_1, x_2, \ldots, x_n) = 0,$$
$$\cdots\cdots$$
$$r_n(x_1, x_2, \ldots, x_n) = 0,$$

where $r_1(x_1, x_2, \ldots, x_n)$ indicates a polynomial in $x_1, x_2, \ldots, x_n$, and the left-hand sides of the following equations must be similarly interpreted. Descartes advises us further to reduce this system of equations to one final equation.]

It seems that Descartes was absolutely absurd in reducing any kind of problem to one of polynomial equations-solving. In Descartes' time differential calculus had not yet been discovered so it is natural that Descartes had no idea about differential equations and other kinds of equations but the polynomial ones. However, even a problem can be so reduced to polynomial equations, the number of equations may not be the same as the number of unknowns: it may be smaller or greater. Even the numbers of equations and unknowns are the same, how to reduce such a system of polynomial equations to the solving of equations involving a single unknown is not at all evident and is actually entirely non-trivial. Even if it can be so reduced, there may occur redundant solutions or loss of solutions. As pointed out by Polya in the above-mentioned classic, the reduced equations in single unknowns may furnish eventually the whole set of solutions only "in general," e.g. in the "regular case." However, the clarification of "in general" or "regular case" will be far from being trivial, as may be seen already from the simplest case in which r_i are all linear in the x's. In fact, this is what we have tried to achieve in the form of Zero-Decomposition Theorems as described in Section 1.

However, in spite of the apparent absurdity of Descartes' Program, the idea of Descartes is of highest significance to the development of mathematics. In this respect let us reproduce some paragraph again from the same classic of Polya:

[Descartes' project failed, but it was a great project and even in its failure it influenced science much more than a thousand and one little projects which happened to succeed. Although Descartes' scheme does not work in all cases, it does work in an inexhaustible variety of cases, among which there is an inexhaustible variety of *important* cases.]

In fact, now-a-days we see that there are really inexhaustible problems from mathematics, from sciences, from technologies, etc. which will ultimately be reduced to polynomial equations-solving. A few of such particular problems have been indicated in Section 2 above. Now, among all kinds of systems of equations to be solved, those of polynomial forms are clearly the simplest, and the most fundamental ones, but at the same time a problem entirely non-trivial at all. Among all kinds of problems in mathematics, the problem of polynomial equations-solving should, therefore, be considered as one of primary importance to be settled with highest priority.

Let us consider now the state of geometry problem-solving and geometry problem-solving in our ancient China. For this sake let us reproduce some paragraphs from a previous paper of the present author appeared in 《 *Proc. ATCM '95* 》 (Singapore, 1996), pp. 66–72:

[The geometry theorem-proving in ancient Greece is well-known. As for in ancient China, one considered usually geometry problem-solving rather than geometry theorem-proving. Most of geometry problems to be solved arose from practical requirements. We may cite in particular:

(I) Problems arising from measurements, ...

(II) Problems about area and volume determination, ...

(III) Problems involving Gou-Gu Form (i.e. a right-angled triangle), ... e.g. the determination of such forms with 3 sides in *rational* ratios.

(IV) Problems arising from actual lifes, ...

A lot of such problems with their solutions appeared already in the two earliest classics fortunately preserved up to the present day, viz. 《 *Nine Chapters of Arithmetic* 》 completed in 1c B.C., and 《 *Zhou Bi Mathematics Manuel* 》, believed to be no later than 100 B.C. To these two classics we may add the more important classic 《 *Annotations to Nine Chapters* 》 due to Liu Hui in 263 A.D.

The results of the geometry problems solved were usually expressed in the form of *Shui* which may sometimes be interpreted as *theorems* in the form of explicit formulas.

Instead of set of axioms the ancient Chinese formulated rather some simple and plausible principles and then proved the geometry theorems in the form of formulas or solved the geometry problems by logical reasoning.

The simplest one of such principles is perhaps the *Out-In Complementary Principle*: When a planar (or a solid) figure be cut into pieces, moved to somewhere else, and combined together, the area (or volume) of the figure will remain the same.

Diverse applications including geometry formula-proving and geometry problem-solving have been made by means of this seemingly trivial principle in an elegant and sometimes unbelievably striking manner. ... We may cite also the proof of the general Gou-Gu formula (i.e. the Pythagorous Theorem) based on the above Principle. The proof appeared in 《 *Zhou Bi* 》 and was attributed to some scholar Shang Gao in year around 1122 B.C.

......

A turning point about geometry problem-solving occurred in Song Dynasty (+960, +1274) and Yuan Dynasty (+1271, +1368). During that period the notions of *Heaven Element, Earth Element*, etc. (equivalent to the present day *unknowns* x, y, etc.), together with allied notions of polynomials and methods of elimination, were introduced. This furnishes a systematic way of solving geometry problems in the following way. Designate the unknowns of the geometry problems to be determined by Heaven Element, Earth Element, etc. The geometry conditions about the geometry entities involved will be turned into some algebraic relations between them which are usually equivalent to polynomial equations in modern form. By eliminating the unknown elements in succession we

12

get finally equations each time in one alone of the unknown elements. When the known data are in numerical values, what is usually the case, we may solve these equations in succession by means of methods of numerical solving developed in Song Dynasty. The results will give then the required solutions of the geometry problems in question. We remark in passing that polynomial equations-solving instead of geometry theorem-proving was the main theme of studies throughout the whole history of mathematics in ancient China.

The method created in that period was called the *Heaven Element Method*. It was in essence a method of algebrization of geometry which permits to reduce in relatively trivial manner geometry problems to the algebraic problems of solving of systems of polynomial equations. Elimination method, which appeared already in 《 *Nine Chapters* 》 for the solving of systems of *linear* equations (equivalent to present day Gaussian elimination) was also developed to the solving of general systems of polynomial equations in the period of Song-Yuan Dynasties. Numerous examples may be found in some still existent classics. Unfortunately the development of Chinese ancient mathematics stopped at this crucial moment during the period of Yuan and Ming Dynasties.

......

The algebrization of geometry, created in ancient China and stopped in the 14th century, was revived in Europe in the 17th century. It was even systematized to the form of analytic geometry owing to the creation of R. Descartes. Descartes emphasized in his 1637 classic on geometry the geometry problem-solving rather than geometry theorem-proving. This was just in the same spirit as in ancient China. The analytic geometry also permits to prove geometry theorems by mere computations in contrast to the proving by purely geometrical reasoning of Euclid. The classic of Descartes showed also the way of reducing geometry problems-solving to polynomial equations-solving, which was just what the Chinese scholars in Song and Yuan Dynasties had tried to do.

In a posthumous work Descartes had founded some doctrine in saying that all problems can be reduced to the solving of problems in mathematics, then to problems in algebra, then to problems of polynomial equations-solving, and finally to problems of solving algebraic equations in single unknowns. In comparing with works of our ancestors, we see that our ancestors and Descartes were on the same lines of thought in the development of mathematics.

The computational method of solving geometry problems (including proving geometry theorems) in reducing them to the solving of polynomial equations lacked however a general way of dealing with the associated system of equations usually in embarrassing confused form.

In recent years the Chinese mathematicians, mainly in MMRC (Math. Mech. Res. Center, Institute of Systems Science, Chinese Academy of Sciences), have developed a general *mechanization method* for solving of arbitrary systems of polynomial equations which permits to be applied to geometry problem-solving, in particular geometry theorem-proving. Our achievements may be considered as a continuation of what our ancestors had founded as well as partial accomplishments of Descartes doctrine.

......]

13

In conclusion, it would be interesting in comparing the way of Descartes in establishing equations for problem-solving with the way of our ancestors in introducing unknowns called Heaven Element, etc. and then establishing equations for problem-solving as described above. In a word, it may be said that Chinese ancient mathematics in the main was developed along the way as indicated in Descartes' Program, and conversely, Descartes' Program may be considered as an overview of the way of developments of Chinese ancient mathematics.

Solving Geometric Problems
with Real Quantifier Elimination

Andreas Dolzmann

Fakultät für Mathematik und Informatik
Universität Passau, D-94030 Passau, Germany
dolzmann@uni-passau.de
http://www.fmi.uni-passau.de/~dolzmann/

Abstract. Many problems arising in real geometry can be formulated as first-order formulas. Thus quantifier elimination can be used to solve these problems. In this note, we discuss the applicability of implemented quantifier elimination algorithms for solving geometrical problems. In particular, we demonstrate how the tools of REDLOG can be applied to solve a real implicitization problem, namely the Enneper surface.

1 Introduction

Since Tarski introduced the first quantifier elimination algorithm for the real numbers in the 1930's (cf. [21]), several other algorithms have been developed. Only some of them are implemented and widely available. Among them there are, for instance, the quantifier elimination by *partial cylindrical algebraic decomposition* implemented in the C-program QEPCAD, the quantifier elimination by *real root counting* implemented in the module QERRC within the Modula II computer algebra system MAS and the quantifier elimination by *virtual substitution* implemented in the REDUCE-package REDLOG. Whereas QEPCAD and QERRC provide a complete quantifier elimination procedure, the quantifier elimination of REDLOG is restricted in the degree of the quantified variables.

The naive approach to use one of the implemented quantifier elimination algorithms for solving a problem often fails in practice. Firstly, the programs can fail due to the limitations of computing time and memory. Secondly, the chosen quantifier elimination algorithm may not be able to handle the wanted class of problems.

Nevertheless it is possible to solve non-trivial problems using the implemented quantifier elimination algorithms. In general this requires both a combination of the available algorithms and a careful formulation of the problem as a first-order formula according to the chosen quantifier elimination method.

We give solutions of problems cited in the literature and compare the results of the different elimination procedures. In particular we discuss the real implicitization of the Enneper surface.

2 The General Framework

We consider multivariate polynomials $f(u, x)$ with rational coefficients, where $u = (u_1 \ldots, u_m)$ and $x = (x_1, \ldots, x_n)$. We call u *parameters* and we call x *main variables*. *Atomic formulas* are polynomial equations $f = 0$, polynomial inequalities $f \geqslant 0$, $f \leqslant 0$, $f > 0$, $f < 0$, and $f \neq 0$. *Quantifier-free* formulas are built from atomic formulas by combining them with the boolean connectors "¬," "∧," and "∨." *Existential formulas* are of the form $\exists x_1 \ldots \exists x_n \psi(u, x)$, where ψ is a quantifier-free formula. Similarly, *universal formulas* are of the form $\forall x_1 \ldots \forall x_n \psi(u, x)$. A *prenex first-order formula* has several alternating blocks of existential and universal quantifiers in front of a quantifier-free formula, the *matrix* of the prenex formula.

The real *quantifier elimination problem* can be phrased as follows: Given a formula φ, find a quantifier-free formula φ' such that both φ and φ' are equivalent in the domain of the real numbers. A procedure computing such a φ' from φ is called a real *quantifier elimination procedure*.

A *background theory* is a set of atomic formulas considered conjunctive. Two formulas φ and φ' are equivalent wrt. a background theory Θ, if and only if

$$\underline{\forall}\left(\bigwedge \Theta \longrightarrow (\varphi \longleftrightarrow \varphi')\right),$$

where $\underline{\forall}$ denotes the universal closure. Notice, that the formula $x^2 - x = 0$ contained in a background theory, does not imply a multiplicative idempotency law, but only describes an equation for the variable x.

A *generic* quantifier elimination procedure assigns to a formula φ a quantifier-free formula φ' and a background theory Θ', such that φ and φ' are equivalent wrt. the background theory Θ'. The computed background theory contains only negated equations. Hence the set

$$\left\{ a \in \mathbb{R}^m \mid \neg(\varphi'(a) \longleftrightarrow \exists x \varphi(a, x)) \right\}$$

has measure zero. Notice, that each quantifier elimination procedure provides also a generic quantifier elimination assigning $\emptyset$ to Θ'. However, the computation of an appropriate Θ' leads in many cases to simpler quantifier-free equivalents and results in a considerable speed-up of the running times. Moreover, in the framework of automatic proving in real geometry generic quantifier elimination allows us to find sufficient conditions automatically, cf. [13] for details.

3 Quantifier Elimination Methods

In this section we sketch the three most important implemented quantifier-elimination algorithms. For a more detailed overview over these algorithms, cf. [14].

3.1 Quantifier Elimination by Partial Cylindrical Algebraic Decomposition

Cylindrical algebraic decomposition (CAD), cf. [6,1], is the oldest and most elaborate implemented real quantifier elimination method. It has been developed by Collins and his students starting in 1974. During the last 10 years particularly Hong made very significant theoretical contributions that improved the performance of the method dramatically, cf. [16,17], resulting in *partial* CAD, cf. [7]. Hong has implemented partial CAD in his program QEPCAD based on the computer algebra C-library SACLIB. QEPCAD is not officially published but available from Hong on request. Brown has contributed the latest very successful improvements to the algorithm, cf. [4]. Unfortunately, the improved version was not available when we made the computations for this article.

We sketch the basic ideas of CAD: Suppose we are given an input formula

$$\varphi(u_1, \ldots, u_m) \equiv Q_1 x_1 \ldots Q_n x_n \psi(u_1, \ldots, u_m, x_1, \ldots, x_n), \quad Q_i \in \{\exists, \forall\}.$$

Let F be the set of all polynomials occurring in ψ as left hand sides of atomic formulas. Call $C \subseteq \mathbb{R}^{m+n}$ *sign invariant* for F, if every polynomial in F has constant sign on all points in C. Then $\psi(c)$ is either "true" or "false" for all $c \in C$.

Suppose we have a finite sequence $\Pi_1, \ldots, \Pi_{m+n}$ with the following properties:

1. Each Π_i is a finite partition of $\mathbb{R}^i$ into connected semi-algebraic cells. For $1 \leqslant j \leqslant n$ each Π_{m+j} is labeled with Q_j.
2. Π_{i-1} consists for $1 < i \leqslant m + n$ exactly of the projections of all cells in Π_i along the coordinate of the i-th variable in $(u_1, \ldots, u_m, x_1, \ldots, x_n)$. For each cell C in Π_{i-1} we can determine the preimage $S(C) \subseteq \Pi_i$ under the projection.
3. For each cell C in Π_m we know a quantifier-free formula $\delta_C(u_1, \ldots, u_m)$ describing this cell.
4. Each cell C in Π_{m+n} is sign invariant for F. Moreover for each cell C in Π_{m+n}, we are given a *test point* t_C in such a form that we can determine the sign of $f(t_C)$ for each $f \in F$ and thus evaluate $\psi(t_C)$.

A quantifier-free equivalent for φ is obtained as the disjunction of all δ_C for which C in Π_m is *valid* in the following recursively defined sense:

1. For $m \leqslant i < m + n$, we have that Π_{i+1} is labeled:
    (a) If Π_{i+1} is labeled "$\exists$," then C in Π_i is valid if at least one $C' \in S(C)$ is valid.
    (b) If Π_{i+1} is labeled "$\forall$," then C in Π_i is valid if all $C' \in S(C)$ are valid.
2. A cell C in Π_{m+n} is valid if $\psi(t_C)$ is "true."

Partial cylindrical algebraic decomposition (PCAD) is an improved version of CAD. It takes the boolean structure of the input formula into account. CAD and also PCAD is doubly exponential in the number of all variables.

3.2 Quantifier Elimination by Virtual Term Substitution

The virtual substitution method dates back to a theoretical paper by Weispfenning in 1988, cf. [22]. During the last five years a lot of theoretical work has been done to improve the method, cf. [18, 23, 11]. The method is implemented within the REDUCE package REDLOG by the author and Sturm, cf. [10].

The applicability of the method in the form described here is restricted to formulas in which the quantified variables occur at most quadratically. Moreover quantifiers are eliminated one by one, and the elimination of one quantifier can increase the degree of other quantified variables. On the other hand there are various heuristic methods included for decreasing the degrees during elimination. One obvious example for such methods is polynomial factorization.

For eliminating the quantifiers from an input formula

$$\varphi(u_1, \ldots, u_m) \equiv Q_1 x_1 \ldots Q_n x_n \psi(u_1, \ldots, u_m, x_1, \ldots, x_n), \quad Q_i \in \{\exists, \forall\}$$

the elimination starts with the innermost quantifier regarding the other quantified variables within ψ as extra parameters. Universal quantifiers are handled by means of the equivalence $\forall x \psi \longleftrightarrow \neg \exists x \neg \psi$. We may thus restrict our attention to a formula of the form

$$\varphi^*(u_1, \ldots, u_k) \equiv \exists x \psi^*(u_1, \ldots, u_k, x),$$

where the $u_{m+1}, \ldots, u_k$ are actually x_i quantified from further outside.

We fix real values $a_1, \ldots, a_k$ for the parameters $u_1, \ldots, u_k$. Then all polynomials occurring in ψ^* become linear or quadratic univariate polynomials in x with real coefficients. So the set

$$M = \{\, b \in \mathbb{R} \mid \psi^*(a_1, \ldots, a_k, b) \,\}$$

of all real values b of x satisfying ψ^* is a finite union of closed, open, and half-open intervals on the real line. The endpoints of these intervals are among $\pm\infty$ together with the real zeros of the linear and quadratic polynomials occurring in ψ^*.

Candidate terms $\alpha_1, \ldots, \alpha_r$ for the zeros can be computed uniformly in u_1, $\ldots, u_k$ by the solution formulas for linear and quadratic equations. For open and half-open intervals, we add expressions of the form $\alpha \pm \varepsilon$, where α is a candidate solution for some left-hand side polynomial. The symbol ε stands for a positive infinitesimal number. Together with the formal expressions ∞ and $-\infty$ all these candidate terms form an *elimination set*. This means M is non-empty if and only if the substitution of at least one element of the elimination set satisfies ψ^*. After substitution of formal expressions possibly involving square roots, ε, or $\pm\infty$, we rewrite the substitution result in such a way that it does not contain any fractions nor one of these special symbols. This process of substituting a term and rewriting the formula is called *virtual substitution*. By disjunctively substituting all candidates into ψ^* we obtain a quantifier-free formula equivalent to $\exists x \psi^*$.

For practical applications this method, of course, has to be refined by a careful selection of smaller elimination sets and by a combination with powerful simplification techniques for quantifier-free formulas, cf. [11] for details. There is a variant of the virtual substitution method for generic quantifier elimination, cf. [13].

Quantifier elimination by virtual substitution is doubly exponential in the number of quantifier blocks but only singly exponential in the number of quantified variables. The number of parameters only plays a minor role in the complexity.

3.3 Quantifier Elimination by Real Root Counting

The basis for this quantifier elimination method is a theorem on real root counting found independently by Becker and Wörmann, cf. [2, 3], and Pedersen, Roy, and Szpirglas, cf. [19, 20]. It is based on a result for counting real zeros of univariate polynomials found by Hermite, cf. [15].

Let $I \subseteq \mathbb{R}[x_1, \ldots, x_n]$ be a zero-dimensional ideal. For $g \in \mathbb{R}[x_1, \ldots, x_n]$ consider the symmetric quadratic form $Q_g = (\text{trace}(m_{gb_i,b_j}))_{1 \leqslant i,j \leqslant d}$ on the linear $\mathbb{R}$-space $S = \mathbb{R}[x_1, \ldots, x_n]/I$, where $\{b_1, \ldots, b_d\} \subseteq S$ is a basis, and the $m_h : S \to S$ are linear maps defined by $m_h(f + I) = (hf) + I$. Let s be the signature of Q_g, and denote by n_+ and n_- the number of real roots of I at which g is positive or negative, respectively. Then $n_+ - n_- = s$.

The use of a Gröbner basis of the ideal I allows to obtain a basis of S, and to perform arithmetic there, cf. [5], thus obtaining the matrix Q_g.

This approach can be extended to obtain the exact number of roots under a side condition, and can be moreover extended to several side conditions:

Let $F, \{g_1, \ldots, g_k\} \subseteq \mathbb{R}[x_1, \ldots, x_n]$ be finite, and assume that $I = Id(F)$ is zero-dimensional. Denote by N the number of real roots $a \in \mathbb{R}^n$ of F for which $g_i > 0$ for $1 \leqslant i \leqslant k$. Define

$$\Gamma(\{g_1, \ldots, g_k\}) = \{ g_1^{e_1} \cdots g_k^{e_k} \mid (e_1, \ldots, e_k) \in \{1, 2\}^k \}.$$

Then defining Q_g as above we have $2^k N = \sum_{\gamma \in \Gamma(\{g_1, \ldots, g_k\})} \text{sig}(Q_\gamma)$.

For real quantifier elimination, this root counting has to be further extended to multivariate polynomials with parametric coefficients in such a way that it will remain correct for *every* real specialization of the parameters including specializations to zero. This task has been carried out by Weispfenning using comprehensive Gröbner bases, cf. [24].

The case of ideals that have a dimension greater than 0 is handled by recursive calls of the approach for zero-dimensional ideals: We construct from one polynomial system several polynomial systems in such a way, that each of the systems has either a smaller number of variables or a smaller dimension.

The complete method has been implemented by the author within the package QERRC of the computer algebra system MAS, cf. [9]. A variant for generic quantifier elimination is under development.

4 Simplification Methods

Simplification of formulas means to compute to a given formula an equivalent one, which is simpler. The simplification of the input formulas and intermediate results are very important for a successful application of quantifier elimination by virtual substitution and for quantifier elimination by real root counting. QEPCAD does not depend heavily on simplification of formulas, because formulas are used only for input and output. For a detailed description of the simplification algorithms sketched in this section and a discussion of the notion of *simple* formulas, cf. [11].

4.1 The Standard Simplifier

The *standard simplifier* is a fast, though sophisticated, simplifier for quantifier-free formulas. It was developed for the implementation of the quantifier elimination by virtual term substitution in the REDLOG package. The standard simplifier was designed to be called very often, for instance, after each elimination step. Beside the simple methods to remove boolean constants and to keep only one of a set of identical subformulas, it implements three main strategies:

Firstly, it simplifies the atomic formulas. All right hand side of atomic formulas are normalized to zero. The left hand side polynomials are normalized to be primitive over $\mathbb{Z}$ in such a way that the highest coefficient wrt. a fixed term order is positive. Furthermore, we drop irrelevant factors of the polynomials. Trivial square sums are detected to be greater than zero or not less than zero, respectively. Optionally, we explode atomic formulas by decomposing the polynomial additively or multiplicatively.

Secondly, smart simplification is applied to conjunctions and disjunctions of atomic formulas: Each pair of atomic formulas involving identical left-hand sides, is replaced by "true," "false," or by one atomic formula using an appropriate relation. Similarly, this method can be applied to some pairs of atomic formulas which differs only in the absolute summand of the left-hand side polynomial.

Thirdly, the techniques used for the smart simplification of flat formulas are applied to nested formulas. This is done by constructing an implicit background theory for each boolean level of the formula.

The standard simplifier offers the option to simplify a formula wrt. a background theory. The output formula is then equivalent to the input formula wrt. the given background theory.

4.2 Simplification of Boolean Normal Forms

For the simplification of boolean normal forms we use two further techniques additionally to the techniques used in the standard simplifier.

The *generalized subsumption* allows us to drop conjunctions of a CNF: Let t_i terms and φ_i, $\rho_i, \rho_i' \in \{<, \leqslant, =, \geqslant, \geqslant, >\}$. Then

$$(t_1 \, \rho_1 \, 0 \wedge \ldots \wedge t_n \, \rho_n \, 0) \vee (t_1 \, \rho_1' \, 0 \wedge \ldots \wedge t_n \, \rho_n' \, 0 \wedge \ldots)$$

can be simplified to $(t_1 \ \rho_1 \ 0 \wedge \ldots \wedge t_n \ \rho_n \ 0)$, if $t_i \ \rho_i' \ 0 \longrightarrow t_i \ \rho_i \ 0$.

The *generalized cut* combines two conjunctions combined disjunctively into one conjunction: if $(t_i \ \rho_i \ 0 \vee t_i \ \rho_i' \ 0) \longleftrightarrow t_i \ \sigma_i \ 0$ then the disjunction

$$(t_1 \ \rho_1 \ 0 \wedge t_2 \ \rho_2 \ 0 \wedge \ldots \wedge t_n \ \rho_n \ 0) \vee (t_1 \ \rho_1' \ 0 \wedge t_2 \ \rho_2 \ 0 \wedge \ldots \wedge t_n \ \rho_n \ 0)$$

can be simplified to $(t_1 \ \sigma_1 \ 0 \wedge t_2 \ \rho_2 \ 0 \wedge \ldots \wedge t_n \ \rho_n \ 0)$. In our implementation, not all possible applications of the generalized cut and the generalized subsumption are performed. We only apply the simplifications in cases, where the respective implication can be decided independently of the terms t_i. Analogous simplification rules hold for conjunctions of disjunctions.

4.3 The Gröbner Simplifier

The *Gröbner simplifier* is an advanced method for the simplification of boolean normal forms. The main technique used for this method is the computation of a Gröbner basis for deciding the ideal membership test, cf. [5]. Let F be a set of polynomials and G a Gröbner basis of F. Then the Gröbner simplifier replaces

$$\bigwedge_{f \in F} f = 0 \wedge \bigwedge_{h \in H} h \ \rho_h \ 0 \quad \text{by} \quad \bigwedge_{f \in G} f = 0 \wedge \bigwedge_{h \in H} \mathrm{Nf}_G(h) \ \rho_h \ 0.$$

Using a radical membership test, we can optionally replace the input formula by false, if $h \in \mathrm{Rad}(F)$ and $\rho_h \in \{<, \neq, >\}$.

There is, of course, a variant which simplifies disjunctions of atomic formulas. Conjunctive and disjunctive normal forms are simplified essentially by applying the Gröbner simplifier to each of the constituents. Additionally there are strategies to relate information contained in different constituents. For the simplification of an arbitrary formula, we compute a CNF or a DNF first. We have implemented a variant of the Gröbner simplifier which automatically decides, whether to compute a CNF or a DNF. This decision is based on a heuristic for estimating which of the normal forms is larger. Like the standard simplifier the Gröbner simplifier can simplify a formula wrt. a background theory.

Although the Gröbner simplifier operates on a boolean normal form, we consider it as a general simplification method. It turned out that in many cases the result of the Gröbner simplification is simpler than the input formula.

We illustrate the technique by means of a very simple example: Consider the input formula $xy + 1 \neq 0 \vee yz + 1 \neq 0 \vee x - z = 0$, which can be rewritten as

$$xy + 1 = 0 \wedge yz + 1 = 0 \longrightarrow x - z = 0.$$

Reducing the conclusion modulo the Gröbner basis $\{x - z, yz + 1\}$ of the premises, we obtain the equivalent formula $xy + 1 = 0 \wedge yz + 1 = 0 \longrightarrow 0 = 0$, which can in turn be easily simplified to "true."

4.4 The Tableau Method

Although the standard simplifier combines information located on different levels, it preserves the basic boolean structure of the formula. The Gröbner simplifier computes a boolean normal from and thus it changes the boolean structure completely. The tableau method, in contrast, provides a technique for changing the boolean structure of a formula slightly by constructing case distinctions. There are three variants of tableau simplifiers:

Given a formula φ and a term t the *tableau simplifier* constructs the following case distinction:

$$(t = 0 \wedge \varphi) \vee (t > 0 \wedge \varphi) \vee (t < 0 \wedge \varphi).$$

This formula is then simplified with the standard simplifier. Even though the constructed case distinction is about three times larger than the original formula, in some cases the simplified equivalent is much smaller than the original formula.

The *iterative tableau simplifier* chooses automatically an appropriate term t by trying all terms contained in φ. The result is then either the original formula or the best result obtained by the tableau simplifier, depending on which one is simpler. The *automatic tableau simplifier* applies the iterative tableau as long as the output is simpler than the input.

5 REDLOG

REDLOG is a REDUCE package implementing a *computer logic system*, cf. [10]. It provides algorithms to deal with first-order logic formulas over various languages and theories. Beside the theory of real closed fields there are algorithms for the theory of discretely valued fields [12] and algebraically closed fields. REDLOG provides an implementation of the quantifier elimination by virtual substitution including variants for generic quantifier elimination, extended quantifier elimination, and linear optimization using quantifier elimination techniques. There are also interfaces to QEPCAD and QERRC available such that these packages can be called from REDLOG and the results are available to be further processed. All simplification algorithms discussed in this note are available in REDLOG. Besides these important algorithms REDLOG provides many tools for normal form computations, constructing, analyzing, and decomposing formulas. The REDLOG source code and documentation are freely available on the WWW.[1]

6 Examples

By means of examples we discuss in this section the applicability of quantifier elimination to some problems taken from various scientific research areas. We compare the timings of the three considered implementations. Furthermore

---

[1] http://www.fmi.uni-passau.de/~redlog/

we will give some hints, how to apply quantifier elimination, even if the naive approach fails.

In this section we allow ourselves to use not only prenex formulas but also general first-order formulas, i.e. boolean combinations of prenex formulas. Furthermore, we use the implication $\Rightarrow$ and the equivalence $\Leftrightarrow$ inside formulas. Notice, that these general formulas can be easily transformed in prenex formulas containing only the boolean connectors $\wedge$ and $\vee$.

All example computations mentioned in this section have been performed on a SUN Ultra 1 Model 140 workstation using 32 MB of memory.

6.1 Real Implicitization

For $n < m$ let $f : \mathbb{R}^n \to \mathbb{R}^m$ be a rational map with the component functions p_i/q, where $p_i, q \in \mathbb{R}[x_1, \ldots, x_n]$ for $1 \leqslant i \leqslant m$. The image $f(\mathbb{R}^n)$ of f is a definable and hence semi-algebraic subset of $\mathbb{R}^m$ described by the formula

$$\exists x_1 \ldots \exists x_n \big(q(x_1, \ldots, x_n) \neq 0 \wedge \bigwedge_{i=1}^{m} p_i(x_1, \ldots, x_n) = u_i q(x_1, \ldots, x_n) \big).$$

Our aim is to obtain a quantifier-free description of $f(\mathbb{R}^n)$ in the variables u_1, $\ldots$, u_m, preferably a single equation, which would provide an implicit definition of f.

Example 1. Descartes' folium $d : \mathbb{R} \to \mathbb{R}^2$ is given by the component functions $3x_1/(1 + x_1^3)$ and $3x_1^2/(1 + x_1^3)$ for u_1 and u_2, cf. [8]. For obtaining an implicit form we apply quantifier elimination to

$$\exists x_1 \big(1 + x_1^3 \neq 0 \wedge 3x_1 = u_1(1 + x_1^3) \wedge 3x_1^2 = u_2(1 + x_1^3)\big).$$

QEPCAD obtains after 1 s the result $u_1^3 - 3u_1 u_2 + u_2^3 = 0$. QERRC obtains after 1.6 s an elimination result with 7 atomic formulas. This can be automatically simplified to the QEPCAD result using the Gröbner simplifier. REDLOG fails on this example due to a violation of the degree restrictions. After simplifying the matrix of the formula with the Gröbner simplifier for CNF's we can eliminate the quantifier with REDLOG. The output formula computed in 0.3 s contains 21 atomic formulas.

Example 2. The Whitney Umbrella is given by

$$(x_1, x_2) \mapsto (x_1 x_2, x_2, x_1^2).$$

QEPCAD produces in 1 s the result $u_3 \geqslant 0 \wedge -u_1^2 + u_2^2 u_3 = 0$. Quantifier elimination by virtual substitution (QEVTS) produces in 0.01 s the much longer result

$$u_3 \geqslant 0 \wedge (u_1^2 - u_2^2 u_3 = 0 \wedge u_1 u_2 \geqslant 0 \vee u_1^2 - u_2^2 u_3 = 0 \wedge u_1 u_2 \leqslant 0).$$

However, including a Gröbner simplification we yield in 0.03 s the same result as produced by QEPCAD. Using QERRC we get the following quantifier-free equivalent in 0.8 s:

$$u_2 \neq 0 \wedge -u_1^2 + u_2^2 u_3 = 0 \vee u_1 = 0 \wedge u_2 = 0 \wedge u_3 + 1 \neq 0 \wedge u_3 \geqslant 0.$$

6.2 Automatic Theorem Proving in Geometry

Example 3 (Steiner-Lehmus theorem, variant). Assume that ABC is a triangle such that $AB > AC$. Then the angle bisector from B to AC is longer than the angle bisector from C to AB (i.e., the longer bisector goes to the shorter side).

In its original form, the Steiner-Lehmus theorem states that *any triangle with two equal internal bisectors is isosceles.* Its contrapositive follows immediately from the variant above.

We put $A = (-1, 0)$, $B = (1, 0)$, and $C = (x_0, y_0)$ with $y_0 > 0$. By $M = (0, b)$ we denote the center and by c the radius of the circumcircle.

The bisectors are constructed using the geometrical theorem proved as Example 4: Let $V = (0, b - c)$ be the point below the x-axis on the circumcircle having equal distance to A and B. Then the angle bisector from C to AB is obtained as CX, where $X = (x, 0)$ is the intersection of CV and AB. The angle bisector from B to AC is obtained analogously: Let $W = (x_1, y_1)$ be the point on the circumcircle with equal distance to A and C lying "west" of the line AC. Let $Y = (x_2, y_2)$ be the intersection of BW and AC. Then the angle bisector is BY.

Our algebraic translation obtained this way reads as follows:

$$\forall b \forall c \forall x \forall x_1 \forall y_1 \forall x_2 \forall y_2 \big(y_1(x_0 + 1) > x_1 y_0 \wedge y_0 > 0 \wedge c > 0 \wedge$$
$$c^2 = 1 + b^2 \wedge c^2 = x_0^2 + (y_0 - b)^2 \wedge x(y_0 + (c - b)) = x_0(c - b) \wedge$$
$$x_1^2 + (y_1 - b)^2 = c^2 \wedge (x_1 + 1)^2 + y_1^2 = (x_1 - x_0)^2 + (y_1 - y_0)^2 \wedge$$
$$(x_1 - 1)y_2 = y_1(x_2 - 1) \wedge (x_0 + 1)y_2 = y_0(x_2 + 1) \wedge$$
$$4 > (x_0 + 1)^2 + y_0^2 \longrightarrow (x - x_0)^2 + y_0^2 < (x_2 - 1)^2 + y_2^2\big).$$

Using QEVTS we obtain after 74 s an elimination result φ^* containing 243 atomic formulas together with the subsidiary conditions

$$\vartheta^* \equiv x_0^2 + 2x_0 + y_0^2 + 1 \neq 0 \wedge$$
$$x_0^2 - 2x_0 + y_0^2 - 3 \neq 0 \wedge x_0 + 1 \neq 0 \wedge x_0 \neq 0 \wedge y_0 \neq 0.$$

QEPCAD proves within 250 s that $\vartheta^* \longrightarrow \varphi^*$, while QEVTS, QERRC fail doing so.

Example 4. Let M be the center of the circumcircle of a triangle ABC. Then $\angle ACB = \angle AMB/2$ (see Figure 1). Choosing coordinates $A = (-a, 0)$, $B = (a, 0)$, $C = (x_0, y_0)$, and $M = (0, b)$ and encoding angles into tangents, an algebraic translation of this problem reads as follows:

$$\forall x \forall t_1 \forall t_2 \forall t \forall b \big(c^2 = a^2 + b^2 \wedge c^2 = x_0^2 + (y_0 - b)^2 \wedge$$
$$y_0 t_1 = a + x_0 \wedge y_0 t_2 = a - x_0 \wedge (1 - t_1 t_2)t = t_1 + t_2 \longrightarrow bt = a\big).$$

Both QEPCAD and QERRC fail on this input. QEVTS together with the Gröbner simplifier yields after 0.06 s the quantifier-free equivalent $a \neq 0 \vee x_0 \neq 0 \vee y_0 \neq 0$ containing non-degeneracy conditions for the triangles.

The generic variants of QEVTS and QERRC produce "true" as a quantifier-free equivalent wrt. the background theory $\{y \neq 0\}$. This computation takes 0.02 s with QEVTS and 2 s with QERRC.

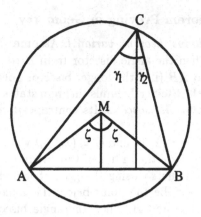

Fig. 1. The angle at circumference is half the angle at center (Example 4).

6.3 The Enneper Surface

The Enneper surface is defined parametrically by

$$x = 3u + 3uv^2 - u^3 \quad y = 3v + 3u^2v - v^3 \quad z = 3u^2 - 3v^2,$$

cf. [8]; in other words as the image of the function

$$f : \mathbb{C}^2 \longrightarrow \mathbb{C}^3 \quad \text{with} \quad f(u,v) = (3u + 3uv^2 - u^3, 3v + 3u^2v - v^3, 3u^2 - 3v^2).$$

The smallest variety V containing the Enneper surface is given by the polynomial

$$\begin{aligned}
p(x,y,z) = \ & 19683x^6 - 59049x^4y^2 + 10935x^4z^3 + 118098x^4z^2 - 59049x^4z + \\
& 59049x^2y^4 + 56862x^2y^2z^3 + 118098x^2y^2z + 1296x^2z^6 + \\
& 34992x^2z^5 + 174960x^2z^4 - 314928x^2z^3 - 19683y^6 + 10935y^4z^3 - \\
& 118098y^4z^2 - 59049y^4z - 1296y^2z^6 + 34992y^2z^5 - 174960y^2z^4 - \\
& 314928y^2z^3 - 64z^9 + 10368z^7 - 419904z^5.
\end{aligned}$$

Using Gröbner bases techniques it is easy to proof that over the complex numbers the image of f is identical to the complete complex variety V. The real Enneper surface is given by the restriction $f \restriction_{\mathbb{R}^2}^{\mathbb{R}^3}$. The question is whether the real Enneper surface is identical to the real variety of p.

This problem can be easily stated as a real quantifier elimination problem. A first-order description of the real image of f is given by

$$\varphi \equiv \exists u \exists v \left(x = 3u + 3uv^2 - u^3 \wedge y = 3v + 3u^2v - v^3 \wedge z = 3u^2 - 3v^2 \right).$$

Using this formula the complete problem can be stated as

$$\forall x \forall y \forall z (\varphi(x,y,z) \Leftrightarrow p(x,y,z) = 0).$$

It turns out, that none of the quantifier elimination algorithms considered in this paper can eliminate all the quantifiers of the above formula. Actually, none of the implementations is able to find a quantifier-free equivalent to φ. In the next paragraphs we will describe how to tackle the problem with the quantifier elimination algorithms and the tools available in the REDLOG system.

In a first step we compute a quantifier-free equivalent to φ. With the implementation of the virtual substitution method we can eliminate one of the existential quantifiers of φ. The solutions of eliminating u or eliminating v are up to a simple substitution identical. We start with the elimination of v and obtain in 0.2 s the result

$$\exists u \left(v_4 \geqslant 0 \wedge \left(v_1 = 0 \wedge v_2 = 0 \wedge v_3 \geqslant 0 \vee v_1 = 0 \wedge v_2 = 0 \wedge v_3 \leqslant 0\right)\right)$$

where

$$v_1 := 108u^6 + 324u^4 - 9u^2z^2 - 54u^2z + 243u^2 - 27y^2 - z^3 - 18z^2 - 81z,$$
$$v_2 := 2u^3 - uz + 3u - x,$$
$$v_3 := 6u^2y + yz + 9y,$$
$$v_4 := 3u^2 - z.$$

Applying the generalized cut to both constituents of the disjunction, we obtain the formula:

$$\exists u \left(v_4 \geqslant 0 \wedge v_1 = 0 \wedge v_2 = 0\right).$$

REDLOG produces this formula by computing a disjunctive normal form. After this simplification only the package QERRC can eliminate the remaining quantifier, but the obtained elimination result φ' is very large. Though it has only 81 atomic formulas a textual representation contains approximately 500 000 characters.

However, we have found a quantifier-free description of the image of the function describing the Enneper surface. One of the atomic formulas contained in the result is the equation $p = 0$. But this fact does not imply any direction of the equivalence we want to prove.

The obtained result is in fact to large, to eliminate the universal quantifiers from

$$\forall x \forall y \forall z \left(\varphi(x, y, z) \Leftrightarrow p(x, y, z) = 0\right)$$

with one of our quantifier elimination procedures. Actually, we are not able to eliminate at least one universal quantifier. Thus we use REDLOG to simplify φ'.

An analysis of the formula yields that the result formula is a disjunction of nine subformulas. Five of these nine are simply conjunctions of atomic formulas and the remaining four subformulas are essentially conjunctive normal forms.

For a simplification of the formula we thus simplify each constituent of the top-level disjunctions with the Gröbner simplifier for conjunctive normal forms. In a first step, we take the Gröbner simplifier without a factorization of atomic formulas. In a second step we use the Gröbner simplifier with the automatic factorization option twice for each constituent of the top-level disjunction. After these simplification we obtain a formula with 50 atomic formulas and about 30000 characters. All these simplifications takes 434 s.

The automatic tableau method can simplify some of the resulting conjunctive normal forms in about 1 s. After these simplifications we get a disjunction of eight conjunctive normal forms. From these eight conjunctive normal forms are six pure conjunctions of atomic formulas. The two other ones are conjunctions containing atomic formulas and only one disjunction of two atomic formulas. Using the equivalents

$$(\alpha \leqslant 0 \vee \beta = 0) \longleftrightarrow (\alpha \cdot \beta^2 \leqslant 0) \quad \text{and} \quad (\alpha = 0 \vee \beta = 0) \longleftrightarrow (\alpha \cdot \beta = 0)$$

we can simplify the formula to a disjunctive normal form ψ with constituents ψ_i.

Let Θ_i be the set of atomic formulas of ψ_i. Using the Gröbner simplifier we can simplify $p = 0$ wrt. the theory Θ_i to "true." This means that $\psi_i \longrightarrow p = 0$ and hence $\psi \longrightarrow p = 0$.

Using a tool of REDLOG, which counts the frequencies of all included atomic formulas we analyze our result. The most frequently occurring atomic formulas in ψ are

$$x = 0, \quad y = 0, \quad z - 3 \neq 0, \quad \text{and} \quad z \neq 0$$

each of them with more than three occurrences.

This observation suggests, that we should study a special case of the implication. For arbitrary formulas Ψ, Γ and formulas $\alpha_1, \ldots, \alpha_n$ with $\alpha_1 \wedge \cdots \wedge \alpha_n$ contradictive the following equivalence holds:

$$(\Psi \Rightarrow \Gamma) \longleftrightarrow (\alpha_1 \wedge \cdots \wedge \alpha_n) \vee (\Psi \Rightarrow \Gamma)$$
$$\longleftrightarrow (\neg \alpha_1 \wedge \Psi) \Rightarrow \Gamma) \wedge \ldots \wedge (\neg \alpha_n \wedge \Psi) \Rightarrow \Gamma)$$
$$\longleftrightarrow (\neg \alpha_1 \Rightarrow (\Psi \Rightarrow \Gamma) \wedge \ldots \wedge (\neg \alpha_n \Rightarrow (\Psi \Rightarrow \Gamma)).$$

Using $x \neq 0$, $y \neq 0$, $z \neq 0$, $z \neq 3$, and $(x = 0 \vee y = 0 \vee z = 0 \vee z = 3)$ for the α_i we can split our implication in a conjunction of five implications. Four of them include simple equations for one of the variables.

For proving the cases including an equation we substitute the values 0 and 3 for the respective variable in the formula $p = 0 \Rightarrow \psi$ and then we eliminate the universal closure of the substitution result. The timings for this elimination are summarized in Table 1. The elimination result is in all cases "true."

Table 1. Elimination times for the special cases.

	$x = 0$	$y = 0$	$z = 0$	$z = 3$
QEVTS	failed	0.7 s	0.1 s	failed
QEPCAD	1.0 s	23.0 s	1.0 s	1.0 s

After we have checked the special cases, we can exclude these cases by simplifying the formula ψ with respect to the theory $\{x \neq 0, y \neq 0, z \neq 0, z \neq 3\}$. This yields a disjunction of two conjunctions. One of the remaining constituents is contradictive as shown by the elimination of the existential closure with QEVTS in 0.8 s. Neither QEPCAD nor QERRC are able to prove this fact. The remaining constituent is a conjunction $\beta_1 \wedge \cdots \wedge \beta_4$ of atomic formulas, where β_1 is the equation $p = 0$. Instead of proving the complete implication, which is not possible with our quantifier elimination procedures, we prove $p = 0 \Rightarrow \beta_i$ for each atomic formula β_i in the conjunction. The first implication holds trivially.

While trying to eliminate one of the quantifiers REDLOG applies a heuristic to decrease the degree of the variables x and y. Namely, it replaces each occurrence of x^2, and y^2 by x and y respectively, adding the additional premise $x > 0 \wedge y > 0$. Finally the quantifier elimination fails. However, after the degree reduction QEPCAD is able to eliminate all universal quantifier. For the elimination it is necessary to give the quantifiers in the order $\forall z \forall y \forall x$. Altogether the three eliminations take 34 s. All results are "true" and thus we have finally proven the equivalence.

7 Conclusions

After approximately 50 years of development, quantifier algorithms can nowaday be used to solve geometrical problems. Even if a first approach fails, the several quantifier elimination methods can be applied in interaction with the user to appropriate subproblems, finally solving the problem. Our results show, that among the three considered algorithms there is no best algorithm, solving all problems. Furthermore they show that not only quantifier elimination algorithms are necessary but even powerful simplification algorithms and a wide spectrum of tools for analyzing and decomposing formulas. Important for the user is a common interface to all algorithms, like REDLOG provides one.

Acknowledgment

We acknowledge the influence of V. Weispfenning on the discussion of the Steiner-Lehmus theorem and the Enneper surface. In particular helpful were his many useful hints for the analysis of the real Enneper surface.

References

1. Dennis S. Arnon, George E. Collins, and Scott McCallum. Cylindrical algebraic decomposition I: The basic algorithm. *SIAM Journal on Computing*, 13(4):865–877, November 1984.
2. Eberhard Becker. Sums of squares and quadratic forms in real algebraic geometry. In *Cahiers du Séminaire d'Historie de Mathématiques*, volume 1, pages 41–57. Université Pierre et Marie Curie, Laboratoire de Mathématiques Fondamentales, Paris, 1991.

3. Eberhard Becker and Thorsten Wörmann. On the trace formula for quadratic forms. In William B. Jacob, Tsit-Yuen Lam, and Robert O. Robson, editors, *Recent Advances in Real Algebraic Geometry and Quadratic Forms*, volume 155 of *Contemporary Mathematics*, pages 271–291. American Mathematical Society, Providence, Rhode Island, 1994. Proceedings of the RAGSQUAD Year, Berkeley, 1990–1991.

4. Christopher W. Brown. Simplification of truth-invariant cylindrical algebraic decompositions. In Oliver Gloor, editor, *Proceedings of the 1998 International Symposium on Symbolic and Algebraic Computation (ISSAC 98)*, pages 295–301, Rostock, Germany, August 1998. ACM Press, New York.

5. Bruno Buchberger. *Ein Algorithmus zum Auffinden der Basiselemente des Restklassenringes nach einem nulldimensionalen Polynomideal*. Doctoral dissertation, Mathematical Institute, University of Innsbruck, Innsbruck, Austria, 1965.

6. George E. Collins. Quantifier elimination for the elementary theory of real closed fields by cylindrical algebraic decomposition. In H. Brakhage, editor, *Automata Theory and Formal Languages. 2nd GI Conference*, volume 33 of *Lecture Notes in Computer Science*, pages 134–183. Gesellschaft für Informatik, Springer-Verlag, Berlin, Heidelberg, New York, 1975.

7. George E. Collins and Hoon Hong. Partial cylindrical algebraic decomposition for quantifier elimination. *Journal of Symbolic Computation*, 12(3):299–328, September 1991.

8. David Cox, John Little, and Donald O'Shea. *Ideals, Varieties and Algorithms*. Undergraduate Texts in Mathematics. Springer-Verlag, New York, Berlin, Heidelberg, 1992.

9. Andreas Dolzmann. Reelle Quantorenelimination durch parametrisches Zählen von Nullstellen. Diploma thesis, Universität Passau, D-94030 Passau, Germany, November 1994.

10. Andreas Dolzmann and Thomas Sturm. Redlog: Computer algebra meets computer logic. *ACM SIGSAM Bulletin*, 31(2):2–9, June 1997.

11. Andreas Dolzmann and Thomas Sturm. Simplification of quantifier-free formulae over ordered fields. *Journal of Symbolic Computation*, 24(2):209–231, August 1997.

12. Andreas Dolzmann and Thomas Sturm. P-adic constraint solving. Technical Report MIP-9901, FMI, Universität Passau, D-94030 Passau, Germany, January 1999. To appear in the proceedings of the ISSAC 99.

13. Andreas Dolzmann, Thomas Sturm, and Volker Weispfenning. A new approach for automatic theorem proving in real geometry. *Journal of Automated Reasoning*, 21(3):357–380, 1998.

14. Andreas Dolzmann, Thomas Sturm, and Volker Weispfenning. Real quantifier elimination in practice. In B. H. Matzat, G.-M. Greuel, and G. Hiss, editors, *Algorithmic Algebra and Number Theory*, pages 221–247. Springer, Berlin, 1998.

15. Charles Hermite. Remarques sur le théorème de M. Sturm. In Emile Picard, editor, *Œuvres des Charles Hermite*, volume 1, pages 284–287. Gauthier-Villars, Paris, 1905.

16. Hoon Hong. An improvement of the projection operator in cylindrical algebraic decomposition. In Shunro Watanabe and Morio Nagata, editors, *Proceedings of the International Symposium on Symbolic and Algebraic Computation (ISSAC 90)*, pages 261–264, Tokyo, Japan, August 1990. ACM Press, New York.

17. Hoon Hong. Simple solution formula construction in cylindrical algebraic decomposition based quantifier elimination. In Paul S. Wang, editor, *Proceedings of the International Symposium on Symbolic and Algebraic Computation (ISSAC 92)*, pages 177–188, Berkeley, CA, July 1992. ACM Press, New York.

18. Rüdiger Loos and Volker Weispfenning. Applying linear quantifier elimination. *The Computer Journal*, 36(5):450–462, 1993. Special issue on computational quantifier elimination.
19. Paul Pedersen. *Counting Real Zeros*. Ph.D. dissertation, Courant Institute of Mathematical Sciences, New York, 1991.
20. Paul Pedersen, Marie-Françoise Roy, and Aviva Szpirglas. Counting real zeroes in the multivariate case. In F. Eysette and A. Galigo, editors, *Computational Algebraic Geometry*, volume 109 of *Progress in Mathematics*, pages 203–224. Birkhäuser, Boston, Basel; Berlin, 1993. Proceedings of the MEGA 92.
21. Alfred Tarski. A decision method for elementary algebra and geometry. Technical report, RAND, Santa Monica, CA, 1948.
22. Volker Weispfenning. The complexity of linear problems in fields. *Journal of Symbolic Computation*, 5(1–2):3–27, February–April 1988.
23. Volker Weispfenning. Quantifier elimination for real algebra—the quadratic case and beyond. *Applicable Algebra in Engineering Communication and Computing*, 8(2):85–101, February 1997.
24. Volker Weispfenning. A new approach to quantifier elimination for real algebra. In B. F. Caviness and J. R. Johnson, editors, *Quantifier Elimination and Cylindrical Algebraic Decomposition*, Texts and Monographs in Symbolic Computation, pages 376–392. Springer, Wien, New York, 1998.

Automated Discovering and Proving for Geometric Inequalities*

Lu Yang, Xiaorong Hou, and Bican Xia

Chengdu Institute of Computer Applications
Academia Sinica, Chengdu 610041, China
luyang@guangztc.edu.cn, xhou@usa.net, bcxia@263.net

Abstract. Automated discovering and proving for geometric inequalities have been considered a difficult topic in the area of automated reasoning for many years. Some well-known algorithms are complete theoretically but inefficient in practice, and cannot verify non-trivial propositions in batches. In this paper, we present an efficient algorithm to discover and prove a class of inequality-type theorems automatically by combining discriminant sequence for polynomials with Wu's elimination and a partial cylindrical algebraic decomposition. Also this algorithm is applied to the classification of the real physical solutions of geometric constraint problems. Many geometric inequalities have been discovered by our program, DISCOVERER, which implements the algorithm in Maple.

1 Introduction

In the last 20 years, the efficiency of computer algorithms for automated reasoning in both algebraic and logic approaches has greatly increased. One has reason to believe that computer will play a much more important role in reasoning sciences in the coming century. People will be able to prove theorems class by class instead of one by one. Since Tarski's well-known work [21], *A Decision Method for Elementary Algebra and Geometry*, published in the early 1950's, the algebraic approaches have made remarkable progress in automated theorem proving. Tarski's decision algorithm, which could not be used to verify any non-trivial algebraic or geometric propositions in practice because of its very high computational complexity, has only got theoretical significance. Some substantial progress was made by Seidenberg [20], Collins [12] and others afterwards, but it was still far away from mechanically proving non-trivial theorems batch by batch, and even class by class. The situation did not change until Wu [25, 26] proposed in 1978 a new decision procedure for proving geometry theorems of "equality type," i.e. the hypotheses and conclusions of the statements consist of polynomial equations only. This is a very efficient method for mechanically proving elementary geometry theorems (of equality type). Chou [8] has successfully implemented Wu's method for 512 examples which include almost all the

---

* This work is supported in part by CAS and CNRS under a cooperative project between CICA and LEIBNIZ, and in part by the National "973" Project of China.

well-known or historically interesting theorems in elementary geometry, and it was reported that for most of the examples the CPU time spent was only few seconds each, or less than 1 second!

The success of Wu's method has inspired in the world the advances of the algebraic approach [18, 22, 23] to automated theorem proving. In the past 20 years, some efficient provers have been developed based on different principles such as Gröbner Bases [5, 6] and Parallel Numerical Method [32]. Especially, Zhang and his colleagues [11, 29] gave algorithms and programs for automatically producing readable proofs of geometry theorems. The achievement has made the studies in automated proving enter a new stage that the proofs created by machines can compare with those by human being, while the decision problem was playing a leading role before. It has also important applications to mathematics mechanization and computer aided instruction (CAI).

Those methods mentioned above are mainly valid to equality-type theorem proving. As for automated inequality discovering and proving, the progress has been slow for many years. Chou and others [9, 10] made helpful approaches in this aspect by combining Wu's method with CAD (Cylindrical Algebraic Decomposition) algorithm or others. Recently, a so-called "dimension-decreasing" algorithm has been introduced by Yang [27]. Based on this algorithm, a program called "BOTTEMA" was implemented on a PC. More than 1000 algebraic and geometric inequalities including hundreds open problems have been verified in this way. The total CPU time spent for proving 120 basic inequalities from Bottema's monograph "Geometric Inequalities," on a Pentium/200, was 20-odd seconds only. To our knowledge, this is the first practical prover which can verify non-trivial inequality-type propositions in batches. As one of the applications, this makes it practical to solve a global-optimization problem by means of verifying finitely many inequalities.

The reason for automated inequality discovering and proving to be a difficult topic is that the concerning algorithms depend fundamentally upon real algebra and real geometry, and some well-known algorithms are complete theoretically but inefficient in practice. In [30, 34, 31], Yang and others introduced a powerful tool, complete discrimination system (CDS) of polynomials, for inequality reasoning. By means of CDS, together with Wu's elimination and a partial CAD algorithm, we present an efficient algorithm to discover and prove a class of inequality-type theorems automatically. A program called "DISCOVERER" was implemented in Maple that is able to discover new inequalities automatically, without requiring us to put forward any conjectures beforehand. For example, by means of this program, we have re-discovered 37 inequalities in the first chapter of the famous monograph [19], "Recent Advances in Geometric Inequalities," and found three mistakes there.

2 The Problem

In general, the problem we study here is:

Give the necessary and sufficient condition which the parameters u must satisfy for the following system TS to have exactly n distinct real solution(s) (or simply, to have real solution(s)):

$$TS : \begin{cases} f_1(u, x_1) = 0, \\ f_2(u, x_1, x_2) = 0, \\ \cdots \cdots \\ f_s(u, x_1, x_2, \ldots, x_s) = 0, \\ g_1(u, X) \geq 0, g_2(u, X) \geq 0, \ldots, g_t(u, X) \geq 0 \end{cases} \tag{1}$$

where $u = (u_1, u_2, \ldots, u_d)$, $X = (x_1, x_2, \ldots, x_s)$,

$$f_i \in Z(u)[x_1, \ldots, x_i], \quad 1 \leq i \leq s,$$
$$g_j \in Z(u)[x_1, \ldots, x_s], \quad 1 \leq j \leq t,$$

and $\{f_1, f_2, \ldots, f_s\}$ is a "normal ascending chain" [33,34] which is "simplicial" [33,34] w.r.t. each g_j ($1 \leq j \leq t$) (also see Section 3 of the present paper for the definitions of these two concepts). Some of the inequalities in (1) may be strict ones.

Before we go further, we illustrate this class of problems by an example:

Give the necessary and sufficient condition for the existence of a triangle with elements s, r, R, where s, r, R are the half perimeter, inradius and circumradius, respectively.

Let a, b, c be the lengths of sides and, without loss of generality, let $s = 1$; we have

$$PS' : \begin{cases} p_1 = a + b + c - 2 = 0, \\ p_2 = r^2 - (1-a)(1-b)(1-c) = 0, \\ p_3 = 4rR - abc = 0, \\ 0 < a < 1, 0 < b < 1, 0 < c < 1, 0 < r, 0 < R. \end{cases}$$

It is easy to see that PS' is equivalent to the following system:

$$TS' : \begin{cases} f_1(r, R, a) = a^3 - 2a^2 + (r^2 + 4rR + 1)a - 4rR = 0, \\ f_2(r, R, a, b) = ab^2 + a(a-2)b + 4rR = 0, \\ f_3(r, R, a, b, c) = a + b + c - 2 = 0, \\ 0 < a < 1, 0 < b < 1, 0 < c < 1, 0 < r, 0 < R. \end{cases}$$

where a, b, c are the variables and r, R the parameters. What we need to do is to find the condition satisfied by r, R, under which TS' has real solution(s). Obviously, this is a problem of the class defined above.

By the algorithm presented in this paper, we can easily find the condition

$$s^4 + 2r^2s^2 - 4R^2s^2 - 20rRs^2 + 12r^3R + 48r^2R^2 + r^4 + 64rR^3 \leq 0,$$

which is called the "Fundamental Inequality" [19] for triangles.

3 Concepts and Notations

In this section, we review some concepts and notations from [33,34] and [30,7, 17].

Given a polynomial g and a triangular set $\{f_1, f_2, \ldots, f_s\}$, let

$$r_s = g, \quad r_{s-i} = \text{resultant}(r_{s-i+1}, f_{s-i+1}, x_{s-i+1}), \quad i = 1, 2, \ldots, s,$$
$$q_s = g, \quad q_{s-i} = \text{prem}(q_{s-i+1}, f_{s-i+1}, x_{s-i+1}), \quad i = 1, 2, \ldots, s;$$

and $\text{res}(g, f_s, \ldots, f_1)$ and $\text{prem}(g, f_s, \ldots, f_1)$ denote r_0 and q_0, respectively.

Definition 3.1 (Normal Ascending Chain [33,34]). Given a triangular set $\{f_1, f_2, \ldots, f_s\}$, by I_i ($i = 1, 2, \ldots, s$) denote the leading coefficient of f_i in x_i. The triangular set $\{f_1, f_2, \ldots, f_s\}$ is called a *normal ascending chain* if

$$I_1 \neq 0, \quad \text{res}(I_i, f_{i-1}, \ldots, f_1) \neq 0, \quad i = 2, \ldots, s.$$

Definition 3.2 (Simplicial [33,34]). A normal ascending chain $\{f_1, f_2, \ldots, f_s\}$ is simplicial with respect to a polynomial g if either $\text{prem}(g, f_s, \ldots, f_1) = 0$ or $\text{res}(g, f_s, \ldots, f_1) \neq 0$.

Definition 3.3 (Discrimination Matrix [30]). Given a polynomial with general symbolic coefficients

$$f(x) = a_0 x^n + a_1 x^{n-1} + \cdots + a_n,$$

the following $2n \times 2n$ matrix in terms of the coefficients

$$\begin{bmatrix}
a_0 & a_1 & a_2 & \cdots & a_n & & & \\
0 & na_0 & (n-1)a_1 & \cdots & a_{n-1} & & & \\
& a_0 & a_1 & \cdots & a_{n-1} & a_n & & \\
& 0 & na_0 & \cdots & 2a_{n-2} & a_{n-1} & & \\
& & \cdots & \cdots & & & & \\
& & \cdots & \cdots & & & & \\
& & & a_0 & a_1 & a_2 & \cdots & a_n \\
& & & 0 & na_0 & (n-1)a_1 & \cdots & a_{n-1}
\end{bmatrix}$$

is called the *discrimination matrix* of $f(x)$, and denoted by $\text{Discr}(f)$. By d_k or $d_k(f)$ denote the determinant of the submatrix of $\text{Discr}(f)$, formed by the first k rows and the first k columns for $k = 1, 2, \ldots, 2n$.

Definition 3.4 (Discriminant Sequence [30]). Let $D_0 = 1$ and $D_k = d_{2k}$, $k = 1, \ldots, n$. We call the $n + 1$-tuple

$$D_0, D_1, D_2, \ldots, D_n$$

the *discriminant sequence* of $f(x)$, and denote it by $\text{DiscrList}(f)$.

The last term D_n is also called the *discriminant* of f with respect to x, and denoted by $\text{Discrim}(f, x)$. It should be noted that the definition of $\text{Discrim}(f, x)$ here is little different from the others which are D_n/a_0.

Definition 3.5 (Sign List [30]). We call the list

$$[\text{sign}(A_0), \text{sign}(A_1), \text{sign}(A_2), \ldots, \text{sign}(A_n)]$$

the *sign list* of a given sequence $A_0, A_1, A_2, \ldots, A_n$, where

$$\text{sign}(x) = \begin{cases} 1, & x > 0, \\ 0, & x = 0, \\ -1, & x < 0. \end{cases}$$

Definition 3.6 (Revised Sign List [30]). Given a sign list $[s_1, s_2, \ldots, s_n]$, we construct a new list

$$[t_1, t_2, \ldots, t_n]$$

(which is called the *revised sign list*) as follows:

- If $[s_i, s_{i+1}, \ldots, s_{i+j}]$ is a section of the given list, where

$$s_i \neq 0, s_{i+1} = \cdots = s_{i+j-1} = 0, s_{i+j} \neq 0,$$

  then, we replace the subsection

$$[s_{i+1}, \ldots, s_{i+j-1}]$$

  with the first $j - 1$ terms of [1]

$$[-s_i, -s_i, \ s_i, \ s_i, -s_i, -s_i, \ s_i, \ s_i, \ldots].$$

- Otherwise, let $t_k = s_k$, i.e. no change for other terms.

Example 3.1. The revision of the sign-list

$$[1, -1, 0, 0, 0, 0, 0, 1, 0, 0, -1, -1, 1, 0, 0, 0]$$

is $[1, -1, 1, 1, -1, -1, 1, 1, -1, -1, -1, -1, 1, 0, 0, 0]$.

Theorem 3.1 ([30]). Given a polynomial $f(x)$ with real coefficients

$$f(x) = a_0 x^n + a_1 x^{n-1} + \cdots + a_n,$$

if the number of the sign changes of the revised sign list of

$$D_0, D_1(f), D_2(f), \ldots, D_n(f)$$

is ν, then the number of the pairs of distinct conjugate imaginary roots of $f(x)$ equals ν. Furthermore, if the number of non-vanishing members of the revised sign list is l, then the number of the distinct real roots of $f(x)$ equals $l - 1 - 2\nu$.

---

[1] that is, let $t_{i+r} = (-1)^{[\frac{r+1}{2}]} \cdot s_i, \ r = 1, 2, \ldots, j - 1.$

Definition 3.7 (Generalized Discrimination Matrix). Given two polynomials $g(x)$ and $f(x)$ where

$$f(x) = a_0 x^n + a_1 x^{n-1} + \cdots + a_n,$$

let [2]

$$r(x) = \text{rem}(f'g, f, x) = b_0 x^{n-1} + b_1 x^{n-2} + \cdots + b_{n-1}.$$

The following $2n \times 2n$ matrix

$$\begin{bmatrix} a_0 \ a_1 \ a_2 \cdots \ a_n & & & \\ 0 \ b_0 \ b_1 \cdots \ b_{n-1} & & & \\ & a_0 \ a_1 \cdots a_{n-1} \ a_n & \\ & 0 \ b_0 \cdots b_{n-2} \ b_{n-1} & \\ & \cdots \ \cdots & \\ & \cdots \ \cdots & \\ & & a_0 \ \ a_1 \ \ a_2 \ \cdots \ a_n \\ & & 0 \ \ b_0 \ \ b_1 \ \cdots b_{n-1} \end{bmatrix}$$

is called the *generalized discrimination matrix* of $f(x)$ with respect to $g(x)$, and denoted by $\text{Discr}(f, g)$.

Definition 3.8 (Generalized Discriminant Sequence [7, 17]). Given two polynomials $f(x)$ and $g(x)$, let the notations be as above, and $D_0 = 1$. Denote by

$$D_1(f, g), D_2(f, g), \ldots, D_n(f, g)$$

the even order principal minors of $\text{Discr}(f, g)$. We call

$$D_0, D_1(f, g), D_2(f, g), \ldots, D_n(f, g)$$

the generalized discriminant sequence of $f(x)$ with respect to $g(x)$, and denote it by $\text{GDL}(f, g)$. Clearly, $\text{GDL}(f, 1) = \text{DiscrList}(f)$.

Theorem 3.2 ([7, 17]). Given two polynomials $f(x)$ and $g(x)$, if the number of the sign changes of the revised sign list of $\text{GDL}(f, g)$ is ν, and the number of non-vanishing members of the revised sign list is l, then

$$l - 1 - 2\nu = c(f, g_+) - c(f, g_-),$$

where [3]

$$c(f, g_+) = \text{card}(\{x \in R | f(x) = 0, g(x) > 0\}),$$

$$c(f, g_-) = \text{card}(\{x \in R | f(x) = 0, g(x) < 0\}).$$

---

[2] By $\text{rem}(a(x), b(x), x)$ denote the remainder of $a(x)$ divided by $b(x)$.
[3] By $\text{card}(\cdot)$ represent the cardinal number of a set.

4 A Theoretical Algorithm

In this section, we give an algorithm which has theoretical significance. The practical algorithm is given in the next section.

Let

$$ps = \{p_i | 1 \leq j \leq n\}$$

be a nonempty, finite set of polynomials. We define

$$\text{mset}(ps) = \{1\} \cup \{p_{i_1} p_{i_2} \cdots p_{i_k} | 1 \leq k \leq n, 1 \leq i_1 < i_2 < \cdots < i_k \leq n\}.$$

For example, if $ps = \{p_1, p_2, p_3\}$, then

$$\text{mset}(ps) = \{1, p_1, p_2, p_3, p_1p_2, p_1p_3, p_2p_3, p_1p_2p_3\}.$$

Given system TS as (1), we define

$$P_{s+1} = \{g_1, g_2, \ldots, g_t\};$$
$$U_i = \bigcup_{q \in \text{mset}(P_{i+1})} \text{GDL}(f_i, q),$$
$$P_i = \{h(u, x_1, \ldots, x_{i-1}) | h \in U_i\}, \quad \text{for } i = s, s-1, \ldots, 2;$$
$$P_1(g_1, g_2, \ldots, g_t) = \{h(u) | h \in U_1\},$$

where U_i means the set consisting of all the polynomials in each $\text{GDL}(f_i, q)$ with q belonging to $\text{mset}(P_{i+1})$. Analogously, we can define $P_1(g_1, \ldots, g_j)$ $(1 \leq j \leq t)$.
Then we have

Theorem 4.1. The necessary and sufficient condition for system TS to have a given number of distinct real solution(s) can be expressed in terms of the signs of the polynomials in $P_1(g_1, g_2, \ldots, g_t)$.

Proof. First of all, we regard f_s and every g_i as polynomials in x_s. By Theorems 3.1 and 3.2 we know that under constraints $g_i \geq 0, 1 \leq i \leq t$, the number of distinct real solutions of $f_s = 0$ can be determined by the signs of polynomials in P_s; let h_j $(1 \leq j \leq l)$ be the polynomials in P_s; we regard every h_j and f_{s-1} as polynomials in x_{s-1}, repeat the same argument as what we did for f_s and g_i's, then we get that, under constraints $g_i \geq 0, 1 \leq i \leq t$, the number of distinct real solutions of $f_s = 0, f_{s-1} = 0$ can be determined by the signs of polynomials in P_{s-1}; do the same argument until $P_1(g_1, g_2, \ldots, g_t)$ is employed. Because the conditions obtained in each step are necessary and sufficient, the theorem holds.

Now, theoretically speaking, we can get the necessary and sufficient condition for system TS to have (exactly n distinct) real solution(s) as follows:

Step 1 Compute $P_1(g_1, g_2, \ldots, g_t)$, the set of polynomials in parameters, which is defined above for TS.

Step 2 By the algorithm of PCAD [2, 3], we can obtain P_1-*invariant cad* D of parameter space $\mathbf{R}^d$ and its *cylindrical algebraic sample (cas)* S [24]. Roughly speaking, D is a finite set of cells that each polynomial of P_1 keeps its sign in each cell; and S is a finit set of points obtained by taking from each cell one point at least, which is called the *sample point* of the cell.

Step 3 For each cell c in D and its sample point $s_c \in S$, substitute s_c into TS (denote it by $TS(s_c)$). Compute the number of distinct real solutions of system $TS(s_c)$, in which polynomials all have constant coefficients now. At the same time, compute the signs of polynomials in $P_1(g_1, g_2, \ldots, g_t)$ on this cell by substituting s_c into them respectively. Record the signs of polynomials in $P_1(g_1, g_2, \ldots, g_t)$ when the number of distinct real solutions of system $TS(s_c)$ equals to the required number n (or when the number > 0, if we are asked to find the condition for TS to have real solution). Obviously, the signs of polynomials in $P_1(g_1, g_2, \ldots, g_t)$ on cell c form a first order formula, denoted by Φ_c.

Step 4 If, in step 3, all we have recorded are $\Phi_{c_1}, \ldots, \Phi_{c_k}$, then $\Phi = \Phi_{c_1} \vee \cdots \vee \Phi_{c_k}$ is what we want.

5 The Practical Algorithm

The algorithm given in Section 4 is not practical in many cases since $P_1(g_1, g_2, \ldots, g_t)$ usually has too many polynomials. Because not all those polynomials are necessary for expressing the condition we want, so we have to give an efficient algorithm to select those necessary polynomials from $P_1(g_1, g_2, \ldots, g_t)$.

Theorem 5.1. Let system TS be given as above. If *PolySet* is a finite set of polynomials in parameters u, e.g.

$$PolySet = \{q_i(u) \in Z[u_1, \ldots, u_d] | 1 \leq i \leq k\},$$

then, by the algorithm of PCAD, we can obtain a *PolySet*-invariant cad D of parameter space $\mathbf{R}^d$ and its cas. If *PolySet* satisfies that

1. the number of distinct real solutions of system TS is invariant in the same cell and
2. the number of distinct real solutions of system TS in two distinct cells C_1, C_2 is the same if *PolySet* has the same sign in C_1, C_2,

then the necessary and sufficient conditions for system TS to have exactly n distinct real solution(s) can be expressed by the signs of the polynomials in *PolySet*. If *PolySet* satisfies (1) only, then some necessary conditions for system TS to have exactly n distinct real solution(s) can be expressed by the signs of the polynomials in *PolySet*.

Proof. We replace parameters u in TS with each sample point respectively. Because D is *PolySet*-invariant and *PolySet* satisfies (1), we can record the signs of polynomials in *PolySet* and the number of distinct real solutions of TS on each cell respectively. Choose all those cells on which TS has n distinct real solution(s). The signs of polynomials in $PolySet$ on those cells form a first order formula, e.g.

$$\Phi = \Phi_1 \vee \Phi_2 \vee \cdots \vee \Phi_l.$$

where each Φ_i represents the signs of polynomials in $PolySet$ on a certain cell on which TS has n distinct real solution(s). We will say that Φ is the condition we want.

Given a parameter $a = (a_1, \ldots, a_d)$, which must fall into a certain cell, if $TS(a)$ has n distinct real solution(s), then a must belong to a cell on which TS has n distinct real solution(s), i.e. a must satisfy a certain formula Φ_i; on the contrary, if a satisfies a certain formula Φ_i, then, because TS has n distinct real solution(s) on the cell represented by Φ_i and $PolySet$ satisfies (2), we thus know that TS must have n distinct real solution(s) on the cell which a belongs to. Thus, the theorem holds.

Given system TS. For every f_i, let

$$R_1 = \mathrm{Discrim}(f_1, x_1),$$
$$R_i = \mathrm{res}(\mathrm{Discrim}(f_i, x_i), f_{i-1}, f_{i-2}, \ldots, f_1), \quad i \geq 2,$$

where the definition of $\mathrm{res}(\cdots\cdots)$ can be found at the beginning of Section 3. For every g_j, let

$$Rg_j = \mathrm{res}(g_j, f_s, f_{s-1}, f_{s-2}, \ldots, f_1).$$

We define

$$BPs = \{R_i | 1 \leq i \leq s\} \bigcup \{Rg_j | 1 \leq j \leq t\}.$$

Clearly, $BPs \subseteq P_1(g_1, g_2, \ldots, g_t)$.

Theorem 5.2. Given system TS, BPs is defined as above. If we consider only those cells which are homeomorphic to $\mathbf{R}^d$ and do not consider those cells which are homeomorphic to $\mathbf{R}^k$ ($k < d$) when use PCAD, then BPs satisfies (1) in theorem 5.1, so the necessary conditions (if we omit the parameters on those cells homeomorphic to $\mathbf{R}^k$ ($k < d$)) for system TS to have n distinct real solution(s) can be expressed by the signs of the polynomials in BPs.

Proof. By PCAD, we can get a BPs-invariant cad of $\mathbf{R}^d$ and its cas. Because we consider only those cells which are homeomorphic to $\mathbf{R}^d$, given a cell C, the signs of each R_i and Rg_j on C are invariant and do not equal to 0.

First of all, by the definition of R_1, the sign of R_1 on C is invariant means that the number of real solutions of $f_1(u, x_1)$ is invariant on C; then we regard $f_2(u, x_1, x_2)$ as a polynomial in x_2, because on C,

$$f_1(u, x_1) = 0, \quad R_2 = \mathrm{res}(\mathrm{Discrim}(f_2, x_2), f_1, x_1) \neq 0,$$

thus $\mathrm{Discrim}(f_2, x_2) \neq 0$ on C, i.e. if we replace x_1 in f_2 with the roots of f_1, the number of real solutions of f_2 is invariant. That is to say, the signs of R_1 and R_2 are invariant on C means the number of real solutions of $f_1 = 0, f_2 = 0$

is invariant on C; now, it's easy to see that the signs of $R_1, \ldots, R_s$ are invariant on C means the number of real solutions of $f_1 = 0, \ldots, f_s = 0$ is invariant on C.

Secondly, by the definition of Rg_j, we know that $Rg_j \neq 0$ means, on C, if we replace $x_1, \ldots, x_s$ in g_j with the roots of $f_1 = 0, \ldots, f_s = 0$, then the sign of g_j is invariant. That completes the proof.

By theorem 5.2, we can start our algorithm from BPs as follows:

Step 1 Let $PolySet = BPs$, $i = 1$.

Step 2 By the algorithm of PCAD [2,3], compute a $PolySet$-invariant cad D of the parameter space $\mathbf{R}^d$ and its cylindrical algebraic sample (cas) S [24]. In this step, we consider only those cells homeomorphic to $\mathbf{R}^d$ and do not consider those homeomorphic to $\mathbf{R}^k$ ($k < d$), i.e., all those cells in D are homeomorphic to $\mathbf{R}^d$ and all sample points in S are taken from cells in D.

Step 3 For each cell c in D and its sample point $s_c \in S$, substitute s_c into TS (denote it by $TS(s_c)$). Compute the number of distinct real solutions of system $TS(s_c)$, in which polynomials all have constant coefficients now. At the same time, compute the signs of polynomials in $PolySet$ on this cell by substituting s_c into them respectively. Obviously, the signs of polynomials in $PolySet$ on cell c form a first order formula, denoted by Φ_c. When all the $TS(s_c)$'s are computed, let

$$set_1 = \{\Phi_c|\ TS \text{ has } n \text{ distinct real solution(s) on } c\},$$

$$set_0 = \{\Phi_c|\ TS \text{ has not } n \text{ distinct real solution(s) on } c\}.$$

Step 4 Decide whether all the recorded Φ_c's can form a necessary and sufficient condition or not by verifying whether $set_1 \cap set_0$ is empty or not (because of Theorem 5.1 and Theorem 5.2). If $set_1 \cap set_0 = \emptyset$, go to Step 5; If $set_1 \cap set_0 \neq \emptyset$, let

$$PolySet = PolySet \cup P_1(g_1, \ldots, g_i), \ i = i + 1,$$

and go back to Step 2.

Step 5 If $set_1 = \{\Phi_{c_1}, \ldots, \Phi_{c_m}\}$, then $\Phi = \Phi_{c_1} \vee \cdots \vee \Phi_{c_m}$ is what we want.

Remark. In order to make our algorithm practical, we do not consider the "boundaries" when use PCAD. So, the condition obtained by this algorithm is a necessary and sufficient one if we omit the parameters on the "boundaries." In another word, all the conditions obtained by our program DISCOVERER, which implements this algorithm, should be understood as follows: if the strict inequalities hold, the conditions hold; if equalities hold, more discussion is needed.

Now, we consider following problem:

Give the necessary and sufficient condition which the parameters u must satisfy for the following system PS to have (exactly n distinct) real solution(s)

$$PS : \begin{cases} h_1(u, X) = 0, h_2(u, X) = 0, \ldots, h_s(u, X) = 0 \\ g_1(u, X) \geq 0, g_2(u, X) \geq 0, \ldots, g_t(u, X) \geq 0 \end{cases} \tag{2}$$

where u means $u_1, u_2, \ldots, u_d$, treated as parameters; X means $x_1, x_2, \ldots, x_s$, treated as variables. That is to say,

$$h_i, g_j \in Z(u_1, \ldots, u_d)[x_1, \ldots, x_s], \quad 1 \leq i \leq s, \; 1 \leq j \leq t.$$

First of all, by Wu's elimination [25, 26], we can reduce the system $h_1(u, X) = 0$, $h_2(u, X) = 0$, $\ldots$, $h_s(u, X) = 0$ to some "triangular sets." Then, if necessary, by WR algorithm [33, 34], make these triangular sets into normal ascending chains in which every chain is simplicial w.r.t. each g_j $(1 \leq j \leq t)$. So, under some nondegenerate conditions, we can reduce PS to some TS's. These nondegenerate conditions, however, do not bring any new limitations on our algorithm because we do not consider "boundaries" when use PCAD and all those parameters which make degenerate conditions true are contained in "boundaries." Another situation we do have to handle is some TS's reduced from PS may have real solutions with dimension greater than 0. In our present algorithm and program, we do not deal with this situation and if it occurs, DISCOVERER outputs a message and does nothing else.

6 Examples

Many problems with various background can be formulated into system PS and can be solved by DISCOVERER automatically.

The calling sequence of DISCOVERER for system PS is:

$$\text{tofind } ([h_1, \ldots, h_s], [g_1, \ldots, g_t], [x_1, \ldots, x_s, u_1, \ldots, u_d], \alpha);$$

where α has following three kind of choices:

- a non-negative integer ν which means the condition for PS to have ν distinct real solution(s) exactly;
- a range $\nu..\mu$ (ν, μ are non-negative integers, $\nu < \mu$) which means the condition for PS to have ν or $\nu + 1$ or $\ldots$ or μ distinct real solutions;
- a range $\nu..n$ (ν is a non-negative integer, n a name) which means the condition for PS to have ν or more than ν distinct real solutions.

Example 6.1. Which triangles can occur as sections of a regular tetrahedron by planes which separate one vertex from the other three?

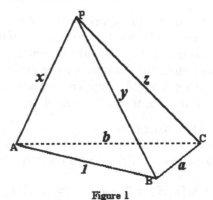

Figure 1

This example appeared as an unsolved problem in the American Mathematical Monthly (Oct. 1994) [15]. In fact, it is a special case of the well-known Perspective-three-Point (P3P) problem (see the following Example 6.2).

If we let $1, a, b$ (assume $b \geq a \geq 1$) be the lengths of three sides of the triangle, and x, y, z the distances from the vertex to the three vertexes of the triangle, respectively (see Figure 1), then, what we need is to find the necessary and sufficient condition that a, b should satisfy for the following system to have real solution(s)

$$\begin{cases} h_1 = x^2 + y^2 - xy - 1 = 0, \\ h_2 = y^2 + z^2 - yz - a^2 = 0, \\ h_3 = z^2 + x^2 - zx - b^2 = 0, \\ x > 0, y > 0, z > 0, a - 1 \geq 0, b - a \geq 0. \end{cases}$$

With our program DISCOVERER, we need only to type in

$$\text{tofind}\,([h_1, h_2, h_3], [x, y, z, a-1, b-a], [x, y, z, a, b], 1..n);$$

DISCOVERER runs 26 seconds on a K6/233 PC with MAPLE 5.3, and outputs

FINAL RESULT:

The system has required real solution(s) IF AND ONLY IF

$$[R1 \geq 0, R2 \geq 0] \quad or \quad [R1 \geq 0, R2 \leq 0, R3 \geq 0]$$

where

$$R1 = a^2 + a + 1 - b^2$$

$$R2 = a^2 - 1 + b - b^2$$

$$\begin{aligned} R3 = 1 &- \frac{8}{3}a^2 - \frac{8}{3}b^2 + \frac{16}{9}a^8 - \frac{68}{27}b^6a^2 + \frac{241}{81}b^4a^4 - \frac{68}{27}b^2a^6 \\ &- \frac{68}{27}b^4a^2 - \frac{68}{27}b^2a^4 - \frac{2}{9}b^6 + \frac{16}{9}b^8 - \frac{2}{9}a^6 + \frac{46}{9}b^2a^2 \\ &+ \frac{16}{9}b^4 + \frac{16}{9}a^4 + \frac{46}{9}b^2a^8 + \frac{46}{9}b^8a^2 - \frac{68}{27}b^6a^4 - \frac{68}{27}b^4a^6 \\ &+ \frac{16}{9}b^4a^8 - \frac{8}{3}b^{10}a^2 + \frac{16}{9}b^8a^4 - \frac{2}{9}b^6a^6 - \frac{8}{3}b^2a^{10} - \frac{8}{3}b^{10} \\ &+ b^{12} - \frac{8}{3}a^{10} + a^{12} \end{aligned}$$

The article [15] has given a sufficient condition that any triangle with two angles $> 60°$ is a possible section. It is easy to see that this condition is equivalent to $[R1 > 0, R2 > 0]$.

Example 6.2. Given the distance between every pair of 3 control points, and given the angle to every pair of the control points from an additional point called the *centre of perspectivity* (say P), find the lengths of the segments joining P and each of the control points. This problem originates from camera calibration and is called *perspective-three-point* (P3P) problem. The corresponding algebraic equation system is called the *P3P equation system*.

So-called a *solution classification* of P3P equation system is to give explicit conditions under which the system has none, one, two, ..., real physical solutions, respectively. This problem had been open for many years [16], until [28] appeared recently.

This example is about a special case of P3P problem and was studied in different way by Gao and Cheng [16]. Suppose the control points A, B, C form an equilateral triangle and $\angle CPA = \angle APB$, give the solution classification. The corresponding equation system is

$$\begin{cases} h_1 = y^2 + z^2 - 2yzp - 1 = 0, \\ h_2 = z^2 + x^2 - 2zxq - 1 = 0, \\ h_3 = x^2 + y^2 - 2zxq - 1 = 0, \\ x > 0,\, y > 0,\, z > 0,\, 1 - p^2 > 0,\, 1 - q^2 > 0,\, p + 1 - 2q^2 \geq 0. \end{cases}$$

where parameters p, q denote the cosines of the angles

$$\angle BPC, \quad \angle CPA\ (= \angle APB),$$

and x, y, z denote the lengths of the segments PA, PB, PC, respectively.

By our program DISCOVERER, we need only type in

tofind $([h_1, h_2, h_3], [x, y, z, 1 - p^2, 1 - q^2, p + 1 - 2q^2], [x, y, z, p, q], 0)$;
tofind $([h_1, h_2, h_3], [x, y, z, 1 - p^2, 1 - q^2, p + 1 - 2q^2], [x, y, z, p, q], 1)$;
tofind $([h_1, h_2, h_3], [x, y, z, 1 - p^2, 1 - q^2, p + 1 - 2q^2], [x, y, z, p, q], 2)$;
tofind $([h_1, h_2, h_3], [x, y, z, 1 - p^2, 1 - q^2, p + 1 - 2q^2], [x, y, z, p, q], 3)$;
tofind $([h_1, h_2, h_3], [x, y, z, 1 - p^2, 1 - q^2, p + 1 - 2q^2], [x, y, z, p, q], 4)$;
tofind $([h_1, h_2, h_3], [x, y, z, 1 - p^2, 1 - q^2, p + 1 - 2q^2], [x, y, z, p, q], 5..n)$;

respectively. After 64 seconds, on a PII/266 PC with MAPLE 5.4, we get the solution classification except some boundaries:

The system has 0 real solution IF AND ONLY IF

$$[R1 > 0,\, R2 > 0,\, R3 < 0,\, R4 > 0]$$

or

$$[R1 < 0,\, R2 > 0,\, R3 < 0]$$

The system has 1 real solution IF AND ONLY IF

$$[R2 < 0,\, R3 < 0,\, R5 > 0]$$

or

$$[R2 < 0,\, R5 < 0]$$

The system has 2 real solutions IF AND ONLY IF

$$[R1 > 0,\, R2 > 0,\, R3 < 0,\, R4 < 0,\, R5 > 0]$$

or

$$[R1 > 0, R2 > 0, R3 > 0, R4 > 0, R5 > 0]$$

or

$$[R1 > 0, R2 > 0, R4 < 0, R5 < 0]$$

The system has 3 real solutions IF AND ONLY IF

$$[R2 < 0, R3 > 0, R4 < 0, R5 > 0]$$

The system has 4 real solutions IF AND ONLY IF

$$[R1 > 0, R2 > 0, R3 > 0, R4 < 0, R5 < 0]$$

The system has 5..n real solutions IF AND ONLY IF

There are not 5..n real solutions in this branch

where

$$R1 = q,$$
$$R2 = 2p - 1,$$
$$R3 = 2q - 1,$$
$$R4 = 2p - 1 - q^2,$$
$$R5 = 2pq^2 - 3q^2 + 1.$$

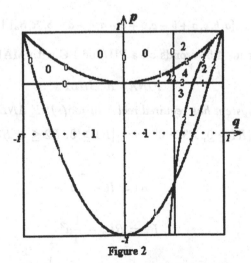

Figure 2

Note that $p + 1 - 2q^2 \geq 0$, the given condition should hold in all cases. And if you want to know the situation when parameters are on a boundary, say $R2$, you need only to type in

$$\text{tofind } ([h_1, h_2, h_3, R2], [x, y, z, 1 - p^2, 1 - q^2, p + 1 - 2q^2], [x, y, z, p, q], 0);$$
$$\text{tofind } ([h_1, h_2, h_3, R2], [x, y, z, 1 - p^2, 1 - q^2, p + 1 - 2q^2], [x, y, z, p, q], 1);$$
$$\text{tofind } ([h_1, h_2, h_3, R2], [x, y, z, 1 - p^2, 1 - q^2, p + 1 - 2q^2], [x, y, z, p, q], 2);$$
$$\text{tofind } ([h_1, h_2, h_3, R2], [x, y, z, 1 - p^2, 1 - q^2, p + 1 - 2q^2], [x, y, z, p, q], 3);$$
$$\text{tofind } ([h_1, h_2, h_3, R2], [x, y, z, 1 - p^2, 1 - q^2, p + 1 - 2q^2], [x, y, z, p, q], 4);$$

In this way, we obtained the complete solution classification, as indicated in Figure 2.

Example 6.3. Solving geometric constraints is the central topic in many current work of developing intelligent Computer Aided Design systems and interactive constraint-based graphic systems. The existing algorithms do not consider whether the given system has real physical solution(s) or not. Our algorithm and program can give the necessary and sufficient condition for the system to have real physical solution(s). For example:

Give the necessary and sufficient condition for the existence of a triangle with elements a, h_a, R, where a, h_a, R means the side-length, altitude, and circumradius, respectively.

Clearly, we need to find the necessary and sufficient condition for the following system to have real solution(s)

$$\begin{cases} f_1 = a^2 h_a^2 - 4s(s-a)(s-b)(s-c) = 0, \\ f_2 = 2Rh_a - bc = 0, \\ f_3 = 2s - a - b - c = 0, \\ a > 0, b > 0, c > 0, a+b-c > 0, b+c-a > 0, \\ c+a-b > 0, R > 0, h_a > 0. \end{cases}$$

We type in

$$\text{tofind}([f_1, f_2, f_3], [a, b, c, a+b-c, b+c-a, c+a-b, R, h_a], [s, b, c, a, R, h_a], 1..n);$$

DISCOVERER runs 4.5 seconds on a PII/266 PC with MAPLE 5.4, and outputs

FINAL RESULT:

The system has required real solution(s) IF AND ONLY IF

$$[R1 \geq 0, R3 \geq 0] \quad or \quad [R1 \geq 0, R2 \leq 0, R3 \leq 0]$$

where

$$R1 = R - \frac{1}{2}a$$

$$R2 = Rh_a - \frac{1}{4}a^2$$

$$R3 = -\frac{1}{2}h_a^2 + Rh_a - \frac{1}{8}a^2$$

In [19], the condition they gave is $R1 \geq 0 \wedge R3 \geq 0$. Now, we know they are wrong and that is only a sufficient condition.

Our program, DISCOVERER, is very efficient for solving this kind of problems. By DISCOVERER, we have discovered or rediscovered about 70 such conditions for the existence of a triangle, and found three mistakes in [19].

References

1. Arnon, D. S., Geometric Reasoning with Logic and Algebra, *Artificial Intelligence*, **37**, 37–60, 1988.
2. Arnon, D. S., Collins, G. E. & McCallum, S., Cylindrical Algebraic Decomposition I: The Basic Algorithm, *SIAM J. Comput.*, **13**:4, 865–877, 1984.
3. Arnon, D. S., Collins, G. E. & McCallum, S., Cylindrical Algebraic Decomposition II: An Adjacency Algorithm for the Plane, *SIAM J.Comput.*, **13**:4, 878–889, 1984.
4. Bottema, O., et al., *Geometric Inequalities*, Wolters-Noordhoff Publ., Groningen, Netherland, 1969.
5. Buchberger, B., Applications of Gröbner Bases in Non-linear Computational Geometry, *Geometric Reasoning*, MIT Press, Cambridge, MA, pp. 413–446, 1989.
6. Buchberger, B., Collins, G. E. & Kutzler, B., Algebraic Methods for Geometric Reasoning, *Annual Review of Computing Science*, **3**, 85–119, 1988.
7. Chen, M., *Generalization of Discrimination System for Polynomials and its Applications*, Ph.D dissertation, Sichuan University, Chengdu, China, 1998. (in Chinese)
8. Chou, S. C., *Mechanical Geometry Theorem Proving*, D. Reidel Publ. Co., Amsterdam, 1988.
9. Chou, S. C., Gao, X. S. & Arnon, D. S., On the Mechanical Proof of Geometry Theorems Involving Inequalities, *Advances in Computing Research*, **6**, JAI Press Inc., pp. 139–181, 1992.
10. Chou, S. C., Gao, X. S. & McPhee, N., A Combination of Ritt-Wu's Method and Collins's Method, *Proc. of CADE-12*, Springer-Verlag, pp. 401–415, 1994.
11. Chou, S. C., Gao, X. S. & Zhang, J. Z., *Machine Proofs in Geometry*, World Scientific Publ. Co., 1994.
12. Collins, G. E., Quantifier Elimination for Real Closed Fields by Cylindrical Algebraic Decomposition, *Lecture Notes in Computer Science*, **33**, Springer-Verlag, pp. 134–183, 1975.
13. Collins, G. E. & Hong, H., Partial Cylindical Algebraic Decomposition for Quantifier Elimination, *J. Symbolic Computation*, **12**, 299–328, 1991.
14. Cox, D., Little, J. & O'Shea, D., *Ideals, Varieties, and Algorithms*, Springer-Verlag, 1992.
15. Folke, E., Which Triangles Are Plane Sections of Regular Tetrahedra? *American Mathematics Monthly*, **101**, 788–789, 1994.
16. Gao, X. S. & Cheng, H. F., On the Solution Classification of the "P3P" Problem, *Proc. of ASCM '98*, Z. Li (ed.), Lanzhou University Press, pp. 185–200, 1998.
17. González-Vega, L., Lombardi, H., Recio, T. & Roy, M.-F., Sturm-Habicht Sequence, *Proc. of ISSAC '89*, ACM Press. pp. 136–146, 1989.
18. Kapur, D., Automated Geometric Reasoning: Dixon Resultants, Gröbner Bases, and Characteristic Sets, *Lecture Notes in Artificial Intelligence*, **1360**, Springer-Verlag, pp. 1–36, 1997.
19. Mitrinovic, D. S., Pecaric, J. E., & Volenec, V., *Recent Advances in Geometric Inequalities*, Kluwer Academic Publ., 1989.
20. Seidenberg A., A New Decision Method for Elementary Algebra, *Annals of Mathematics*, **60**, 365–371, 1954.
21. Tarski, A., *A Decision Method for Elementary Algebra and Geometry*, University of California Press, Berkeley, 1951.
22. Wang, D. M., Elimination Procedures for Mechanical Theorem Proving in Geometry, *Ann. Math. Artif. Intell.*, **13**, 1–24, 1995.

23. Wang, D. M., A Decision Method for Definite Polynomial, MM Research Preprints, No. **2**, Beijing, pp. 68–74, 1987.
24. Winkler, F., *Polynomial Algorithms in Computer Algebra*, Springer-Verlag, Wien, 1996.
25. Wu, W. T., On the Decision Problem and the Mechanization of Theorem-proving in Elementary Geometry, *Scientia Sinica*, **21**, 159–172, 1978.
26. Wu, W. T., *Mechanical Theorem Proving in Geometries: Basic Principles* (translated from the Chinese by X. Jin and D. Wang), Springer-Verlag, Wien, 1994.
27. Yang, L., Practical Automated Reasoning on Inequalities: Generic Programs for Inequality Proving and Discovering, *Proc. of the Third Asian Technology Conference in Mathmatics*, W.-C. Yang et al. (eds.), Springer-Verlag, pp. 24–35, 1998.
28. Yang, L., A Simplified Algorithm for Solution Classification of the Perspective-three-Point Problem, MM Research Preprints, No. **17**, Beijing, pp. 135–145, 1998.
29. Yang, L., Gao, X. S., Chou, S. C. & Zhang, J. Z., Automated Production of Readable Proofs for Theorems in Non-Euclidean Geometries, *Lecture Notes in Artificial Intelligence*, **1360**, Springer-Verlag, pp. 171–188, 1997.
30. Yang, L., Hou, X. R. & Zeng, Z. B., A Complete Discrimination System for Polynomials, *Science in China*, Series **E 39**:6, 628–646, 1996.
31. Yang, L. & Xia, B. C., Explicit Criterion to Determine the Number of Positive Roots of a Polynomial, MM Research Preprints, No. **15**, Beijing, pp. 134–145, 1997.
32. Yang, L., Zhang, J. Z. & Hou, X. R., A Criterion of Dependency between Algebraic Equations and Its Applications, *Proc. of the 1992 International Workshop on Mathematics Mechanization*, International Academic Publ., Beijing, pp. 110–134, 1992.
33. Yang, L., Zhang, J. Z. & Hou, X. R., An Efficient Decomposition Algorithm for Geometry Theorem Proving Without Factorization, *Proc. of Asian Symposium on Computer Mathematics*, H. Shi & H. Kobayashi (eds.), Scientists Incorporated, Japan, pp. 33–41, 1995.
34. Yang, L., Zhang, J. Z. & Hou, X. R., *Nonlinear Algebraic Equation System and Automated Theorem Proving*, Shanghai Scientific and Technological Education Publ. House, Shanghai, 1996. (in Chinese)
35. Zhang, J. Z., Yang, L. & Hou, X. R., A Criterion for Dependency of Algebraic Equations with Applications to Automated Theorem Proving, *Science in China*, Series **A 37**, 547–554, 1994.

Proving Newton's Propositio Kepleriana Using Geometry and Nonstandard Analysis in Isabelle

Jacques D. Fleuriot and Lawrence C. Paulson

Computer Laboratory – University of Cambridge
New Museums Site, Pembroke Street
Cambridge CB2 3QG
{jdf21,lcp}@cl.cam.ac.uk

Abstract. The approach previously used to mechanise lemmas and Kepler's Law of Equal Areas from Newton's **Principia** [13] is here used to mechanically reproduce the famous *Propositio Kepleriana* or Kepler Problem. This is one of the key results of the Principia in which Newton demonstrates that the centripetal force acting on a body moving in an ellipse obeys an inverse square law. As with the previous work, the mechanisation is carried out through a combination of techniques from geometry theorem proving (GTP) and Nonstandard Analysis (NSA) using the theorem prover Isabelle. This work demonstrates the challenge of reproducing mechanically Newton's reasoning and how the combination of methods works together to reveal what we believe to be flaw in Newton's reasoning.

1 Introduction

The reasoning of Newton's *Philosophiæ Naturalis Principia Mathematica* (the *Principia* [14]), as it was originally published, is a mixture of geometric and algebraic arguments together with Newton's own proof techniques. These combine to produce a complex mathematical reasoning that is used to explain the physical world. The demonstrations of *Lemmas* and *Propositions* in the *Principia* are, in fact, proof sketches that require a lot of work on the part of the reader for a detailed understanding. There are several reasons that make the *Principia* a very difficult text to master. First of all, the proofs are very involved and one requires an adequate knowledge of geometry to be able to understand many of the steps made by Newton. Secondly, Newton's exposition can be tedious and difficult to grasp in places. Many mathematicians contemporary to Newton, despite their grounding in ancient Greek geometry and familiarity with the style of the exposition, had difficulties understanding Newton's mathematical reasoning. This gives an indication of the demands that a thorough study of the *Principia* has on the modern reader.

As we mentioned in a previous paper, Newton's geometry is also notable for his use of *limit* or *ultimate* arguments in his proofs [13]. These are implicit notions of differential calculus that are at the core of Newton's treatment and give Newton's geometry an infinitesimal nature. Newton further adds motion to his reasoning and enriches the geometry with various kinematics concepts that enable points to move towards points for example. Thus, Newton's geometry consists in studying the relations, such as ratios, between various parts of the constructed diagrams as certain of its elements tend towards limiting positions or become infinitely small.

In this paper, we build on the tools and techniques that we presented before [13]. In section 2, we review the geometric methods and concepts that we have formalised in this work. We also give examples of theorems proved in Isabelle using these techniques. Section 3 is a brief introduction to the concepts from NSA that we use; it also outlines the infinitesimal aspects of our geometry. We then present in Section 4, as a case study, the proof of the **Kepler Proposition**. This is a key proof of the Book 1 of the *Principia* and our work follows, in its steps, the analysis made by Densmore [9]. This extended case study shows our combination of techniques from geometry and NSA at work to provide a formal proof of a major proposition. The challenge inherent to the mechanisation of Newton's reasoning– especially in an interactive environment such as Isabelle where the user guides the proof– will become obvious as we highlight the steps and difficulties encountered. Section 5 offers our comments and conclusions.

2 Geometry

We use methods that are based on *geometric invariants* [5,6] and high level geometry lemmas about these invariants. A particular property is ideal as an invariant if it ensures that the proofs generated are short. This enables some of the proofs to be derived automatically using the powerful tactics of Isabelle's classical reasoner. Also, the methods should be powerful enough to prove many properties without adding auxiliary points or lines. The other important aspect is to achieve diagram independence for the proofs, that is, the same proof can be applied to several diagrams.

2.1 The Signed Area Method

In this method, there are basic rules about geometric properties called signed areas. These can be used to express various geometric concepts such as collinearity (`coll`), parallelism ($\parallel$) and so on. Moreover, the basic rules can be combined to prove more complex theorems which deal with frequently-used cases and help simplify the search process.

We represent the line from point A to point B by $A—B$, its length by $\text{len}(A—B)$, and the *signed* area $S_{\text{delta}}ABC$ of a triangle is the usual notion of area with its sign depending on how the vertices are ordered. We follow the usual approach of having $S_{\text{delta}}ABC$ as positive if $A—B—C$ is in anti-clockwise direction and negative otherwise. Some of the rules and definitions used are:

$$a - b \parallel c - d \equiv (S_{\text{delta}}\, a\, b\, c = S_{\text{delta}}\, a\, b\, d)$$
$$\text{coll}\, a\, b\, c \implies \text{len}(a - b) \times S_{\text{delta}}\, p\, b\, c = \text{len}(b - c) \times S_{\text{delta}}\, p\, a\, b$$

We can also introduce new points using the following property and define the *signed area of a quadrilateral* $S_{quad}\,a\,b\,c\,d$ in terms of signed areas of triangles:

$$S_{delta}\,a\,b\,c = S_{delta}\,a\,b\,d + S_{delta}\,a\,d\,c + S_{delta}\,d\,b\,c$$
$$S_{quad}\,a\,b\,c\,d \equiv S_{delta}\,a\,b\,c + S_{delta}\,a\,c\,d$$

We have proved a number of theorems about the sign of $S_{quad}\,a\,b\,c\,d$ that depend on the ordering of the vertices, for example $S_{quad}\,a\,b\,c\,d = -S_{quad}\,a\,d\,c\,b$.

When dealing with geometry proofs, we often take for granted conditions that need to be stated explicitly for machine proofs: for example, two points making up a line should not coincide. The machine proofs are valid only if these conditions are met. These are known as *non-degenerate* conditions and are required in many cases to prevent the denominators of fractions from becoming zero in the various algebraic statements.

2.2 The Full-Angle Method

A full-angle $\langle u, v \rangle$ is the angle from line u to line v measured anti-clockwise. We define the equivalence relation of *angular* equality as follows:

$$x =_a y \equiv \exists n \in \mathbb{N}.\ |x - y| = n\pi$$

and can use it to express that two lines are perpendicular

$$a - b \perp c - d \equiv \langle a - b, c - d \rangle =_a \frac{\pi}{2}$$

Other properties of full-angles concern their sign and how they can be split or joined. The same rule therefore either introduces a new line or eliminates a common one from the full angles depending on the direction in which it is used.

$$\langle u, v \rangle =_a -\langle v, u \rangle$$
$$\langle u, v \rangle =_a \langle u, x \rangle + \langle x, v \rangle$$

Full-angles are used instead of traditional angles because their use simplifies many proofs by eliminating case-splits. Moreover, as we have already mentioned, these methods are useful to us since they relate closely to the geometric properties used by Newton [13]. They preserve the intuitive nature of his geometry and can easily be extended with infinitesimal notions, as we will see shortly.

2.3 A Simple Example: Euclid I.29

Euclid's proposition 29 of Book I [10], can be easily proved using the full-angle method. The proposition states that if $A - B \parallel C - D$ and the transversal $P - Q$ intersects $A - B$ and $C - D$ then $\langle A - B, P - Q \rangle = \langle C - D, P - Q \rangle$.

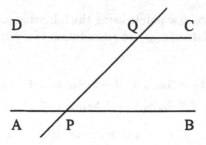

Fig. 1. Euclid Proposition I.29

We prove this theorem easily by using the rules about full-angles given in Section 2.2 and the fact the angle between two parallel lines is zero.

Proof:

$$A - B \parallel C - D \Longrightarrow \langle A - B, C - D \rangle =_a 0$$
$$\langle A - B, P - Q \rangle + \langle P - Q, C - D \rangle =_a 0$$
$$\langle A - B, P - Q \rangle =_a -\langle P - Q, C - D \rangle$$
$$\langle A - B, P - Q \rangle =_a \langle C - D, P - Q \rangle.$$

2.4 Extending the Geometric Theory

The main aim of the *Principia* is to investigate mathematically the motion of bodies such as planets. Thus, we need to have definitions for geometric figures such as the circle, the ellipse and their tangents. The ellipse is especially important for this work since Kepler's Problem is concerned with elliptical motion. The circle can be viewed as a special case of the ellipse where the foci coincide.

$$\mathtt{ellipse}\, f_1\, f_2\, r \equiv \{p.\, |\mathtt{len}(f_1 - p)| + |\mathtt{len}(f_2 - p)| = r\}$$
$$\mathtt{circle}\, x\, r \equiv \mathtt{ellipse}\, x\, x\, (2 \cdot r)$$
$$\mathtt{arc\_len}\, x\, a\, b \equiv |\mathtt{len}(x - a)| \times \langle a - x, x - b \rangle$$
$$\mathtt{e\_tangent}\, (a - b)\, f_1\, f_2\, E \equiv (\mathtt{is\_ellipse}\, f_1\, f_2\, E \,\wedge\, a \in E \,\wedge$$
$$\langle f_1 - a, a - b \rangle =_a \langle b - a, a - f_2 \rangle)$$

We need to prove a number of properties relating to the ellipse such as the one stating that *all parallelograms described around a given ellipse are equal to each other* (Figure 2).

This relationship appears (in slightly different wording) as **Lemma 12** of the *Principia* where it is employed in the solution of Proposition 11, the famous *Propositio Kepleriana* or Kepler problem. Newton refers us to the "writers on the conics sections" for a proof of the lemma. This lemma is demonstrated in

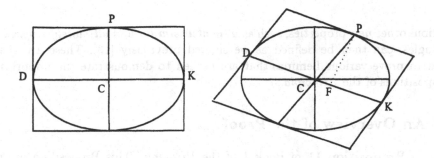

Fig. 2. Circumscribed Parallelograms

Book 7, Proposition 31 in the *Conics* of Apollonius of Perga [1]. Of course, unlike Newton, we have to prove this result explicitly in Isabelle to make it available to any other proof that might use it.

3 Infinitesimal Geometry

In this section, we give a brief overview of our geometry containing infinitesimals. We first give formal definitions for the various types of numbers that exist and which can be used to describe geometric quantities.

3.1 The Nonstandard Universe $\mathbb{R}^*$

Definition 1. *In an ordered field extension $\mathbb{R}^* \supseteq \mathbb{R}$, an element $x \in \mathbb{R}^*$ is said to be an* **infinitesimal** *if $|x| < r$ for all positive $r \in \mathbb{R}$;* **finite** *if $|x| < r$ for some $r \in \mathbb{R}$;* **infinite** *if $|x| > r$ for all $r \in \mathbb{R}$.*

The extended, richer number system $\mathbb{R}^*$ is known as the *hyperreals*. It has been developed in Isabelle through purely definitional means using an *ultrapower* construction. We will not give more details of this substantial construction in the present paper so as to concentrate on the geometric aspects only.

Definition 2. *$x, y \in \mathbb{R}^*$ are said to be* **infinitely close**, *$x \approx y$ if $|x - y|$ is infinitesimal.*

This is an important equivalence relation that will enable us to reason about infinitesimal quantities. For example, we can formalise the notion of two points coinciding by saying that the distance between them is infinitely close to zero. Two geometric quantities that become ultimately equal can also be modelled using it. The relation and its properties are used to formalise ultimate situations that might be considered degenerate by ordinary GTP methods [13].

Using the relation, we can also define the concept of two full-angles being infinitely close:

$$a_1 \approx_a a_2 \equiv \exists n \in \mathbb{N}. \, |a_1 - a_2| \approx n\pi$$

Various other *new* properties, such as *ultimately similar* and *ultimately congruent* triangles, can then be defined as we showed previously [13]. These are then used to prove various Lemmas that are needed to demonstrate the important Propositions of the *Principia*.

4 An Overview of the Proof

This is **Proposition 11** of Book 1 of the *Principia*. This Proposition is important for both mathematical and historical reasons as it lays the foundations for Kepler's first law of Gravitation. It provides the mathematical analysis that could explain and confirm Kepler's guess that planets travelled in ellipses round the sun [15].

The proof of this proposition will be studied in detail as it gives a good overview of the mixture of geometry, algebra and limit reasoning that is so characteristic of Newton's *Principia*. It also gives an idea of the depth and amount of mathematical expertise involved in Newton's proof. The proof that Newton describes, though relatively short on paper, becomes a major demonstration when expanded and reproduced using Isabelle. The elegance of many of the constructions, which could be glossed over, are revealed through the detailed analysis.

We give formal justifications of the steps made by Newton in ultimate situations through our formal and logical use of infinitesimals. Infinitesimal reasoning is notorious for leading to contradictions. However, nonstandard analysis is generally believed to be consistent and hence ensure that our mechanisation is rigorous. We will give the enunciation of the Proposition and the proof (sketch) provided by Newton. We will then expand on the sketch and provide detailed proofs of the steps that are made by Newton. This will require the use of the rules from the geometric and NSA theories developed in Isabelle.

4.1 Proposition 11 and Newton's Proof

Proposition 11 is in fact stated as a problem by Newton at the start of Section 3 of the *Principia*. This section deals with *"the motion of bodies in eccentric conic section"*. Particular orbits and laws governing forces that are relevant to the universe are investigated. The mathematical tools are developed for later use in Book III of the *Principia* when natural phenomena of our world are investigated. Our task consists in expressing Newton's *result* as a goal which is then proved. Figure 3 shows Newton's original diagram used for this Proposition.

Proposition 11 *If a body revolves in an ellipse; it is required to find the law of the centripetal force tending to the focus of the ellipse*
Newton's Proof:
Let S be the focus of the ellipse. Draw SP cutting the diameter DK of the ellipse in E, and the ordinate Qv in x; and complete the parallelogram QxPR. It is evident that EP is equal to the greater semiaxis AC: for drawing HI from the other focus H of the ellipse parallel to EC, because CS, CH are equal, ES,

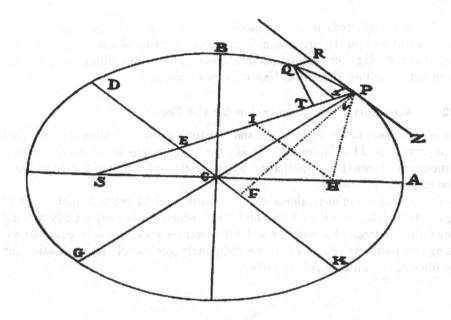

Fig. 3. Newton's Original Diagram for Proposition 11

EI will be also equal; so that EP is the half-sum of PS, PI, that is (because of the parallels HI, PR, and the equal angles IPR, HPZ), of PS, PH, which taken together are equal to the whole axis 2AC. Draw QT perpendicular to SP, and putting L for the principal latus rectum of the ellipse (or for $\frac{2BC^2}{AC}$), we shall have

$$L \cdot QR : L \cdot Pv = QR : Pv = PE : PC = AC : PC,$$
$$\text{also, } L \cdot Pv : Gv \cdot Pv = L : Gv, \text{ and, } Gv \cdot Pv : Qv^2 = PC^2 : CD^2$$

By Corollary 2, Lemma 7, when the points P and Q coincide, $Qv^2 = Qx^2$, and Qx^2 or $Qv^2 : QT^2 = EP^2 : PF^2 = CA^2 : PF^2$, and (by Lemma 12) $= CD^2 : CB^2$. Multiplying together corresponding terms of the four proportions, and by simplifying, we shall have

$$L \cdot QR : QT^2 = AC \cdot L \cdot PC^2 \cdot CD^2 : PC \cdot Gv \cdot CD^2 \cdot CB^2 = 2PC : Gv,$$

since $AC \cdot L = 2BC^2$. But the points Q and P coinciding, 2PC and Gv are equal. And therefore the quantities $L \cdot QR$ and QT^2, proportional to these, will also be equal. Let those equals be multiplied by $\frac{SP^2}{QR}$, and $L \cdot SP^2$ will become equal to $\frac{SP^2 \cdot QT^2}{QR}$. And therefore (by Corollary 1 and 5, Proposition 6) the centripetal force is inversely as $L \cdot SP^2$, that is, inversely as the square of the distance SP. Q.E.I.

Newton's derivation concludes that the centripetal force, for a body moving in an ellipse, is inversely proportional to the square of the distance.

Our proof proceeds in several steps where we set up various relationships that we will need for the conclusion. This involves proving Newton's intermediate results and storing them as intermediate theorems (we avoid calling them lemmas so as not to confuse them with Newton's own Lemmas).

4.2 A Geometric Representation for the Force

An investigation of the Proposition and Newton's result indicates that our goal is to prove that $\exists k \in \mathbb{R}. force \approx k \times \frac{1}{SP^2}$ (i.e. $force \propto \frac{1}{SP^2}$ ultimately). We now demonstrate through a combination of geometric and infinitesimal procedures how to prove the theorem.

Our combination of methods was previously used to prove *Kepler's Law of Equal Areas* [13]. This is an important result which states that a body moving under the influence of a centripetal force describes equal areas in equal times. Using this result we can now derive a completely **geometric** representation for the force acting on the orbiting body.

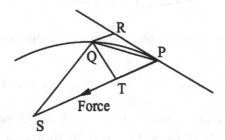

Fig. 4. Diagram for Geometric Representation of Force

Consider Figure 4 in which a point P is moving along an arc of finite curvature under the influence of a centripetal force acting towards S. Let Q be a point infinitely close to P, that is, the length of the arc from P to Q is infinitesimal. QR, parallel to SP, represents the displacement from the rectilinear motion (along the tangent) due to the force acting on P. QT is the perpendicular dropped to SP. From Newton's **Lemma 10, Corollary 3**, we have that displacement "in the very beginning of motion" is proportional to the force and the square of the time, and hence (for some real proportionality constant k_1) that

$$force \approx k_1 \times \frac{\text{len}(Q - R)}{\text{Time}^2} \tag{1}$$

Since the distance between P and Q is infinitesimal, the angle $\langle P - S, S - Q \rangle$ is infinitely small, and hence the area of the sector SPQ ($\text{S}_{\text{arc}}\,\text{S P Q}$) is infinitely close to that of the triangle SPQ:

$$\langle P - S, S - Q \rangle \approx_a 0 \Longrightarrow \text{S}_{\text{arc}}\,\text{S P Q} \approx \text{S}_{\text{delta}}\,\text{S P Q}$$
$$\Longrightarrow \text{S}_{\text{arc}}\,\text{S P Q} \approx 1/2 \times \text{len}(Q - T) \times \text{len}(S - P) \tag{2}$$

From Kepler's Law of Equal Areas, we can replace **Time** by $S_{arc}SPQ$ [13] and, hence, using (1) and (2), we have the following geometric representation for the force (for some new proportionality constant k)

$$force \approx k \times \frac{\text{len}(Q-R)}{\text{len}(Q-T)^2 \times \text{len}(S-P)^2} \tag{3}$$

This is a general result (**Proposition 6** of the *Principia*) that applies to any motion along an arc under the influence of a central force. We justify the use of a circular arc for the general arc by the fact that it is possible to construct a circle at the point P that represents the best approximation to the curvature there. This circle, sometimes known as the *osculating circle*[1], has the same first and second derivative as the curve at the given point P. Thus, the osculating circle has the same curvature and tangent at P as the general curve and an infinitesimal arc will also be same. We refer the reader to Brackenridge for more details on the technique [2, 3].

With this result set up, to prove the Kepler Problem, we need to show that the ratio involving the infinitesimal quantities QR and QT is equal or infinitely close to some constant (finite) quantity. Thus, the proof of Proposition 11 involves, in essence, eliminating the infinitesimals from relation (3) above. This relation is transformed using the geometry of the ellipse to one involving only macroscopic (i.e. non-infinitesimals) aspects of the orbit. We show next how the various GTP and NSA techniques are applied to the analysis of an elliptical orbit to determine the nature of the centripetal force.

4.3 Expanding Newton's Proof

A detailed account of our mechanisation of Newton's argument for Proposition 11 would take several pages since the proof sketch given by Newton is complex and we would have to present a large number of derivations. We will highlight the main results that were proved and, in some cases, details of the properties that needed to be set up first. We will also mention the constraints that needed to be satisfied within our framework before the various ratios that were proved could be combined. Our mechanisation was broken down into several steps that roughly followed from Newton's original proof. The main results that are set up are as follows (see Fig. 3):

- $\text{len}(E-P) = \text{len}(A-C)$
- $\text{len}(A-C)/\text{len}(P-C) = L \times \text{len}(Q-R)/L \times \text{len}(P-v)$
- $L \times \text{len}(P-v)/(\text{len}(G-v) \times \text{len}(P-v)) = L/\text{len}(G-v)$
- $\text{len}(G-v) \times \text{len}(P-v)/\text{len}(Q-v)^2 = \text{len}(P-C)^2/\text{len}(C-D)^2$
- $\text{len}(Q-v)^2/\text{len}(Q-T)^2 \approx \text{len}(C-D)^2/\text{len}(C-B)^2$

Step 1: Proving $\text{len}(E-P) = \text{len}(A-C)$

---

[1] from the Latin *osculare* meaning to kiss– the term was first used by Leibniz

This result shows that the length of EP is independent of P and Newton's proof uses several properties of the ellipse. We will give a rather detailed overview of this particular proof as it gives an idea of the amount of work involved in mechanising Newton's geometric reasoning. Moreover, the reader can then compare Newton's proof style and prose with our own proof and see the GTP methods we have formalised in action.

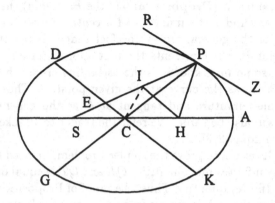

Fig. 5. Construction for Step 1 of Proposition 11

In Figure 5, the following holds

- C is the centre of the ellipse with S and H the foci
- P is a point of the curve
- RZ is the tangent at P
- the conjugate diameter $D - K \parallel P - Z$
- $P - S$ intersects $D - K$ at E
- $H - I \parallel E - C$ and $H - I$ intersects $P - S$ at I

Since $\mathtt{H - I} \parallel \mathtt{E - C}$, the following theorem holds,

$$\mathtt{H - I} \parallel \mathtt{E - C} \Longrightarrow \mathtt{S_{delta}\, C\, E\, I = S_{delta}\, C\, E\, H} \tag{4}$$

But the foci are collinear with and (by Apollonius III.45 [1]) equidistant from the centre of the ellipse; so the following can be derived using the signed-area method,

$$\mathtt{coll\ S\, C\, H} \Longrightarrow \mathtt{len(S - C) \times S_{delta}\, C\, E\, H = len(C - H) \times S_{delta}\, C\, S\, E}$$
$$\Longrightarrow \mathtt{S_{delta}\, C\, E\, H = S_{delta}\, C\, S\, E} \tag{5}$$

Also, points S, E and I are collinear and therefore combining with (4) and (5) above, we verify Newton's "ES, EI **will also be equal**"

$$\mathtt{coll\ S\, E\, I} \Longrightarrow \mathtt{len(S - E) \times S_{delta}\, C\, E\, I = len(E - I) \times S_{delta}\, C\, S\, E}$$
$$\Longrightarrow \mathtt{len(S - E) = len(E - I)} \tag{6}$$

Next, the following derivations can be made, with the help of the last result proving Newton's "*EP* is the **half-sum** of *PS, PI*"

$$\text{coll } E\,I\,P \Longrightarrow \text{len}(E-P) = \text{len}(E-I) + \text{len}(I-P)$$
$$\Longrightarrow \text{len}(E-P) = \text{len}(S-E) + \text{len}(I-P)$$
$$\Longrightarrow 2 \times \text{len}(E-P) = \text{len}(E-P) + \text{len}(S-E) + \text{len}(I-P)$$
$$\Longrightarrow 2 \times \text{len}(E-P) = \text{len}(S-P) + \text{len}(I-P)$$
$$\Longrightarrow \text{len}(E-P) = \frac{\text{len}(S-P) + \text{len}(I-P)}{2} \tag{7}$$

Note the use of the following theorem in the derivation above

$$\text{coll } S\,E\,P \Longrightarrow \text{len}(S-E) + \text{len}(E-P) = \text{len}(S-P)$$

Next, Newton argues that in fact (7) can be written as

$$\text{len}(E-P) = \frac{\text{len}(S-P) + \text{len}(H-P)}{2} \tag{8}$$

So, a proof of $\text{len}(I-P) = \text{len}(H-P)$ is needed to progress further. This will follow if it can be shown that $\triangle\text{PHI}$ is an isosceles, that is

$$\langle P-H, H-I \rangle = \langle H-I, I-P \rangle \tag{9}$$

To prove (9), both $H-I \parallel P-Z$ and $H-I \parallel P-R$ are derived first using

$$H-I \parallel E-C \wedge E-C \parallel P-Z \Longrightarrow H-I \parallel P-Z \tag{10}$$
$$H-I \parallel P-Z \wedge \text{coll } P\,Z\,R \Longrightarrow H-I \parallel P-R \tag{11}$$

From (10), (11), and the proof of Euclid I.29 given in Section 2.3

$$H-I \parallel P-Z \Longrightarrow \langle P-H, H-I \rangle = \langle H-P, P-Z \rangle \tag{12}$$
$$H-I \parallel P-R \Longrightarrow \langle H-I, I-P \rangle = \langle R-P, P-I \rangle$$
$$\Longrightarrow \langle H-I, I-P \rangle = \langle R-P, P-S \rangle \tag{13}$$

From the definition of the tangent to an ellipse and the collinearity of P, I, and S (also recall that full-angles are angles between *lines* rather than rays and are measured anti-clockwise),

$$\text{e\_tangent } (P-Z) \text{ S H Ellipse} \Longrightarrow \langle H-P, P-Z \rangle = \langle R-P, P-I \rangle$$
$$\Longrightarrow \langle H-P, P-Z \rangle = \langle R-P, P-S \rangle \tag{14}$$

From (12), (13) and (14), the following is deduced as required

$$\langle P - H, H - I \rangle = \langle H - I, I - P \rangle$$

Thus, we have $\texttt{len}(I - P) = \texttt{len}(H - P)$ (Euclid I.6 [10]), and hence (8) is proved, that is, Newton's assertion that "[EP is the half sum of] PS, PH".

Next, it follows from the definition of an ellipse that the sum of $\texttt{len}(S - P)$ and $\texttt{len}(P - H)$ is equal to the length of the major axis, that is,

$$P \in \texttt{Ellipse} \Longrightarrow \texttt{len}(S - P) + \texttt{len}(P - H) = 2 \times \texttt{len}(A - C) \qquad (15)$$

From (15) and (8), we can finally derive the property that Newton states as being evident: "*EP is equal to the greater semiaxis AC*"

$$\texttt{len}(E - P) = \texttt{len}(A - C) \qquad (16)$$

The first step has shown Newton's geometric reasoning in action. For the next steps, as the various ratios are derived, we will not always show the detailed derivations of the geometric theorems. We will concentrate on the setting up of the proportions and how everything is put together to get the final result. We will state Newton's Lemmas when they are used and theorems about infinitesimals that we use.

Step 2: Showing $\frac{L \cdot QR}{L \cdot Pv} = \frac{QR}{Pv} = \frac{PE}{PC} = \frac{AC}{PC}$

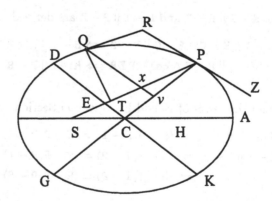

Fig. 6. Construction for Steps 2—4 of Proposition 11

In Figure 6, in addition to properties already mentioned, the following holds

- $QT \perp SP$
- $QxPR$ is a parallelogram
- Q, x, and v are collinear
- Q is infinitely close to P

It is easily proved that $v - x \parallel C - E$ and so the following theorem follows

$$v - x \parallel C - E \Longrightarrow \langle P - v, v - x \rangle = \langle P - C, C - E \rangle \tag{17}$$

From (17) and the fact that $\triangle Pvx$ and $\triangle PCE$ share P as a common vertex, it follows that they are *similar*. Also, since $QxPR$ is a parallelogram, we have $\text{len}(Q - R) = \text{len}(P - x)$. Thus, the following derivations follow

$$\text{SIM } PVxPCE \Longrightarrow \frac{\text{len}(P - E)}{\text{len}(P - C)} = \frac{\text{len}(P - x)}{\text{len}(P - v)} = \frac{\text{len}(Q - R)}{\text{len}(P - v)} = \frac{\text{len}(A - C)}{\text{len}(P - C)} \tag{18}$$

One of the substitution used in (18) follows from (16) proved in the previous step. The equations above verify Newton's ratios.

Step 3: Showing $\frac{L \cdot Pv}{Gv \cdot Pv} = \frac{L}{Gv}$

The proof of the ratio

$$\frac{L \times \text{len}(P - v)}{\text{len}(G - v) \times \text{len}(P - v)} = \frac{L}{\text{len}(G - v)} \tag{19}$$

is trivial and we will not expand on it. We only note that the constant L is known as the *latus rectum*[2] of the ellipse at A.

Step 4: Showing $\frac{Gv \cdot Pv}{Qv^2} = \frac{PC^2}{CD^2}$

By Apollonius I.21 [1], *if the lines DC and Qv are dropped ordinatewise to the diameter PG, the squares on them DC^2 and Qv^2 will be to each other as the areas contained by the straight lines cut off GC, CP, and Gv, vP on diameter PG.* Algebraically, we proved the following property of the ellipse,

$$\frac{\text{len}(D - C)^2}{\text{len}(Q - v)^2} = \frac{\text{len}(G - C) \times \text{len}(C - P)}{\text{len}(G - v) \times \text{len}(P - v)}$$
$$= \frac{\text{len}(P - C)^2}{\text{len}(G - v) \times \text{len}(P - v)}$$

Rearranging the terms, we get the required ratio,

$$\frac{\text{len}(G - v) \times \text{len}(P - v)}{\text{len}(Q - v)^2} = \frac{\text{len}(P - C)^2}{\text{len}(D - C)^2} \tag{20}$$

Step 5: Showing $\frac{Qv^2}{QT^3} \approx \frac{CD^2}{CB^2}$ and intermediate ratios

In Figure 7, we have the additional property,

---

[2] The latus rectum is defined as $L = 2 \times \text{len}(B - C)^2 / \text{len}(A - C)$

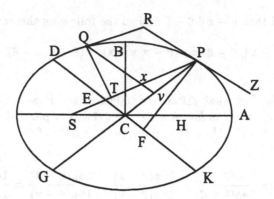

Fig. 7. Construction for Step 5 of Proposition 11

$- PF \perp DK$

Again, it can be easily proved that $Qx \parallel EF$. The following theorem then follows from Euclid I.29 as given in Section 2.3

$$Q - x \parallel E - F \Longrightarrow \langle Q - x, x - E \rangle = \langle F - E, E - x \rangle$$
$$\Longrightarrow \langle Q - x, x - T \rangle = \langle F - E, E - P \rangle \qquad (21)$$

Since $\langle P - F, F - E, \rangle = \langle x - T, T - Q \rangle = \pi/2$ and (21), it follows that $\triangle PEF$ and $\triangle QxT$ are similar. The next theorems (using (16) where needed) then hold and verify Newton's intermediate results for the current Step.

$$\textbf{SIM PEFQxT} \Longrightarrow \frac{\text{len}(Q - x)^2}{\text{len}(Q - T)^2} = \frac{\text{len}(P - E)^2}{\text{len}(P - F)^2} = \frac{\text{len}(C - A)^2}{\text{len}(P - F)^2} \qquad (22)$$

Newton's **Lemma 12** (See Figure 2) is now needed for the next result. According to the Lemma, the parallelogram circumscribed about DK and PG is equal to the parallelogram circumscribed about the major and minor axes of the ellipse. Thence, we have the following theorem

$$\text{len}(C - A) \times \text{len}(C - B) = \text{len}(C - D) \times \text{len}(P - F) \qquad (23)$$

Rearranging (23), we have $\text{len}(C - A)/\text{len}(P - F) = \text{len}(C - D)/\text{len}(C - B)$ and substituting in (22), leads to

$$\frac{\text{len}(Q - x)^2}{\text{len}(Q - T)^2} = \frac{\text{len}(C - D)^2}{\text{len}(C - B)^2} \qquad (24)$$

By Newton's **Lemma 7, Corollary 2**, when the distance between Q and P becomes infinitesimal as they coincide, we have the following result [13]:

$$\frac{\text{len}(Q-v)}{\text{len}(Q-x)} \approx 1 \tag{25}$$

Now, to reach the final result for this step, we need to substitute $\text{len}(Q-v)$ for $\text{len}(Q-x)$ in (22). However, we cannot simply carry out the substitution even though the quantities are infinitely close. Indeed, one has to be careful when multiplying the quantities on both sides of the $\approx$ relation because they might no longer be infinitely close after the multiplication. Consider, the non-zero infinitesimal ϵ,

$$\epsilon \approx \epsilon^2 \text{ but } \epsilon \times 1/\epsilon \not\approx \epsilon^2 \times 1/\epsilon$$

It is possible, however, to multiply two infinitely close quantities by any finite quantity; the results are still infinitely close. This follows from the theorem:

$$x \approx y \wedge u \in \text{Finite} \implies x \times u \approx y \times u \tag{26}$$

Now, assuming that $\text{len}(C-D)$ and $\text{len}(C-B)$ are both finite but not infinitesimal (for example, $\text{len}(C-D), \text{len}(C-B) \in \mathbb{R}$), then $\text{len}(C-D)/\text{len}(C-B)$ is **Finite**. Hence, the ratio of *infinitesimals* $\text{len}(Q-x)/\text{len}(Q-T)$ is **Finite**. Therefore, from (25), (26), and using (24) the following theorem is derived:

$$\frac{\text{len}(Q-v)^2}{\text{len}(Q-T)^2} \approx \frac{\text{len}(C-D)^2}{\text{len}(C-B)^2} \tag{27}$$

This gives the result that we wanted for the fifth step of the proof of Proposition 11. We are now ready for putting all the various results together in the next and final step. This will then conclude the formal proof of the Proposition.

Step 6: Putting the ratios together

Combining (20) and (27), with the help of theorem (26) and some algebra yields,

$$\frac{\text{len}(Q-v)^2}{\text{len}(Q-T)^2} \approx \frac{\text{len}(C-D)^2}{\text{len}(C-B)^2} \wedge \frac{\text{len}(G-v) \times \text{len}(P-v)}{\text{len}(Q-v)^2} \in \text{Finite}$$

$$\implies \frac{\text{len}(G-v) \times \text{len}(P-v)}{\text{len}(Q-T)^2} \approx \frac{\text{len}(P-C)^2}{\text{len}(C-B)^2} \tag{28}$$

which is combined with (19) to derive the next relation between ratios. The reader can check that both sides of the $\approx$ relation are multiplied by finite quantities ensuring the results are infinitely close:

$$\frac{L \times \text{len}(P - v)}{\text{len}(Q - T)^2} \approx \frac{\text{len}(P - C)^2 \times L}{\text{len}(C - B)^2 \times \text{len}(G - v)} \tag{29}$$

The next task is to combine the last result (29) with (18) to yield the following ratio which is equivalent to Newton's "$L \cdot QR : QT^2 = AC \cdot L \cdot PC^2 \cdot CD^2 : PC \cdot Gv \cdot CD^2 \cdot CB^2$"

$$\frac{L \times \text{len}(Q - R)}{\text{len}(Q - T)^2} \approx \frac{\text{len}(P - C) \times L \times \text{len}(A - C)}{\text{len}(C - B)^2 \times \text{len}(G - v)} \tag{30}$$

But, we know that $L = 2 \times \text{len}(B - C)^2/\text{len}(A - C)$, so (30) can be further simplified to give Newton's other ratio "$L \cdot QR : QT^2 = 2PC : Gv$"

$$\frac{L \times \text{len}(Q - R)}{\text{len}(Q - T)^2} \approx \frac{2 \times \text{len}(P - C)}{\text{len}(G - v)} \tag{31}$$

Once these ratios have been derived, Newton says **"But the points Q and P coinciding, $2PC$ and Gv are equal. And therefore the quantities $L \cdot QR$ and QT^2, proportional to these are also equal."**

We formalise this by showing that $\text{len}(P - v) \approx 0$ as the distance between Q and P becomes infinitesimal; thus, it follows that $2 \times \text{len}(P - C)/\text{len}(G - v) \approx 1$ and so, using (31) and the transitivity of $\approx$, we have the result

$$\frac{L \times \text{len}(Q - R)}{\text{len}(Q - T)^2} \approx 1 \tag{32}$$

The final step in Newton's derivation is **"Let those equals be multiplied by $\frac{SP^2}{QR}$ and $L \cdot SP^2$ will become equal to $\frac{SP^2 \cdot QT^2}{QR}$"**. This final ratio gives the geometric representation for the force, as we showed in Section 4.2, and hence enables Newton to deduce immediately that the centripetal force obeys an inverse square law.

We would like to derive Newton's result in the *same way*, but remark that

$$\text{len}(S - P) \in \text{Finite} - \text{Infinitesimal} \;\wedge\; \text{len}(Q - R) \in \text{Infinitesimal}$$
$$\implies \frac{\text{len}(S - P)^2}{\text{len}(Q - R)} \in \text{Infinite} \tag{33}$$

as Q and P become coincident. So, there seems to be a problem with simply multiplying (32) by Newton's ratio SP^2/QR since we cannot ensure that the results are infinitely close. Our formal framework *forbids* the multiplication that Newton does as the result is not necessarily a theorem!

Therefore, we need to find an **alternative** way of arriving at the same result as Newton. Recall from Section 4.2, that we have proved the following geometric representation for the centripetal force:

$$\textit{force} \approx k \times \frac{\text{len}(Q - R)}{\text{len}(Q - T)^2} \times \frac{1}{\text{len}(S - P)^2} \tag{34}$$

Now from (32), we can deduce that since $L \in$ Finite $-$ Infinitesimal, the following theorems hold

$$\frac{\text{len}(Q-R)}{\text{len}(Q-T)^2} \in \text{Finite} - \text{Infinitesimal} \tag{35}$$

$$\frac{\text{len}(Q-T)^2}{\text{len}(Q-R)} \approx L \tag{36}$$

$$\frac{\text{len}(Q-T)^2}{\text{len}(Q-R)} \in \text{Finite} \tag{37}$$

Since (35) holds and $1/\text{len}(S-P)^2 \in$ Finite, it follows that *force* $\in$ Finite and so we can now use the following theorem about the product of *finite*, infinitely close quantities

$$a \approx b \wedge c \approx d \wedge a \in \text{Finite} \wedge c \in \text{Finite} \Longrightarrow a \times c \approx b \times d$$

with (34), (36), and (37) to yield

$$force \times L \approx k \times \frac{\text{len}(Q-R)}{\text{len}(Q-T)^2} \times \frac{1}{\text{len}(S-P)^2} \times \frac{\text{len}(Q-T)^2}{\text{len}(Q-R)}$$

$$\approx k \times \frac{1}{\text{len}(S-P)^2} \tag{38}$$

Note that we also used the fact that $\approx$ is symmetric in the derivation above. Finally from (38), we get to the celebrated result since L is finite (real) and constant for a given ellipse,

$$force \quad \approx \quad \frac{k}{L} \times \frac{1}{\text{len}(S-P)^2}$$

$$force \propto_{ultimate} \frac{1}{\text{len}(S-P)^2} \tag{39}$$

5   Final Comments

We would like to conclude by mentioning some important aspects of this mechanisation and possible changes to the geometry theory that could improve automation. We also briefly review what we have achieved.

5.1 On Finite Geometric Witnesses

We have made an important remark about steps involving infinitesimals, ratios of infinitesimals and the infinitely close relation. Whenever we are dealing with such ratios, care needs to be exercised as we cannot be sure what the result of dividing two infinitesimals is: it can be infinitesimal, finite or infinite. We notice, when carrying out our formalisation, that whenever Newton is manipulating the ratio of vanishing quantities, he usually makes sure that this can be expressed in terms of some finite quantity as in the proof for Step 5 of Section 4.3. Thus, the ratio of infinitesimals is shown to be infinitely close or even equal to some finite quantity. This ensures that such a finite ratio can be used safely and soundly within our framework. The importance of setting up such *finite geometric witnesses* cannot be under-stated since the rigour of NSA might prevent steps involving ratios of infinitesimals from being carried out otherwise. Indeed, we have seen that the lack of a finite witness in the last step of Newton's original argument prevents us from deriving the final result in the same way as he does. The alternative way we went about deriving the result, however, is sound and follows from rules that have been proved within our framework.

5.2 Further Work

In our previous work [13], we mentioned the existence of other methods, such as the Clifford algebra, that provide short and readable proofs [4, 11]. Although these algebraic techniques are more difficult to relate with the geometric concepts that are actually used in Newton's reasoning, interesting work done by Wang et al. has come to our attention in which powerful sets of rewrite rules have been derived to carry out proofs in Euclidean geometry [12, 16]. It would be interesting to see how these could be integrated with Isabelle's powerful simplifier to provide a greater degree of automation in some of our proofs. In a sense, such an approach would match in the level of details some of the results that Newton states (as obvious) and does not prove in depth.

As an interesting observation, it is worth noting that the Kepler Problem can be proved, or even discovered, using algebraic computations. This has been demonstrated through the work on mechanics done by Wu [17], and also by Chou and Gao [7, 8], in the early nineties.

5.3 Conclusions

We have described in detail the machine proof of Proposition 11 of the *Principia* and shown how the theories developed in Isabelle can be used to derive Newton's geometric representations for physical concepts. We have used the same combination of geometry and NSA rules introduced in our previous work to confirm, through a study of one of the most important Propositions of the Principia, that Newton's geometric and ultimate procedures can be cast within the rigour of our formal framework. The discovery of a step in Newton's reasoning that could not be justified formally– in contrast with other ones where Newton explicitly sets up finite witnesses– is an important one. The alternative derivation presented in this work shows how to use our rules to deduce the same result soundly.

Once again, the mechanisation of results from the *Principia* has been an interesting and challenging exercise. Newton's original reasoning, though complex and often hard to follow, displays the impressive deductive power of geometry. The addition of infinitesimal notions results in a richer, more powerful geometry in which new properties can emerge in ultimate situation. Moreover, we now have new, powerful tools to study the model built on Newton's exposition of the physical world.

Acknowledgements

Support from the ORS and Cambridge Commonwealth Trusts is gratefully acknowledged. This research was also funded by EPSRC grant GR/K57381 'Mechanizing Temporal Reasoning'. We also wish to thank Data Connection for supporting us with a Postgraduate Travel Award.

References

1. Apollonius. *Conics*. Translation by R. Catesby Taliaferro. In *Great Books of the Western World*, Vol. 11, Chicago, Encyclopedia Britannica (1939)
2. J. B. Brackenridge. Newton's mature dynamics and the Principia: A simplified solution to the Kepler Problem. Historia Mathematica **16** (1989), 36—45.
3. J. B. Brackenridge. *The Key to Newton's Dynamics: The Kepler Problem and Newton's Principia*. University of California Press, 1995.
4. H.-B. Li and M.-T. Cheng. Clifford algebraic reduction method for mechanical theorem proving in differential geometry. *J. Automated Reasoning* **21** (1998), 1–21.
5. S. C. Chou, X. S. Gao, and J. Z. Zhang. Automated generation of readable proofs with geometric invariants, I. Multiple and shortest proof generation. *J. Automated Reasoning* **17** (1996), 325–347.
6. S. C. Chou, X. S. Gao, and J. Z. Zhang. Automated generation of readable proofs with geometric invariants, II. Theorem proving with full-angles. *J. Automated Reasoning* **17** (1996), 349–370.
7. S. C. Chou, X. S. Gao. Automated reasoning in differential geometry and mechanics: II. Mechanical theorem proving. *J. Automated Reasoning* **10** (1993), 173–189.
8. S. C. Chou, X. S. Gao. Automated reasoning in differential geometry and mechanics: III. Mechanical formula derivation. *IFIP Transaction on Automated Reasoning* 1–12, North-Holland, 1993.
9. D. Densmore. *Newton's Principia: The Central Argument*. Green Lion Press, Santa Fe, New Mexico, 1996.
10. Euclid. *Elements*. Translation by Thomas L. Heath. Dover Publications Inc., 1956.
11. D. Wang. Geometry Machines: From AI to SMC. Proceedings of the 3rd International Conference on Artificial Intelligence and Symbolic Mathematical Computation (Stey, Austria, September 1996), LNCS 1138, 213–239.
12. S. Fevre and D. Wang. Proving geometric theorems using Clifford algebra and rewrite rules. Proceedings of the 15th International Conference on Automated Deduction (CADE-15) (Lindau, Germany, July 5-10, 1998), LNAI 1421, 17–32.

13. J. D. Fleuriot and L. C. Paulson. A combination of nonstandard analysis and geometry theorem proving, with application to Newton's Principia. Proceedings of the 15th International Conference on Automated Deduction (CADE-15) (Lindau, Germany, July 5-10, 1998), LNAI 1421, 3-16.

14. I. Newton. *The Mathematical Principles of Natural Philosophy*. Third edition, 1726. Translation by A. Motte (1729). Revised by F. Cajory (1934). University of California Press.

15. D. T. Whiteside. The mathematical principles underlying Newton's *Principia Mathematica*. Glasgow University Publication 138, 1970.

16. D. Wang. Clifford algebraic calculus for geometric reasoning with application to computer vision. Automated Deduction in Geometry (D. Wang et al., Eds.), LNAI 1360, Springer-Verlag, Berlin Heidelberg, 1997, 115-140.

17. W.-t. Wu. Mechanical theorem proving of differential geometries and some of its applications in mechanics. *J. Automated Reasoning* **7** (1991), 171-191.

Readable Machine Solving in Geometry and ICAI Software MSG *

Chuan-Zhong Li[1] and Jing-Zhong Zhang[2]

[1] Institute for Educational Software, Guangzhou Normal University
[2] Joint Laboratory for Auto-Reasoning, CICA and GNU

Abstract. Readability is a fuzzy concept and there are several levels of readability. Readable machine solving avails the development of mathematics culture and is useful to education. There are several methods of readable machine solving, such as the logic method, points elimination method, geometry information searching system or deductive database method. Based on these methods, some new types of educational software have been developed. As an example, the intelligent educational software "Mathematics Lab: Solid Geometry" (MSG) is introduced in this paper.

1 Introduction

Because of various reasons, machine proving of geometry theorems is one of the main focuses of automated reasoning. Since the pioneering work of Wu [28] in 1977, highly successful algebraic methods for solving geometry problems have been developed. For most geometry statements, one can assert whether it is true or false by a computer system using these methods within seconds [1].

These algebraic methods generally can only tell whether a statement is true or not. If one wants to know the proving process, one usually has to look at tedious computations of polynomials. In other words, the solution produced by algebraic methods is generally not readable.

Researchers have been studying automated generation of readable solving using computer systems since the work by Gelernter and others [20]. With a great amount of efforts in the past 40 years, especially with the new techniques developed in recent years [31, 2, 29, 10], the advance in this direction is noticeable. There are now two classes of methods, the *points elimination* methods and the geometry deductive database methods (abbr. GDD method), to produce readable solutions for geometric problems. Using these methods, about one thousand non-trivial geometric problems were solved by computer programs and most of the solutions are short and readable.

The research on *readable machine solving* is useful in education. Two pieces of ICAI software named *Geometry Experts* (abbr. GEX) [17, 18, 16] and *Mathematics Lab: Solid Geometry* (MSG) are developed based on this work. Now both

---

* This work was supported in part by the Chinese National Science Foundation, the 863 Foundation and DOES company.

are used in about 400 middle schools in China, and were warmly welcomed by teachers and students.

This paper reviews recent advances on readable machine solving and then introduces the ICAI software MSG. In Section 2, we give an introduction to the readable machine solving methods. In Section 3, we introduce the points elimination methods with some examples. In Section 4, we explain GDD (GISS) methods briefly. Section 5 introduces the structure and working principle of software MSG, and lists its functions. In the last Section, we give some comments on future development and propose some problems for future research.

2 What Is Readable Solving and Why Readability Is Wanted

By *readable solving* we mean algorithms that not only find the solutions of certain problems but also generate a human readable solving process such that one can check why the solution is true and how to get to the conclusion.

Certainly, *readability* is a fuzzy concept. Generally, if a solving process is very long, it will be difficult to check and understand. So we may say that a very long solution is *not readable*. But we cannot determine exactly how long is very long or how short is readable. Moreover, there are several levels of *readability*. A proof for a geometry theorem may be *readable* to a mathematician but not to a student in middle school. Therefore, we use the word *readable* with a different meaning in different situations. For instance, while talking about the education, *readable solving* means that it can be checked and understood easily by most students.

There are two obvious reasons for seeking readability in problem solving:

1 Readability avails the development of mathematics culture. Mathematics is not only a science field but also a *culture*. The development of mathematical culture needs the support from most people. If people were told only the mathematical conclusion but can not check and understand the process, they will lose their interest in mathematics.
2 Readability is useful to education. While the students study mathematics, they mainly learn the ideas, principles and methods. So they should understand why a conclusion is true and how to get it. In our viewpoint, readable machine solving can play an important role in ICAI (Intelligent Computer Aided Instruction).

Because it is very difficult to write programs to produce readable solutions for various mathematics problems, most mathematics software can only give the final conclusion. The research of readable machine solving, especially in geometry, is a very attractive field for a long time to come.

3 A Sketch of the Point Elimination Method

In 1992, a new idea for automated geometry solving was proposed and implemented [31, 2]. It is called the *point elimination method*. The program based on

the method can produce short and readable proofs for most non-trivial geometry theorems. Roughly speaking, this approach is to represent the hypotheses in
a statement in a *constructive way* and to describe the conclusion of the statement by equations of *geometric quantities*. The proving process is to eliminate
geometry objects from the conclusion by using a few *formulae according to the
constructions and the geometric quantities*. We will use two examples to illustrate
this idea.

We will use capital English letters to denote points. We denote by $\overline{AB}$ the
signed length of the oriented segment from A to B and its absolute value by AB.
The *signed area* S_{ABC} of triangle ABC is the usual area with a sign depending
on the order of the three vertices of the triangle. About the relationship between
the area and the length, there is a very useful tool:

Proposition 1 (The Co-side Theorem). *Let M be the intersection of two
non-parallel lines AB and PQ and $Q \neq M$. Then $\frac{\overline{PM}}{\overline{QM}} = \frac{S_{PAB}}{S_{QAB}}$.*

The co-side theorem is often used to eliminate the intersecting point of two
lines. This will be shown clearly in the following example.

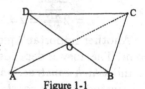

Figure 1-1

*Example 1. Let O be the intersection of the two diago-
nals AC and BD of a parallelogram $ABCD$. Show that
O is the midpoint of AC, or $\frac{\overline{AO}}{\overline{OC}} = 1$.*

Analysis. To show $\frac{\overline{AO}}{\overline{OC}} = 1$, by using the Co-side Theorem we have

$$\frac{\overline{AO}}{\overline{OC}} = \frac{S_{ABD}}{S_{BCD}}.$$

Point O is thus eliminated from the conclusion. Now the problem becomes to
show $S_{ABD} = S_{BCD}$. Because $AB \parallel CD$ and $AD \parallel BC$, we have

$$S_{ABD} = S_{ABC}, S_{BCD} = S_{ABC}.$$

So point D is also eliminated and now it is evident that the conclusion is valid.

It is now possible to get some basic idea about the Points Elimination method
from the above example:

(1) The *geometric quantities.* Here they are ratios of signed lengths and signed
areas. We use the algebraic expressions in geometric quantities to represent
the goals of problems. For more problems, more geometric quantities would
be introduced.
(2) The *constructions.* In example 1, we can use the constructive process to
represent the conditions of the problem. Points are introduced by constructions in the order: A, B, C, D, O; and are eliminated in the reverse order:
O, D, A, B, C. Of course, we can use more kinds of constructions to enlarge
the class of problems.

3) The *formulae*. To eliminate a point, formulae are needed. The main task in building a proving system is to choose the proper formulae for eliminating points.

Let us use the idea of Eliminating Points to solve another problem:

Example 2. M is the midpoint of the side BC of triangle ABC and N is on the side AC such that $\overline{AC} = 3\overline{NC}$. Point P is the intersection of lines AM and BN. Find the ratio $\frac{\overline{AP}}{\overline{PM}}$.

For the problem, the constructive process can be:

C1 Take any three non-collinear points A,B and C.
C2 Draw point M as the midpoint of segment BC.
C3 Take point N on segment AC such that $\overline{AC} = 3\overline{NC}$.
C4 Draw point P as the intersection of two lines AM and BN.

Solution by Points Elimination method

$$\frac{\overline{AP}}{\overline{PM}} = \frac{S_{ABN}}{S_{BMN}} \quad \text{(To eliminate } P \text{ by Co-side Theorem)}$$
$$= \frac{S_{ABN}}{\frac{S_{BCN}}{2}} \quad \text{(To eliminate } M \text{)}$$
$$= 2\frac{\overline{AN}}{\overline{NC}} = 2*2 = 4 \quad \text{(To eliminate } N \text{)}.$$

Another important question is how to select the formula to eliminate a given point. The answer depends on *what kind of geometric quantity the point appeared in* and *which construction the point was introduced by*. As an instance, if point P is introduced by construction *intersection of two lines* and appeared in geometric quantity *ratio of two segments*, then it can be eliminated by using the co-side theorem. Therefore, if m kinds of geometric quantities and n types of constructions are used, the number of formulae needed for eliminating points should be about the nm.

Based on the idea of *eliminating points*, we can develop various methods for geometry solving by choosing different *geometric quantities, constructions* and *formulae* for eliminating points or other geometry objects. In fact, a series of algorithms based on this idea have been developed. The *area method* for Euclidean plane geometry was first proposed in [31] and improved in [3, 2, 6]; the *volume method* for solid geometry was given in [8]. To get more readable solutions for some classes of problems, the full-angle method [5], the vector method and complex number method [4, 7, 6, 26] were developed. The eliminating method for non-Euclidean geometry, by which many new theorems have been discovered, was established later in [29]. In the following, we sketch only the methods for Euclidean plane geometry. They are the area method, the vector or complex number method and the full angle method.

More recently, the points elimination method has been extended to the classes of geometry problems in which the geometric quantities can be represented as elements in the Clifford Algebra [22, 27, 14, 30]. Using the extended Points Elimination method, a computer program has been developed, by which one can solve problems involving lines and conics and the solutions are generally short and readable [30].

Among all the above methods, the area method is the most useful one. It is a complete method for constructive propositions in Euclidean plane geometry. In the area method, we need three geometric quantities: the sign areas, ratios of collinear or parallel oriented segments and a new quantity named *Pythagorean Difference*. The *Pythagorean Difference* is used to treat geometry problems involving perpendiculars or angles. For points A, B, and C, the *Pythagorean difference* P_{ABC} is defined to be $P_{ABC} = \overline{AB}^2 + \overline{CB}^2 - \overline{AC}^2$.

For four points A, B, C and D, the notation $AB \perp CD$ implies that one of the following conditions is true: $A = B$, or $C = D$, or line AB is perpendicular to line CD. For a quadrilateral $ABCD$, let $P_{ABCD} = P_{ABD} - P_{CBD} = \overline{AB}^2 + \overline{CD}^2 - \overline{BC}^2 - \overline{DA}^2$. Then, the Pythagorean theorem can be generalized as follow.

Proposition 2. $AC \perp BD$ iff $P_{ABD} = P_{CBD}$ or $P_{ABCD} = 0$.

As an application of the above proposition, we give:

Example 3 (The Orthocenter Theorem). Let the two altitudes AF and BE of triangle ABC meet in H. Show that $CH \perp AB$.

Figure 1-2

Proof. By Proposition 2, we need only to show $P_{ACH} = P_{BCH}$. Since $BH \perp AC$ and $AH \perp BC$, by Proposition 2 we can eliminate point H: $P_{ACH} = P_{ACB}$; $P_{BCH} = P_{BCA}$. Thus we have the proof:

$$\frac{P_{ACH}}{P_{BCH}} = \frac{P_{ACB}}{P_{BCA}} = \frac{P_{ACB}}{P_{ACB}} = 1.$$

In the construction process, we always use several introduced points to decide a new point. Such as, if points A,B,C,D are known, we can get the intersection of lines AB and CD. All of the secrets to eliminate points are how to represent the position of a new point with that of the old ones. By the co-side theorem, we use the ratio of areas to represent the position of intersection.

Now we are going to explain the vector method or complex number method briefly. For each point P in the plane, let $\overrightarrow{P}$ be the corresponding vector from the origin to P or complex number, and $\tilde{P}$ the conjugate number of $\overrightarrow{P}$. Then we denote by $\overrightarrow{AB} = \overrightarrow{B} - \overrightarrow{A}$ the vector from point A to point B.

The *inner product* of vectors $\overrightarrow{AB}$ and $\overrightarrow{CD}$ is denoted by $\langle \overrightarrow{AB}, \overrightarrow{CD} \rangle$, and the *exterior product* by $[\overrightarrow{AB}, \overrightarrow{CD}]$.

We can define the *signed area of a triangle ABC* as

$$S_{ABC} = \frac{[\overrightarrow{AB}, \overrightarrow{AC}]}{2}.$$

Then the exterior product $[\overrightarrow{AB}, \overrightarrow{CD}]$ is actually the *signed area or the quadrilateral $ACBD$*, i.e.,

$$S_{ACBD} = S_{ACD} - S_{BCD} = \frac{[\overrightarrow{AB}, \overrightarrow{CD}]}{2}.$$

Then we have

$$S_{ABC} = \frac{[\overrightarrow{AB}, \overrightarrow{AC}]}{2} = \frac{1}{4i}((\tilde{B} - \tilde{A})(\overrightarrow{C} - \overrightarrow{A}) - (\overrightarrow{B} - \overrightarrow{A})(\tilde{C} - \tilde{A})).$$
$$P_{ABC} = 2\langle \overrightarrow{BA}, \overrightarrow{BC} \rangle = (\overrightarrow{B} - \overrightarrow{A})(\tilde{C} - \tilde{A}) + (\tilde{B} - \tilde{A})(\overrightarrow{C} - \overrightarrow{A}).$$

Therefore, the area method for Euclidean geometry can be translated into the language of vectors or complex numbers. The solutions thus produced are generally longer than those produced by the area method. But the vector or complex method is still useful for solving the problems in some classes. Because this method has two advantages:

(1) The vectors or complex numbers themselves can be eliminated as geometric quantities directly.
(2) The vectors or complex numbers are useful to represent the relationships about angles or similar triangles.

We give an example for vector method:

Example 4 (Cantor's Theorem). Suppose A, B, C, D are on a circle with center O and E, F, G be the midpoints of CD, AB and AD respectively. If N be the intersection of the lines through E parallel to OF and through F parallel to OE, then the conclusion can be represented by

$$((\langle \overrightarrow{GB}, \overrightarrow{BC} \rangle = \langle \overrightarrow{NB}, \overrightarrow{BC} \rangle))$$

(i.e. $GN \perp BC$)

Here is the machine proof by the vector method.

$$\frac{\langle \overrightarrow{BC}, \overrightarrow{BG} \rangle}{\langle \overrightarrow{BC}, \overrightarrow{BN} \rangle}$$ The Conclusion.

$$= \frac{\langle \overrightarrow{BC}, \overrightarrow{BG} \rangle \cdot 1}{(\langle \overrightarrow{BC}, \overrightarrow{BE} \rangle + \langle \overrightarrow{BC}, \overrightarrow{BF} \rangle - \langle \overrightarrow{BC}, \overrightarrow{BO} \rangle) \cdot 1}$$ S1. Elim. point N.

$$= \frac{\langle \overrightarrow{BC}, \overrightarrow{BG} \rangle \cdot 1}{(\langle \overrightarrow{BC}, \overrightarrow{BF} \rangle - \langle \overrightarrow{BC}, \overrightarrow{BO} \rangle + 1/2\langle \overrightarrow{BC}, \overrightarrow{BD} \rangle + 1/2\langle \overrightarrow{CB}, \overrightarrow{CB} \rangle) \cdot 1}$$ S2. Elim. point E.

$$= \frac{(2) \cdot \langle \overrightarrow{BC}, \overrightarrow{BG} \rangle \cdot 1}{(-2\langle \overrightarrow{BC}, \overrightarrow{BO} \rangle + \langle \overrightarrow{BC}, \overrightarrow{BD} \rangle + \langle \overrightarrow{CB}, \overrightarrow{CB} \rangle + \langle \overrightarrow{BA}, \overrightarrow{BC} \rangle) \cdot 1}$$ S3. Elim. point F.

$$= \frac{(-2) \cdot (1/2\langle \overrightarrow{BC}, \overrightarrow{BD} \rangle + 1/2\langle \overrightarrow{BA}, \overrightarrow{BC} \rangle) \cdot 1}{(2\langle \overrightarrow{BC}, \overrightarrow{BO} \rangle - \langle \overrightarrow{BC}, \overrightarrow{BD} \rangle - \langle \overrightarrow{CB}, \overrightarrow{CB} \rangle - \langle \overrightarrow{BA}, \overrightarrow{BC} \rangle) \cdot 1}$$ S4. Elim. point G.

$$= \frac{-(\langle \overrightarrow{BC}, \overrightarrow{BD} \rangle + \langle \overrightarrow{BA}, \overrightarrow{BC} \rangle) \cdot (2)}{(-2\langle \overrightarrow{BC}, \overrightarrow{BD} \rangle - 2\langle \overrightarrow{BA}, \overrightarrow{BC} \rangle) \cdot 1}$$ S5. Elim. point O.

$$= 1$$ S6. Simplify.

For solving geometric problems about angles, using *angle* as the main geometric quantity directly will be better than using both *area* and *vector*. We use a new kind of angle called full-angle, which is more convenient to machine solving. Intuitively, a full-angle $\angle[u, v]$ is the directed angle from line u to line v. Two full-angles $\angle[l, m]$ and $\angle[u, v]$ are equal if there exists a rotation K such that

$K(l) \parallel u$ and $K(m) \parallel v$. For the geometric meaning of the addition of full-angles, let l, m, u, and v be four lines and K be a rotation such that $K(l) \parallel v$. Then $\angle[u, v] + \angle[l, m] = \angle[u, K(m)]$.

For all parallel lines $AB \parallel PQ$, $\angle[0] = \angle[AB, PQ]$ is a constant. For all perpendicular lines $AB \perp PQ$, $\angle[1] = \angle[AB, PQ]$ is a constant. It is clear that $\angle[1] + \angle[1] = \angle[0]$ and $\angle[u, v] + \angle[0] = \angle[u, v]$.

Using full-angle method, many geometry properties can be described concisely. For instance, we have:

Proposition 3. *F1 For any line UV, $\angle[AB, CD] = \angle[AB, UV] + \angle[UV, CD]$.*
F2 PQ is parallel to UV iff $\angle[AB, PQ] = \angle[AB, UV]$ for any point $A \neq B$.
F3 PQ is perpendicular to UV iff $\angle[AB, PQ] = \angle[1] + \angle[AB, UV]$.
F4 For three non-collinear points X, A, and B, $XA = XB$ iff $\angle[AX, AB] = \angle[AB, XB]$.
*F5 (**The Inscribed Angle Theorem**) Four non-collinear points A, B, C, and D are cyclic iff $\angle[AD, CD] = \angle[AB, CB]$.*

For ordinary angles, above propositions are true only under certain additional conditions like "a line is inside an angle", "two points are on the same side of a line", etc.

Even using the above simple properties of full-angles, we can produce short solutions for many difficult geometry problems automatically. The technique of eliminating points is now extended to that of *eliminating lines*. For instance, if PQ is perpendicular to UV, by F3 we can eliminate line PQ from full-angle $\angle[AB, PQ]$: $\angle[AB, PQ] = \angle[1] + \angle[AB, UV]$. We illustrate the method with an example.

*Example 5 (*Miquel's Axiom*).* [1] If four circles are arranged in sequence, each two successive circles intersecting, and a circle pass through one pair of each such pair of intersections, then the remaining intersections lie on another circle.

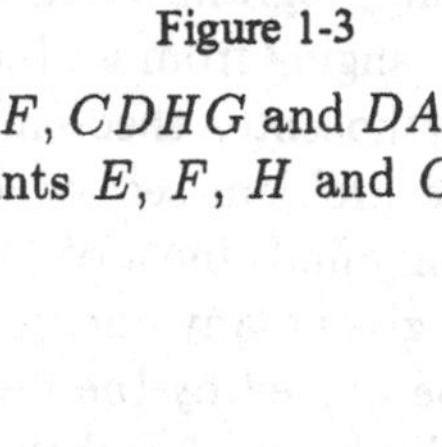

Figure 1-3

Let us assume that the four circles are $ABFE, BCGF, CDHG$ and $DAEH$. Show that if points A, B, C and D are cyclic then points E, F, H and G are also cyclic.

The Machine Proof
$\angle[HG, HE] - \angle[FG, FE]$

  ($\angle[HG, HE] = \angle[HG, HD] + \angle[HD, HE] = \angle[GC, DC] - \angle[EA, DA]$,
  because cyclic(G, H, D, C), cyclic(E, H, D, A). (F1, F5))

$= -\angle[FG, FE] + \angle[GC, DC] - \angle[EA, DA]$

  ($\angle[FG, FE] = \angle[FG, FB] + \angle[FB, FE] = \angle[GC, CB] - \angle[EA, BA]$,

---

[1] All previous methods failed to prove this theorem automatically by far due to computer system memory limitation.

because cyclic(G, F, B, C), cyclic(E, F, B, A). (F1, F5))

$= \angle[GC,DC] - \angle[GC,CB] - \angle[EA,DA] + \angle[EA,BA]$

$\quad$ ($\angle[GC,DC] - \angle[GC,CB] = -\angle[DC,CB]$. (F1))

$= -\angle[EA,DA] + \angle[EA,BA] - \angle[DC,CB]$

$\quad$ ($-\angle[EA,DA] + \angle[EA,BA] = \angle[DA,BA]$. (F1))

$= -\angle[DC,CB] + \angle[DA,BA]$

$\quad$ ($\angle[DC,CB] = \angle[DA,BA]$,

$\quad$ because cyclic (C, D, B, A). (F5))

$= \angle[0]$

The full-angle method as reported here is not complete. By using the area method, we can give a complete method for proving theorems involved full-angles. Full-angles are also useful in GDD method, which will be mentioned in next Section.

4 The Geometric Information Searching System

To obtain proofs in a more traditional style, the algorithm based on the search method has been proposed in [9, 10]. It is called GDD (Geometry Deductive Database) method or GISS (Geometry Information Searching System) method. For a given geometric configuration, using GDD method, one can get the information bank containing all the messages about the figure that can be deduced from the built-in reasoning rules. It is meaning that the fixpoint

The work uses the *synthetic approaches* are closely related to GDD method. In [23], Nevins used a combination of forward chaining and backward chaining with emphasis on the forward chaining. But before [9], all previous synthetic work deals with theorems involving straight lines only and the fixpoint is not reached. The idea of adding auxiliary points and the problem of including circles into the program are discussed respectively in [25, 24] and [23] but both not implemented.

The program based on GDD has been tested with 160 geometry configurations ranging from well-known geometry theorems such as the centroid theorem, the orthocenter theorem, and Simson's theorem to problems recently proposed in the problem section of the *American Mathematical Monthly*. The program not only finds most of the well-known properties of these configurations but also often gives many unexpected results. As is known, most of the 160 theorems can not be proved by the previous programs based on synthetic approaches.

The basic ideas behind the GDD method, such as fixpoints and the treatment of negative clauses, come from the deductive database theory [15]. But the strength of GDD method is mainly based on the following improvements:

(1) We propose the idea of *structured deductive database*. In traditional deductive databases, each n-ary predicate is associated with an n-dimensional relation. This is not suitable for geometry for two reasons: the excessively large database and repetitive representation of information. On average, the

databases for the 160 tested geometry configurations would be of size 242,117 if using the traditional representation. This problem was solved in GDD by using some simple mathematical structures such as sequences and equivalent classes to represent facts in the database. The size of the structured database is one thousand times smaller.

(2) We propose a new *data-based search strategy* . GDD keeps a list of "new data" and for each new data the system searches the rule set (or intensional database) to find and apply the rules using this data. If using the data-based strategy, the redundant deductions caused by repeated application of a rule to the same facts will be automatically eliminated.

(3) Selecting appropriate rules. All the previous work based on the synthetic approach [20, 23] uses geometric rules about the congruent triangles as its basic geometric rules. Most of the basic results in geometry such as the centroid theorem, the orthocenter theorem, and Simson's theorem are beyond the scope of these rules without adding auxiliary points.

(4) Adding of auxiliary points. In GDD, about twenty rules of adding auxiliary points are implemented and thirty nine of the one hundred sixty configurations solved by the program need auxiliary points. This is the first implementation of a nontrivial set of rules of adding auxiliary points.

(5) Numerical diagrams of geometry statements are used as models to deal with the negative information in the rules. The program can construct the diagram automatically for a class of constructive geometry statements. In all the previous work using diagrams as models [20], the diagrams are constructed by the user.

Here we give an example for the GDD method:

Example 6. Let the three altitudes of a triangle ABC are AD,BE,CF. Also point D is on the segment BC. Show $\angle[FDA] = \angle[ADE]$.

The conclusion can be represented by full-angles in form

$$\angle[DF, DA] = \angle[DA, DE]$$

The machine proof by GDD method is like the following:

(1) $\angle[DF, DA] = \angle[DA, DE]$ (by (2) and $DB \perp DA$).
(2) $\angle[DE, DB] = \angle[DB, DF]$ (by (3)+(4)).
(3) $\angle[BD, BF] = \angle[ED, EC]$ (by (5)).
(4) $\angle[BF, DF] = \angle[EC, CD]$ (by (6)).
(5) A, D, B, E are concyclic (by $DB \perp DA$ and $EB \perp EA$).
(6) A, D, C, F are concyclic (by $DC \perp DA$ and $FC \perp FA$).

By the GDD method, the computer system not only obtains the conclusion, but also produces a lot of informations of the figure, such as: 24 pairs of equal angles, 7 sets of similar triangles and 30 pairs of equal ratio of segments.

5 Educational Software MSG

There are several methods to produce readable solving for geometry problems. Every method has specific advantages. For the same theorem, one can get proofs with different styles by using different methods. Using the above method, about one thousand non-trivial geometry problems (most of them can be found in [2, 12, 13, 11]) were solved by computer programs and most of the solutions are readable and short.

These automated reasoning methods for readable machine solving are mature enough to be used in ICAI. Using these methods, we can make our educational software more intelligent and attractive. How to use automated reasoning methods to develop educational software? What would be the educational software with automated reasoning like? The intellective educational software MSG [21] may give us some explanation.

The full name of MSG is *Mathematics Lab: Solid Geometry*. It is one of a series educational software named *Mathematics Laboratory*. To fit the request of teachers and students in middle schools and high schools, the working style of MSG is more traditional than GEX. It means that, unlike the volume method for solid geometry in [8], we use only quantities that appeared in the geometry textbook. The program of MSG consists mainly of four parts. They are *constructing tools, information boxes, reasoning rules* and the *solving controls*.

The *constructing tools* allow users not only to draw solid geometric figure, but also send the constructing process to information boxes. Based on the information, MSG can do graphic translating and reasoning for solving.

A deductive database consists of *information boxes*. Every box is for saving one kind of information. There are four kinds of angles, six kinds of distances, three kinds of perpendicular and three kinds of parallel in solid geometry. So the database is more complex than that in plane geometry.

We use about 220 *reasoning rules*. A rule may be one axiom, lemma, theorem, formula, definition or skill for solving. They are chosen from plane and solid geometric textbooks in middle schools and high schools in China. When a solid figure was drawn and a problem was proposed, MSG can do reasoning by using part or all of the rules to get new information about the figure. Of course, new information will be put into the database. The process will stop until a solution is obtained.

The *solving controls* can select or remove rules used in reasoning. This is mainly according to the knowledge level of students or the request by user. When automated solving failed, the controls may give, by an expert knowledge bank, some suggestions about how to add auxiliary points and auxiliary lines.

We will show the functions of MSG using the following example.

Example 7. ABCD is a square, $EA \perp planeABCD$, $AF \perp BE$, F is the foot from A to line BE and G the foot from F to line CE. Show that $AG \perp EC$.

FUNCTION 1. Dynamical Construct and Generating Problems

To solve a problem by computer system, it is necessary to make the environment in which the user can describe his problems conveniently. After the software MSG was started, the construction menu or toolbar can help the user to draw various solid figures by mouse and to prepare proposing problems. This is the construction of our example:

1. Take two points A, B in space.
2. Draw a square $ABCD$.
3. Add segments EA, EB, EC.
4. Take a point E at the line through A and perpendicular to plane ABC.
5. Draw the foot F from point A to line BE.
6. Draw the foot G from point F to line EC.
7. Add segment AG.

Now the figure you constructed is as the following figure.

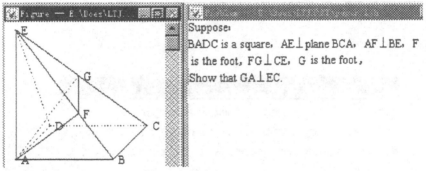

Figure 2-1

While you drag a free point by mouse, the graph would change, but it still keeps the geometry properties given during constructing. The program will automatically remember the constructing procedure. These graphs are dynamical and can be changed. They can always keep their properties given by the constructing procedure. In other words, they have intelligence and know the characteristics they should have. When a graph is constructed, you can add a conclusion such that some problems will be generated. In the above example, we should add the conclusion: $AG \perp EC$ (Figure 2-1).

FUNCTION 2. Automated Reasoning and Alternate Solving

Based on the basic properties of the graph and the additional conditions, using a series of rules (axioms, definitions, theorems and formulae), the program can find many geometric properties of the graph and solve the problems proposed by user. For this example, the program will automatically get the conclusion $AG \perp EC$ (Figure 2-2)

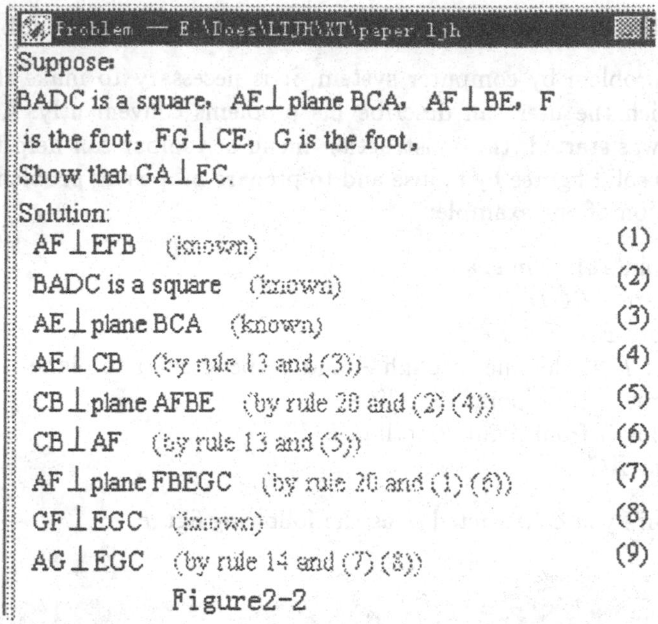

Problem — E:\Docs\LTJH\XT\paper.ljh

Suppose:
BADC is a square, AE⊥plane BCA, AF⊥BE, F is the foot, FG⊥CE, G is the foot,
Show that GA⊥EC.

Solution:

AF⊥EFB (known)	(1)
BADC is a square (known)	(2)
AE⊥plane BCA (known)	(3)
AE⊥CB (by rule 13 and (3))	(4)
CB⊥plane AFBE (by rule 20 and (2) (4))	(5)
CB⊥AF (by rule 13 and (5))	(6)
AF⊥plane FBEGC (by rule 20 and (1) (6))	(7)
GF⊥EGC (known)	(8)
AG⊥EGC (by rule 14 and (7) (8))	(9)

Figure2-2

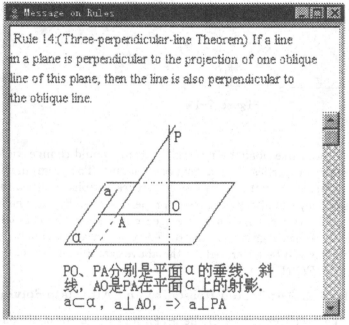

Message on Rules
Rule 14:(Three-perpendicular-line Theorem) If a line in a plane is perpendicular to the projection of one oblique line of this plane, then the line is also perpendicular to the oblique line.
P
a
O
A
α
PO、PA分别是平面α的垂线、斜线，AO是PA在平面α上的射影.
a⊂α, a⊥AO, => a⊥PA

Figure 2-3

and list the reasons step by step. If one asks for the rule used in a step, the program can give details of the rule (Figure 2-3).

The procedure of automated reasoning is closely related to the figures. If point D is between point A and B, $AB = AD + BD$, otherwise $AB = BD - AD$. Users can do exercises themselves by using the tools of "Manual Solving". When you give yourself a reasoning step by step, the program will judge each step and give

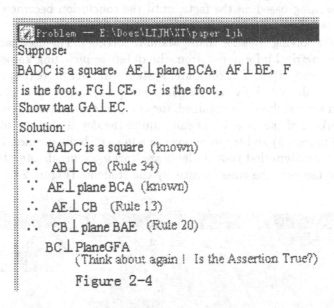

Figure 2-4

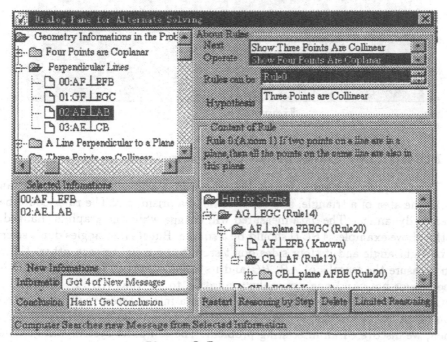

Figure 2-5

comments on your reasons. The program will keep the reason step if it thinks it is right (Figure 2-4).

Alternatively, the user may choose some facts from the information provided by computer system as the presupposition. Then the computer system will do a step of reasoning based on the facts, until the conclusion become clear (Figure 2-5). This is called "Alternate Solving".

Alternate solving is easier than "Manual Solving". The user can choose one or more geometrical information from the database including known conditions and deduced properties. The computer system will infer more properties or conclusions from them and give the new information for the user to choose (Figure 2-5). When the conclusion is obtained, the computer system will generate a proof by the thinking of the user. Users can inquire the detail of every reasoning rule anytime (Figure 2-3) and try to remove or add some built-in rules. If you inform the computer system that your students are learning some chapter, the computer system will use only the rules learned by the students (Figure 2-6).

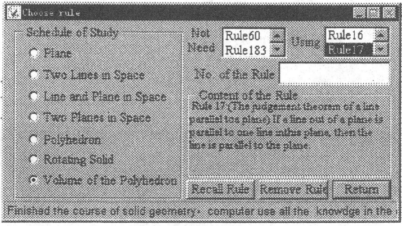

Figure 2-6

FUNCTION 3. Measure and Rotation in Space

The computer system can find the length of a segment, the degree of an angle, the area of a triangle, and the volume of a prism, etc. The ruler could be fixed or dynamical. The data measured will change while the graph is changed. In the above example, $ABCD$ should be a square. But its four angles don't seem to be right angle and its four sides don't seem to be equal to each other. Users can measure the lengths of AB, BC and its angles to check whether $ABCD$ is a square or not. And one can make a rotation to see that $ABCD$ is indeed a square at some position (Figure 2-7). The data in the screen will be changed along with points moving. When $EA = AB$, $\angle GAF$ is of 30 degree (Figure 2-8). Then, we discovered an interesting problem. The computer system will give its proof using the function of reasoning (Figure 2-9).

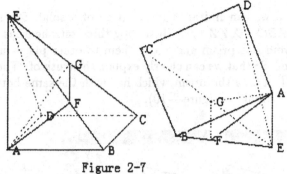

Figure 2-7

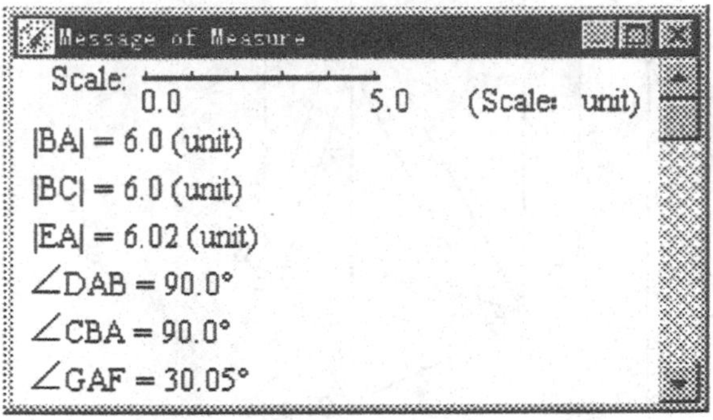

Scale: |———————————|
     0.0   5.0   (Scale: unit)

$|BA| = 6.0$ (unit)

$|BC| = 6.0$ (unit)

$|EA| = 6.02$ (unit)

$\angle DAB = 90.0°$

$\angle CBA = 90.0°$

$\angle GAF = 30.05°$

Figure 2-8

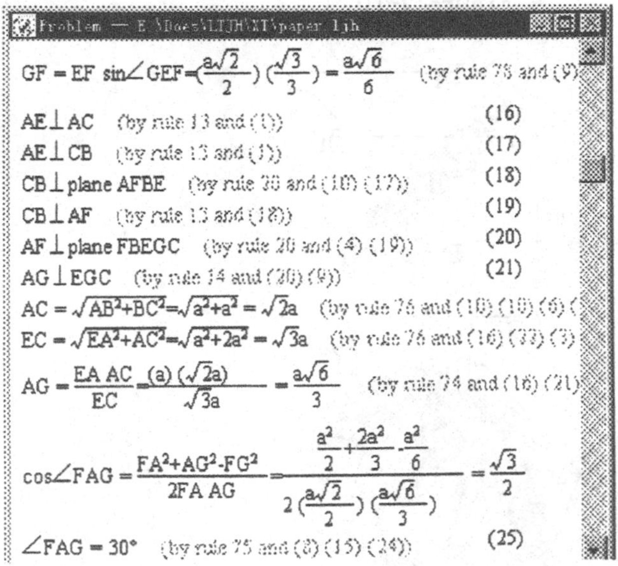

Problem — E:\Docs\LIJH\XT\paper_ljh

$$GF = EF \sin\angle GEF = \left(\frac{a\sqrt{2}}{2}\right)\left(\frac{\sqrt{3}}{3}\right) = \frac{a\sqrt{6}}{6} \quad \text{(by rule 78 and (9))}$$

$AE \perp AC$ (by rule 13 and (1))   (16)

$AE \perp CB$ (by rule 13 and (1))   (17)

$CB \perp$ plane $AFBE$ (by rule 20 and (16) (17))   (18)

$CB \perp AF$ (by rule 13 and (18))   (19)

$AF \perp$ plane $FBEGC$ (by rule 20 and (4) (19))   (20)

$AG \perp EGC$ (by rule 14 and (20) (9))   (21)

$AC = \sqrt{AB^2 + BC^2} = \sqrt{a^2 + a^2} = \sqrt{2}a$ (by rule 78 and (10) (10) (6) (

$EC = \sqrt{EA^2 + AC^2} = \sqrt{a^2 + 2a^2} = \sqrt{3}a$ (by rule 78 and (16) (77) (3)

$$AG = \frac{EA\ AC}{EC} = \frac{(a)(\sqrt{2}a)}{\sqrt{3}a} = \frac{a\sqrt{6}}{3} \quad \text{(by rule 74 and (16) (21))}$$

$$\cos\angle FAG = \frac{FA^2 + AG^2 - FG^2}{2FA\ AG} = \frac{\dfrac{a^2}{2} + \dfrac{2a^2}{3} - \dfrac{a^2}{6}}{2\left(\dfrac{a\sqrt{2}}{2}\right)\left(\dfrac{a\sqrt{6}}{3}\right)} = \frac{\sqrt{3}}{2}$$

$\angle FAG = 30°$ (by rule 75 and (3) (15) (24))   (25)

Figure 2-9

FUNCTION 4. Part Copy and Translation of Figure

Using this function, we can imitate the partition of a solid. For instance, we can draw a prism $ABC - XYZ$ first, then copy three tetrahedrons $AXYZ$, $ABCZ$ and $ABYZ$ from this prism and move them to other places using the parallel moving function. So that we can clearly explain the truth of "The volume of a pyramid is equal to 1/3 of the prism which has with the same bottom and equal height with the pyramid" (Figure 2-10).

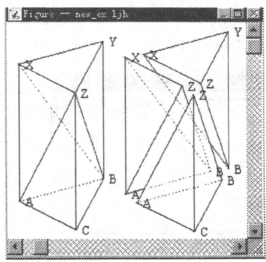

Figure 2-10

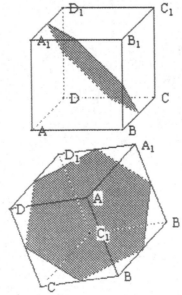

Figure 2-11

FUNCTION 5. Section of a Solid

The user can cut a prism, a pyramid, etc. with arbitrary assigned plane. Such as the section of cube which is cut by its main diagonals mid-perpendicular-plane is a regular hexagon. And we can see its exact shape after rotating the cube to an adequate position (Figure 2-11).

6 Concluding Remarks

We have seen that *machine readable solving* in geometry is useful for developing the ICAI software. The ICAI software based on automated reasoning can provide a *micro-world*, which represents the knowledge structure of real solids. The micro-world has at least three advantages for teaching and learning:

(1) In the micro-world, the knowledge about the solids is divided into some elements and relationships, and these basic parts can also be combined to make meaningful structures. User can control and operate on them with both graphic translating and symbolical reasoning. So it is a full-scale model of real solids.

(2) The dynamical construction provides an environment for students to study. They can change the figure on screen continuously by hand and observe it to find geometric conjectures.

(3) It can be a stage on which teachers can build their knowledge bank by saving solved problems. The bank may help teachers to prepare their lessons more easily.

Under the teacher's instruction, students can do constructing, reasoning, computing, measuring and figure translating in the micro-world. With the computer system, they may ask, answer, guess and create problems. Learning will become more interesting.

But this is just a starting point. There remains a lot of work to do. Such as:

(1) To develop similar software for physics and chemistry. Problems in these courses may be easier than in geometry. But representing knowledge is more complex, and there would be more kinds of problems.

(2) To improve the program such that users can add reasoning rules proved by themselves.

(3) In machine readable solving for geometry, it is faster to use signed area and full-angle. But solutions using traditional geometric quantities would be more welcomed. An idea is to first reason with any effective tool, and then translate the solution into a more traditional style.

We hope that more work will appear in this direction.

References

1. Chou, S.C., *Mechanical geometry theorem proving*, D.Reidel Publishing Company, Dordrecht, Netherland, 1988.

2. Chou, S.C., Gao, X.S. and Zhang, J.Z., *Machine proofs in geometry*, World Scientific, Singapore, 1994.

3. Chou, S.C., Gao, X.S. and Zhang, J.Z., Automated production of traditional proofs for constructive geometry theorems, *Proc. of Eighth IEEE Symposium on Logic in Computer Science*, pp. 48–56, IEEE Computer Society Press, 1993.

4. Chou, S.C., Gao, X.S. and Zhang, J.Z., Mechanical geometry theorem proving by vector calculation, *Proc. of ISSAC-93*, ACM Press, Kiev, pp. 284–291.

5. Chou, S.C., Gao, X.S. and Zhang, J.Z., Automated generation of readable proofs with geometric invariants, II. proving theorems with full-angles, *J. of Automated Reasoning*, 17, 349–370, 1996.

6. Chou, S.C., Gao, X.S. and Zhang, J.Z., Recent advances of automated geometry theorem proving with high level geometry invariants, *Proc. of Asian Symposium on Computer Mathematics*, Kobe, Japan, 1996, pp. 173–186.

7. Chou, S.C., Gao, X.S. and Zhang, J.Z., Vectors and automated geometry reasoning, TR-94-3, CS Dept., WSU, Jan. 1994.

8. Chou, S.C., Gao, X.S. and Zhang, J.Z., Automated production of traditional proofs in solid geometry, *J. Automated Reasoning*, 14, 257–291, 1995.

9. Chou, S.C., Gao, X.S. and Zhang, J.Z., A fixpoint approach to automated geometry theorem proving, WSUCS-95-2, CS Dept, Wichita State University, 1995 (to appear in J. of Automated Reasoning).

10. Chou, S.C., Gao, X.S. and Zhang, J.Z., The geometry information searching system by forward reasoning, *Proc. of Inter. Symposium on Logic and Software Engineering*, Beijing, World Scientific, Singapore, 1996.

11. Chou, S.C., Gao, X.S. and Zhang, J.Z., A collection of 110 geometry theorems and their machine produced proofs using full-angles, TR-94-4, CS Dept., WSU, March 1994.

12. Chou, S.C., Gao, X.S. and Zhang, J.Z., Automated solution of 135 geometry problems from American Mathematical Monthly, TR-94-8, CS Dept., WSU, August 1994.

13. Chou, S.C., Gao, X.S. and Zhang, J.Z., A collection of 90 mechanically solved geometry problems from non-Euclidean geometries, TR-94-10, CS Dept., WSU, Oct. 1994.

14. Fevre, S. and Wang, D., Proving geometric theorems using Clifford algebra and rewrite rules, *Proc. of CADE-15*, LNAI 1421, pp. 17–32, Springer, 1998.

15. Gallaire, H., Minker, J. and Nicola, J.M., Logic and databases: a deductive approach, *ACM Computing Surveys*, 16(2), 153–185, 1984.

16. Gao, X.S., Building dynamic mathematical models with Geometry Expert, III. a deductive dabase, Preprint, 1998.

17. Gao, X.S., Zhu, C.C. and Huang, Y., Building dynamic mathematical models with Geometry Expert, I. geometric transformations, functions and plane curves, *Proc. of the Third Asian Technology Conference in Mathematics*, Ed. W.C. Yang, pp. 216–224, Springer, 1998.

18. Gao, X.S., Zhu, C.C. and Huang, Y., Building dynamic mathematical models with Geometry Expert, II. linkages. *Proc. of the Third Asian Symposium on Computer Mathematics*, Ed. Z.B. Li, pp. 15–22, Lanzhou University Press, 1998.

19. Gao, X.S., Zhang, J.Z. and Chou, S.C., *Geometry Expert* (Chinese book with software), Chiu Chang Mathematics Publishers, Taipei, 1998.

20. Gelernter, H., Hanson, J. R. and Loveland, D. W., Empirical explorations of the geometry-theorem proving machine, *Proc. West. Joint Computer Conf.*, pp. 143–147, 1960.

21. Li, C.Z. and Zhang, J.Z., *MathLab: solid geometry* (Chinese book with software), China Juvenile Publishers, Beijing, 1998.
22. Li, H. (1996): Clifford algebra and area method. MM-Research Preprints No. bf 14, pp. 37–69, Institute of Systems Science, Academia Sinica, 1996.
23. Nevins, A.J., Plane geometry theorem proving using forward chaining, *Artificial Intelligence*, **6**, 1–23, 1975.
24. Reiter, R., A semantically guided deductive system for automatic theorem proving, *IEEE Tras. on Computers*, **C-25**(4), 328–334, 1976.
25. Robinson, A., Proving a theorem (as done by man, logician, or machine), *Automation of Reasoning*, Ed. J. Siekmann and G. Wrightson, pp. 74–78, Springer, 1983.
26. Stifter, S. (1993): Geometry theorem proving in vector spaces by means of gröbner bases, *Proc. of ISSAC-93*, Kiev, pp. 301–310, ACM Press, 1993.
27. Wang, D., Clifford algebraic calculus for geometric reasoning with application to computer system vision, *Automated Deduction in Geometry*, Ed. D. Wang, LNAI **1360**, pp. 115–140, Springer, 1997.
28. Wu, W.T., On the decision problem and the mechanization of theorem-proving in elementary geometry, *Scientia Sinica*, **21**, 159–172. Re-published in *Automated Theorem Proving: after 25 Years*, Ed. W.W. Bledsoe and D.W. Loveland, pp. 235–242, 1984.
29. Yang, L., Gao, X.S., Chou, S.C. and Zhang, J.Z., Automated production of readable proofs for theorems in non-Euclidean geometries, *Automated Deduction in Geometry*, Ed. D. Wang, LNAI **1360**, pp. 171–188, Springer, 1997.
30. Yang, H.Q., Zhang, S.G. and Feng, G.C., Clifford algebra and mechanical geometry theorem proving, *Proc. of the Third Asian Symposium on Computer Mathematics*, Ed. Z.B. Li, pp. 49–64, Lanzhou University Press, 1998.
31. Zhang, J.Z., Chou, S.C. and Gao, X.S., Automated production of traditional proofs for theorems in Euclidean geometry, I. the Hilbert intersection point theorems, TR-92-3,CS Dept., WSU, 1992. Also in *Annals of Mathematics and AI*, **13**, 109–137, 1995.

Plane Euclidean Reasoning

Desmond Fearnley-Sander

Department of Mathematics, University of Tasmania
GPO Box 252-37, Hobart, Tasmania 7001, Australia
URL: http://www.maths.utas.edu.au/People/dfs/dfs.html
EMAIL: dfs@hilbert.maths.utas.edu.au

Abstract. An automatic reasoning system for plane Euclidean geometry should handle the wide variety of geometric concepts: points, vectors, angles, triangles, rectangles, circles, lines, parallelism, perpendicularity, area, orientation, inside and outside, similitudes, isometries, sine, cosine, ... It should be able to construct and transform geometric objects, to compute geometric quantities and to prove geometric theorems. It should be able to call upon geometric knowledge transparently when it is needed. In this paper a type of ring generated by points and numbers is presented which may provide a formal basis for reasoning systems that meet these requirements. The claim is that this simple algebraic structure embodies all the concepts and properties that are investigated in the many different theories of the Euclidean plane.

A reasoner for plane Euclidean geometry would

- know basic geometric facts;
- know basic trigonometry;
- handle complex numbers;
- compute geometric quantities, such as areas of polygons and cosines of angles;
- work with diagrams;
- interact transparently with reasoners that need geometry;
- prove geometry theorems.

The best current computational systems are far from meeting these requirements. On the one hand, while computer algebra systems such as Mathematica and Maple, have some capability of handling local aspects of this program, such as deriving trigonometric formulae, their reliance on *ad hoc* rules makes the problem of integration of these fragmentary capabilities quite intractable. On the other hand, the wonderful geometry theorem provers based on the methods of Wu and Buchberger, are dependent upon introduction of coordinates, a feature that in its own way is *ad hoc*, since we know that coordinates are in fact substantially irrelevant. Though one might hope that Euclid's original axiomatic approach would be the way forward, attempts to automate it have had scant success.

In my opinion what is needed for the automation of reasoners with the indicated capabilities is an expressive, computationally tractable, simple uniform formalism. In this paper a candidate for that role is presented. An algebraic structure, called a *Euclid ring*, is introduced that is generated by points and real numbers, that fully supports Euclidean plane geometry, trigonometry, transformations and complex numbers, and that is coordinate-free. The vector fragment of this structure is the well-known two-dimensional positive-definite Clifford algebra introduced by W. K. Clifford. To use this algebra, many authors identify points with "vectors from the origin"; this is ungeometrical and avoidable, as our treatment shows.

No new geometric results are presented in this paper. Its thrust is rather to demonstrate the richness of Euclid ring semantics by showing how concepts such as angle and area arise naturally, and to demonstrate the algebraic power of the Euclid ring formalism by giving a range of elementary computations and proofs that may be obtainable automatically by methods currently under development.

1 Mathematical Structure

A *Euclid ring* is a ring Ω, with unit element 1, and subsets $\mathcal{R}$ and $\mathcal{P}$ such that

1. $\mathcal{R} \cup \mathcal{P}$ generates Ω;
2. $\mathcal{R}$ is the field of real numbers and is a subring of Ω;
3. $\mathcal{P}$ is a 2-dimensional affine space over $\mathcal{R}$;
4. $\alpha A = A\alpha$ for $\alpha \in \mathcal{R}$, $A \in \mathcal{P}$;
5. $(B - A)^2 > 0$ for $A \neq B \in \mathcal{P}$;
6. $0 \notin \mathcal{P}$.

Elements of $\mathcal{P}$ are called *points* and denoted by Roman capitals A, B, C, $\ldots$, Q; note that U, V, W, X, Y and Z do not denote points. *Real numbers* are denoted by Greek letters, $\alpha, \beta, \gamma, \ldots$. The number 0 is the additive identity of the ring and the number 1 is its multiplicative identity. $\mathcal{P}$ is closed under affine combinations (meaning that $A, B \in \mathcal{P}$, $\alpha + \beta = 1 \Rightarrow \alpha A + \beta B \in \mathcal{P}$) and is generated by three linearly independent points. Indeed, thanks to items 1 and 3, Ω itself is generated by the real numbers and three linearly independent points.

All elements of a Euclid ring are sums of products of numbers and points and it is easy to calculate with them; for example, to verify that if $P = \frac{1}{3}(A + B + C)$ then

$$(A - B)^2 + (B - C)^2 + (C - A)^2 = 3\left[(A - P)^2 + (B - P)^2 + (C - P)^2\right]$$

is a matter of high-school algebra (though one must not assume that multiplication is commutative).

We will progressively show how various types of *geometric objects*, including vector and angle, *geometric quantities*, including length and area, *geometric predicates*, including perpendicularity and parallelism, and *geometric transformations*, including translation and similitude, are represented in Euclid ring terms. Note, however, that, just as in natural language, it is not the case that every well-formed term has an interpretation.

2 Rules of Interpretation and Physical Facts

A good way to approach the semantics of a Euclid ring is to view it as providing an algebraic theory of a physical domain, the drawing board of the engineer or architect. The link between Ω and the physical drawing board requires two *rules of interpretation*: the elements of $\mathcal{P}$ are to be interpreted geometrically, as their name suggests, as *points*, and the number $(B - A)^2$ is to be interpreted as the square of the *distance* (in some fixed unit) between the points A and B. We shall write $|B - A| = \sqrt{(B - A)^2}$.

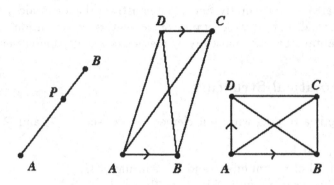

Fig. 1. Facts 1, 2 and 3.

The interpretation of other algebraic entities will, as we shall see, be forced by these rules of interpretation and by certain fundamental *physical facts* like (see Figure 1) the following:

Fact 1. A point P lies on the line segment between points A and B if and only if the distance from A to B is the same as the sum of the distances from A to P and from P to B;

Fact 2. Points A, B, C and D are successive vertices of a parallelogram if and only if the point mid-way between A and C is the same as the point mid-way between B and D;

Fact 3. A parallelogram with vertices A, B, C and D is a rectangle if and only if the distance from A to C is the same as the distance from B to D;

Fact 4. The area of a square with side-length 1 is 1.

It is to be emphasized that these are statements about the physical world of the drawing board which we seek to model. Although, for example, a rectangle may be defined in various ways, we always find, upon measurement, that Fact 2 and Fact 3 are confirmed, and in any model we shall certainly insist that the corresponding statements, formulated mathematically, be true.

Fact 1 forces the interpretation of $P = \alpha A + \beta B$ where $\alpha + \beta = 1$, $\alpha \geq 0$ and $\beta \geq 0$; for then we have

$$P = \alpha A + \beta B \Leftrightarrow P - A = \beta(B - A) \text{ and } B - P = \alpha(B - A),$$

and these equations imply that $|P - A| = \beta|B - A|$ and $|B - P| = \alpha|B - A|$ and hence that $|B - A| = |B - P| + |P - A|$. If (1) is to hold we must interpret P as the point which divides the line segment from A to B in the ratio $\beta : \alpha$. (See Figure 1.) The interpretation of a general affine combination (for which $\alpha + \beta = 1$ but either α or β may be negative) now follows easily. In particular, $\frac{1}{2}(A + B)$ is the *mid-point* of the line segment between A and B.

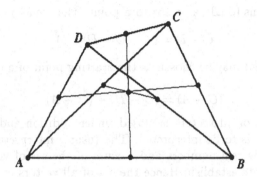

Fig. 2. A thorem on quadrilaterals.

We now find that even such a simple algebraic identity as

$$\frac{1}{2}\left[\frac{1}{2}(A + B) + \frac{1}{2}(C + D)\right] = \frac{1}{2}\left[\frac{1}{2}(A + C) + \frac{1}{2}(B + D)\right]$$

$$= \frac{1}{2}\left[\frac{1}{2}(A + D) + \frac{1}{2}(B + C)\right]$$

has a geometric interpretation (see Figure 2) which, though obvious, is a nontrivial fact. Having shown that $P = \frac{1}{3}(A + B + C)$ is the centroid of the triangle with vertices A, B and C we see that the algebraic identity stated in Section 1 is a metric theorem which can be checked experimentally by making careful drawings and measurements. Each interpretable theorem of our algebra affords a further test of the efficacy of our model — if ever such a theorem, in its physical interpretation, fails to hold, we must reject the model. Every model is tentative, even Euclid's.

3 Vectors

Fact 2 (see Figure 1), together with the equivalence

$$B - A = C - D \Leftrightarrow \frac{1}{2}(A + C) = \frac{1}{2}(B + D)$$

forces us to interpret the difference of two points $B - A$ as the *vector* from A to B. We will use letters $U, V, W, X, \ldots$ from the end of the alphabet to denote vectors.

If P is a point then $1P = P$, a point, but no other real multiple of P can be a point; for if αP, with $\alpha \neq 1$, was a point Q, we would have $0 = \alpha/(\alpha - 1)P - 1/(\alpha - 1)Q \in \mathcal{P}$. We do not need to interpret αP for α different from 0 and 1. Also a vector $A - B$ cannot be a point P, since that would entail that $\frac{1}{2}A = \frac{1}{2}(B + P) \in \mathcal{P}$, contrary to the above.

The equivalence

$$B = A + X \Leftrightarrow X = B - A$$

shows what it means to add a vector to a point. The identity

$$C - D = (A + C - D) - A$$

shows that any point may be chosen as the starting point of a given vector. The identity

$$(C - A) = (C - B) + (B - A)$$

shows that the set of all vectors is closed under addition and also shows how addition of vectors is to be interpreted. The (usual) interpretation of multiplication of vectors by real numbers and closure of the set of vectors under this operation are easy to establish. Hence the set of all vectors is a linear space; it is two-dimensional, having as basis the pair of vectors $B - A$ and $C - A$ where A, B and C are linearly independent points. Note, though, that some of what follows is independent of dimension.

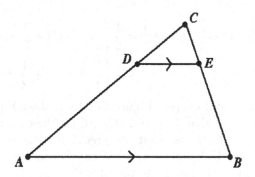

Fig. 3. A theorem on ratios of lengths.

Example. A familiar classical affine theorem (see Figure 3) is expressed by the implication:

$$D = \alpha A + \beta B, \; E = \alpha A + \beta C \Rightarrow D - E = \beta(B - C). \quad \Box$$

Example. As an example of a modern affine theorem (involving the notion of vector in an essential way) we invite the reader to interpret and generalize the identity:

$$\left[\frac{1}{2}(A_1 + A_2) - \frac{1}{2}(A_6 + A_1)\right] + \left[\frac{1}{2}(A_3 + A_4) - \frac{1}{2}(A_2 + A_3)\right]$$
$$+ \left[\frac{1}{2}(A_5 + A_6) - \frac{1}{2}(A_4 + A_5)\right] = 0. \quad \square$$

The square of the vector X from A to B is the positive real number $X^2 = (B - A)^2$, interpreted, of course, as the square of its length, and we shall write $|X| = \sqrt{X^2}$.

4 Perpendicularity

From an algebraic point of view the fact that

$$(X + Y)^2 = (X - Y)^2 \Leftrightarrow YX = -XY \Leftrightarrow (X + Y)^2 = X^2 + Y^2$$

is trivial, but its geometric interpretation is subtle and important. (It will be made completely clear when, in a moment, we show how a product XY is to be interpreted.) The empirical Fact 3 (see Figure 1) forces us to define perpendicularity of vectors X and Y in such a way that $X \perp Y$ if and only if $(X + Y)^2 = (X - Y)^2$; or, equivalently, via

$$X \perp Y \Leftrightarrow YX = -XY.$$

Then one of the above equivalences is the *theorem of Pythagoras*.

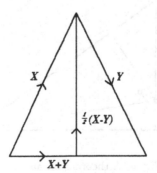

Fig. 4. An isosceles triangle theorem.

Example. The theorems suggested by Figures 4 and 5 are proved as follows:

$$X^2 = Y^2 \Rightarrow (X + Y)(X - Y) = -(X - Y)(X + Y)$$
$$\Rightarrow X + Y \perp X - Y. \quad \square$$

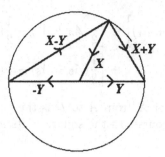

Fig. 5. A circle theorem.

Example. (See Figure 6.) The ring identity

$$(B - C)(A - P) + (C - A)(B - P) + (A - B)(C - P)$$
$$= -(A - P)(B - C) - (B - P)(C - A) - (C - P)(A - B)$$

implies that the altitudes of a triangle intersect; for it shows that if $B - C \perp A - P$ (that is, if $(B - C)(A - P) = -(A - P)(B - C)$) and $C - A \perp B - P$ then $A - B \perp C - P$. It is easy to devise a similar proof that the perpendicular bisectors of the sides intersect. Theorems arising in this way from identities may be said to be *embodied* in a Euclid ring. So far as I know the problem of characterising those theorems that are embodied is open. □

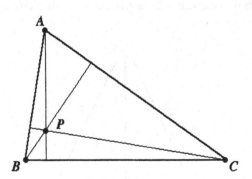

Fig. 6. A theorem on altitudes.

A vector Y with $Y^2 = 1$ is called a *direction*. The *direction of a non-zero vector X* is $\frac{X}{|X|}$.

Example. (See Figure 7). The line with direction Y through a point $B = A + X$ on the circle with centre A and radius $|X|$ consists of all points $P = A + X + \tau Y$ with $\tau \in \mathbf{R}$. The condition for P to lie on the circle is that $(X + \tau Y)^2 = X^2$, or, equivalently, that

$$\tau = 0 \quad \text{or} \quad \tau = -(XY + YX);$$

hence there is a unique intersection point if and only if $Y \perp X$. Observe that $XY + YX$ is a real number, since it equals $(X+Y)^2 - X^2 - Y^2$. $\square$

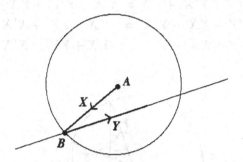

Fig. 7. A theorem on secants.

5 Parallelism

Vectors X and Y are *parallel*, and we write $X \| Y$, if one is a multiple of the other. It is clear that if $X \| Y$ then $XY = YX$. Let us prove the converse. Assuming without loss of generality that X and Y have equal lengths, we have

$$XY = YX \Rightarrow (X - Y)(X + Y) = 0$$
$$\Rightarrow Y = \pm X$$
$$\Rightarrow Y \| X.$$

We have here used the fact that any non-zero vector V is invertible (with inverse $V^{-1} = \frac{V}{V^2}$) and hence

$$VW = 0 \Leftrightarrow V = 0 \text{ or } W = 0.$$

6 Angles

Whatever definition one adopts for the (oriented) angle $\widehat{X, Y}$ between directions X and Y one would like it to be such that whenever $\widehat{X, Y} = \widehat{X', Y'}$ one has (as Figure 8 suggests)

$$X - Y' \parallel Y - X',$$
$$X + Y' \parallel X' + Y,$$
$$X + Y' \perp Y - X'$$
$$\text{and } X' + Y \perp X - Y';$$

equivalently, the following four equations should hold:

$$XY - Y'Y - XX' + Y'X' = YX - YY' - X'X + Y'Y'$$
$$XX' + XY + Y'X' + Y'Y = X'X + Y'Y' + YX + YY'$$
$$XY + Y'Y - XX' - Y'X' = -YX - YY' + X'X + Y'Y'$$
$$XY - Y'Y + XX' - Y'X' = -YX + YY' - X'X + X'Y'.$$

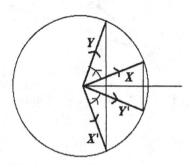

Fig. 8. Equality of angles.

Adding, we see that these equations imply that

$$XY = X'Y'.$$

Accordingly, we define the *angle* $\widehat{X,Y}$ *between directions* X and Y to be the product XY.

As it is sometimes convenient to talk about the *angle* $\widehat{X,Y}$ *between non-zero vectors* X and Y we define this to be the angle between their directions:

$$\widehat{X,Y} = \frac{XY}{|X||Y|}.$$

Note that it is only for *directions* X and Y that the angle $\widehat{X,Y}$ is the product XY.

Example. (See Figure 9.) Let $X^2 = 1 = X_1^2$, $Y \neq 0$. Then

$$\widehat{X,Y} = \widehat{X_1,Y} \Rightarrow XY = X_1 Y$$
$$\Rightarrow X = X_1$$
$$\Rightarrow X_1 \| X.$$

Here we have multiplied both sides of one equation by Y^{-1}. $\quad\square$

If X, Y and Z are directions then

$$(XY)(YZ) = XY^2 Z = XZ.$$

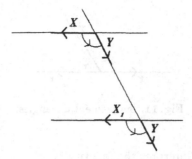

Fig. 9. A theorem on parallel lines.

Note that an arbitrary angle VW between directions V and W has the form YZ for a suitable Z, namely $Z = YVW$; it will be shown in a moment that Z is a direction. Assuming this, we see that the set of all angles is closed under multiplication. Since $1 = X^2$ is an angle and

$$XYYX = 1 = YXXY,$$

angles form a group under multiplication, with $(XY)^{-1} = YX$. In particular XY determines YX.

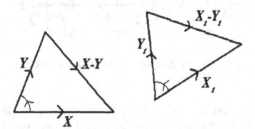

Fig. 10. A theorem on congruent triangles.

Example. (See Figure 10.)

$$X^2 = X_1^2,\ Y^2 = Y_1^2,\ XY = X_1 Y_1 \Rightarrow (X - Y)^2 = (X_1 - Y_1)^2.$$

This is a familiar result about congruence of triangles. The theory of congruence may be developed without introducing transformations and the same applies to similarity. □

We call $1 = XX$ the *identity angle*, $-1 = -XX$ the *straight angle*, the *complement* of XY and $-YX$ its *supplement*. Then, for example (see Figure 11)

$$XYY(-X) = X(-X) = -1.$$

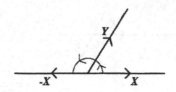

Fig. 11. Supplementary angles.

Readers may like to interpret the identity

$$(X_1X_2)(X_2X_3)...(X_{n-1}X_n)(X_nX_1) = 1$$

where the X_j are directions.

The geometric meaning of the conditions for perpendicularity and parallelism now become clear (see Figure 12, (a) and (b)):

$$X \perp Y \Leftrightarrow \widehat{X,Y} = \widehat{X,-Y},$$
$$X \| Y \Leftrightarrow \widehat{X,Y} = \widehat{Y,X}.$$

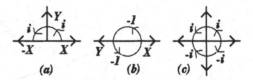

Fig. 12. Right angles and straight angles.

Let us now prove that if X, Y and Z are vectors then so is XYZ. This holds trivially if X and Y are linearly dependent, and otherwise it holds for $Z = X$ by virtue of the identity

$$XYX = \left[(X+Y)^2 - X^2 - Y^2 \right] X - Y$$

and for an arbitrary $Z = \alpha X + \beta Y$ because

$$XYZ = \alpha XYX + \beta XY^2.$$

Moreover if X and Y are directions then

$$(XYZ)^2 = \alpha^2 + \alpha\beta(XY + YX) + \beta^2 = Z^2;$$

in particular, (a result that was needed above) if Z is also a direction then so is XYZ. This argument also entails that the group of angles is Abelian, since for directions X, Y and Z we have

$$XYZXYZ = 1$$

and hence
$$ZXYZ = YX = YZZX.$$

If X and U are arbitrary directions one easily checks that the vector $U - XUX$ is perpendicular to X; hence there exists a direction Y perpendicular to X.

7 Orientation

Let X and Y be a pair of perpendicular directions. If V and W are any other such pair, then $XY = -YX$ and $VW = -WV$ and so
$$(XY)^2 = -1 = (VW)^2.$$

We know that $VW = XZ$ for a suitable direction Z, and hence that
$$(XY)^2 - (XZ)^2 = 0;$$

because multiplication of angles is commutative this is equivalent to
$$(XY - XZ)(XY + XZ) = 0$$

or, indeed, to
$$X(Y - Z)X(Y + Z) = 0.$$

It follows immediately that either $Z = Y$ or $Z = -Y$ and hence that either $VW = XY$ or $VW = X(-Y) = -XY$.

When we model a line as a one-dimensional affine space a choice of orientation (left and right) must be made. The situation in two dimensions is similar. Write $i = XY$. Then i and $-i$ are called *right angles*, and one of them, say, i, is called *positive* and the other *negative*.

Note that
$$VW = \pm i \Rightarrow (VW)^2 = i^2 = -1 \Rightarrow VW = -WV \Rightarrow V \perp W.$$

Multiplication of positive and negative right angles has the expected properties (see Figure 12 (c)):
$$i^2 = (-i)^2 = -1,$$
$$i(-i) = (-i)i = 1,$$
$$(-1)^2 = 1.$$

For any direction V the unique direction W such that VW is a positive right angle is Vi :
$$VW = i \Leftrightarrow W = Vi.$$

Hence i anti-commutes with vectors:
$$Vi = V(VVi) = -V(ViV) = -iV.$$

The directions V and Vi are linearly independent (or not parallel) since
$$VVi = ViV \Rightarrow V^2 i = -V^2 i \Rightarrow V = 0.$$

Hence they form a basis for the space of all vectors.

8 Trigonometry

Consider the angle VW between directions V and W. Since (see Figure 13)

$$W = \alpha V + \beta V i \Leftrightarrow VW = \alpha + \beta i$$

we see that VW is uniquely expressible as a linear combination of the angles 1 and i. We define the *cosine* and *sine* of the angle $VW = \alpha + \beta i$ to be the real numbers

$$\cos VW = \alpha \quad \text{and} \quad \sin VW = \beta.$$

Despite appearances these definitions are genuinely coordinate-free.

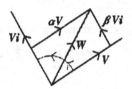

Fig. 13. Sine and cosine of the angle VW.

Since also $WV = \alpha - \beta i$ we have

$$1 = VWWV = (\alpha + \beta i)(\alpha - \beta i) = \alpha^2 + \beta^2 = \cos^2 VW + \sin^2 VW.$$

Hence $-1 \le \cos VW \le 1$ and $-1 \le \sin VW \le 1$; the first of these is the *Cauchy-Schwartz inequality*. Of course, $\cos 1 = 1$, $\sin 1 = 0$, $\cos i = 0$, $\sin i = 1$, $\cos(-1) = -1$, $\sin(-1) = 0$, $\cos(-i) = 0$ and $\sin(-i) = -1$.

Also, for example (see Figure 13),

$$\cos W(Vi) = \cos(\alpha - \beta i)i = \cos(\beta + \alpha i) = \sin VW.$$

The *"addition formulae"* are obtained by taking real and imaginary parts in a product

$$(VW)(XY) = (\alpha + \beta i)(\gamma + \delta i)$$
$$= (\alpha\gamma - \beta\delta) + (\alpha\delta + \beta\gamma)i$$

to get

$$\cos(VW)(XY) = \cos VW \cos XY - \sin VW \sin XY$$

and

$$\sin(VW)(XY) = \cos VW \sin XY + \sin VW \cos XY.$$

If X and Y are arbitrary non-zero vectors, we have

$$\cos \widehat{X,Y} = \frac{XY + YX}{2\,|X|\,|Y|} \quad \text{and} \quad \sin \widehat{X,Y} = \frac{i(YX - XY)}{2\,|X|\,|Y|}.$$

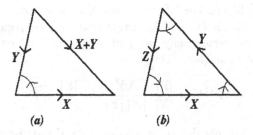

Fig. 14. The cosine and sine rules.

Immediate consequences are the *cosine rule* (see Figure 14 (a)),

$$(X + Y)^2 = X^2 + Y^2 + XY + YX$$
$$= X^2 + Y^2 - 2\,|X|\,|Y|\cos\widehat{X,Y}$$

and the *sine rule* (see Figure 14 (b)): if $X + Y + Z = 0$, then

$$XY - YX = YZ - ZY = ZX - XZ$$

and hence

$$\frac{\sin\widehat{X,-Y}}{|Z|} = \frac{\sin\widehat{Y,-Z}}{|X|} = \frac{\sin\widehat{Z,-X}}{|Y|}.$$

The cosine rule and the Cauchy-Schwartz inequality together entail the *triangle inequality*,

$$|X + Y| \le |X| + |Y|.$$

An angle is called *acute* if its cosine is positive and *obtuse* otherwise; and it is called *positive* if its sine is positive and *negative* otherwise. It is also convenient to call the sign (1 or 0 or -1) of $\sin\widehat{X,Y}$ the *orientation* of $\widehat{X,Y}$.

It is now easy to prove, for example, that at least two angles of any triangle are acute and to prove that the product of two acute positive angles is positive. Also one may relate the present definitions to properties of triangles by showing that for a triangle $\langle A, B, C \rangle$ with a right angle at B the cosine of the angle between the vectors $B - A$ and $C - A$ is $\frac{|B-A|}{|C-A|}$ while its sine has absolute value $\frac{|C-B|}{|C-A|}$.

9   Complex Numbers

Let U, V, U_1 and V_1 be directions, with $\xi UV = \eta U_1 V_1$, where $\xi > 0$ and $\eta > 0$ are real numbers. We will prove that $\xi = \eta$ and (hence) $UV = U_1 V_1$. Indeed, we may write $UV' = U_1 V_1$ for a suitable direction V' and we have

$$\xi UV = \eta UV' \Rightarrow \xi V = \eta V'$$
$$\Rightarrow \xi^2 = \eta^2$$
$$\Rightarrow \xi = \eta;$$

here we first multiplied on the left by U and then squared.

Any positive multiple ξUV of an angle may also be regarded as a ratio of vectors $U^{-1}(\xi V)$; conversely, any ratio of vectors $X^{-1}Y$ (where $X \neq 0$) is a positive multiple of an angle:

$$X^{-1}Y = \frac{|Y|}{|X|}\left(\frac{XY}{|X||Y|}\right) = \frac{|Y|}{|X|}\widehat{X,Y}.$$

Clearly $X^{-1}Y$ does not determine either X or Y but, by what has just been shown, it does determine both the angle $\widehat{X,Y}$ and the ratio $\frac{|Y|}{|X|}$. Equivalently,

$$X^{-1}Y = X_1^{-1}Y_1 \Leftrightarrow \widehat{X,Y} = \widehat{X_1,Y_1} \quad \text{and} \quad \frac{|Y|}{|X|} = \frac{|Y_1|}{|X_1|};$$

Thus $X^{-1}Y = X_1^{-1}Y_1$ if and only if the triangles with edge vectors X and Y, X_1 and Y_1 are directly similar.

As we shall show in a moment, it is reasonable to call ratios of vectors $X^{-1}Y$ *complex numbers*. (Note that $X^{-1}Y = 0 \Leftrightarrow Y = 0$.) The number $\frac{|Y|}{|X|}$ is the *absolute value* of $X^{-1}Y$ and $\widehat{X,Y}$ is its *angle*. The complex number $X^{-1}Y$ also determines another complex number YX^{-1}, its *conjugate*, given by

$$YX^{-1} = \frac{|Y|}{|X|}\widehat{Y,X},$$

since $\widehat{Y,X} = (\widehat{X,Y})^{-1}$.

Real numbers are ratios of parallel vectors:

$$Y\|X \Leftrightarrow Y = \alpha X \text{ where } \alpha \in \mathcal{R} \Leftrightarrow X^{-1}Y = \alpha \in \mathcal{R}.$$

Imaginary numbers are ratios of perpendicular vectors:

$$Y\perp X \Leftrightarrow Y = \beta Xi \text{ where } \beta \in \mathcal{R} \Leftrightarrow X^{-1}Y = \beta i \in \mathcal{R}i.$$

And complex numbers of absolute value 1 are ratios of vectors of equal length, or what we have been calling angles:

$$|Y| = |X| \Leftrightarrow \frac{|Y|}{|X|} = 1 \Leftrightarrow X^{-1}Y = \widehat{X,Y}.$$

It follows immediately from the corresponding facts for angles that every complex number may be written uniquely in the form $X^{-1}Y$, where X is a given non-zero vector, and that the non-zero complex numbers form an Abelian group under multiplication. (See Figure 15.) Also, since

$$X^{-1}Y + X^{-1}Z = X^{-1}(Y + Z),$$

the complex numbers form an Abelian group under addition. Thus the set $\mathcal{C}$ of all ratios of vectors is a field.

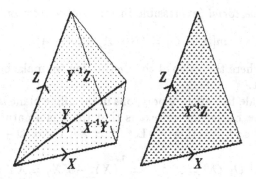

Fig. 15. Complex multiplication.

For $X \neq 0$ we have

$$Y = \alpha X + \beta X i \Leftrightarrow X^{-1}Y = \alpha + \beta i;$$

because X and xi are linearly independent, the pair of complex numbers, 1 and i, form a basis for $\mathcal{C}$ as a linear space over $\mathcal{R}$. The coordinates, α and β, are the so-called *real* and *imaginary parts*, $\Re[\zeta]$ and $\Im[\zeta]$, of the complex number $\zeta = X^{-1}Y$.

Observe that

$$YX^{-1} = \alpha - \beta i$$

and

$$\left|X^{-1}Y\right|^2 = \frac{Y^2}{X^2} = X^{-1}YYX^{-1} = \alpha^2 + \beta^2.$$

Since these properties characterize the complex field, our terminology is appropriate.

Conjugation, $X^{-1}Y \mapsto YX^{-1}$, is an automorphism of $\mathcal{C}$ that carries i to $-i$.

10 Area

If we define the (oriented) *scale-free area* of a polygon $\langle A_1, A_2, \ldots, A_k \rangle$ to be the quantity

$$m[A_1, A_2, \ldots, A_k] := A_1 \wedge A_2 + A_2 \wedge A_3 + \cdots + A_k \wedge A_1$$

where $A \wedge B := \frac{1}{2}(AB - BA)$, then we get a full theory of plane area, as expounded in [4]. The main properties are that

- area is additive;
- area is translation-invariant;
- area of triangles is characterised, up to choice of scale, by three basic theorems of Euclid.

Note that area is *vectorial* (expressible in terms of vectors only), since

$$m[A, B, C] = (B - A) \wedge (C - A);$$

this is one point where the value of the distinction we make between points and vectors is manifest.

The space of scale-free areas of polygons in the affine plane is one-dimensional. What is new in the Euclidean context is that there is a natural choice of scale. Choose a point O and direction X. Let $Y = Xi$, $O_1 = O + X$, $O_2 = O + Y$. Then:

$$m[O, O_1, O_2] = X \wedge Y = \frac{1}{2}(XY - YX) = XY = i$$

and $m[O, O_1, O_1 + O_2, O_2] = 2i$. According to Fact 4 (of Section 2), the area of the unit square should be 1. We therefore define the (oriented, scaled) *area* of a polygon $\langle A_1, A_2, \ldots, A_k \rangle$ to be the real number

$$\mu[A_1, A_2, \ldots, A_k] := \frac{1}{2i} m[A_1, A_2, \ldots, A_k].$$

For the area of a triangle $\langle A, B, C \rangle$ with $V = B - A$, $W = C - A$ we have a familiar formula:

$$\begin{aligned}
\mu[A, B, C] &= \frac{1}{2i} V \wedge W \\
&= \frac{1}{2}|V|.|W|(\frac{VW - WV}{2i|V|.|W|}) \\
&= \frac{1}{2}|V|.|W|\sin \widehat{V, W} \\
&= \frac{1}{2}\text{base}[A, B, C].\text{height}[A, B, C].
\end{aligned}$$

An arbitrary square with vertices $A_1, A_2 = A_1 + X$, $A_3 = A_1 + X + Xi$, $X_4 = A_1 + Xi$ has area

$$\begin{aligned}
\mu[A_1, A_2, A_3, A_4] &= \mu[A_1, A_2, A_3] + \mu[A_1, A_3, A_4] \\
&= \frac{1}{2} \, \Im[X(X + Xi)] + \frac{1}{2} \, \Im[(X + Xi)(-X)] \\
&= \Im[X^2 i] \\
&= X^2.
\end{aligned}$$

11   Affine Maps

Denote by $\mathcal{P}$ the set of all points, by $\mathcal{L}$ the (real) linear space spanned by $\mathcal{P}$, and by $\mathcal{V}$ the set of all vectors (a proper linear subspace of $\mathcal{L}$).

A function $f : \mathcal{P} \to \mathcal{P}$ is called an *affine map* if it preserves lines and ratios of distances along lines, or, equivalently, if

$$f[\alpha A + \beta B] = \alpha f[A] + \beta f[B]$$

for all $A, B \in \mathcal{P}$ and $\alpha, \beta \in \mathcal{R}$ with $\alpha + \beta = 1$. (Actually it is only necessary to assume this for α and β both positive.)

An affine map preserves parallelism of lines since $\langle A, B, C, D \rangle$ is a parallelogram if and only if $\frac{1}{2}(A + C) = \frac{1}{2}(B + D)$.

Any affine map $f{:}\mathcal{P} \to \mathcal{P}$ can be extended naturally to a linear function $f : \mathcal{L} \to \mathcal{L}$. To achieve this one must set

$$f[\alpha A] = \alpha f[A] \quad \text{for all} \quad \alpha \in \mathcal{R}, \ A \in \mathcal{P},$$

and

$$f[B - A] = f[B] - f[A] \quad \text{for all } A, B \in \mathcal{P}.$$

Since every element of $\mathcal{L}$ can be shown to be either a unique multiple of a unique point, αA, or a vector $B - A$, and since

$$
\begin{aligned}
B - A = C - D &\Rightarrow \tfrac{1}{2}(A + C) = \tfrac{1}{2}(B + D) \\
&\Rightarrow \tfrac{1}{2}(f[A] + f[C]) = \tfrac{1}{2}(f[B] + f[D]) \\
&\Rightarrow f[B] - f[A] = f[C] - f[D],
\end{aligned}
$$

this determines f on $\mathcal{L}$ in a well-defined fashion. Now for arbitrary points A and B and real numbers α and β, one has, in the case where $\alpha + \beta \neq 0$,

$$
\begin{aligned}
f[\alpha A + \beta B] &= f[(\alpha + \beta)(\frac{\alpha}{\alpha + \beta} A + \frac{\beta}{\alpha + \beta} B)] \\
&= (\alpha + \beta) f[\frac{\alpha}{\alpha + \beta} A + \frac{\beta}{\alpha + \beta} B] \\
&= \alpha f[A] + \beta f[B],
\end{aligned}
$$

and, in the case where $\alpha + \beta = 0$,

$$
\begin{aligned}
f[\alpha A + \beta B] &= f[((\alpha + 1)A + \beta B) - A] \\
&= f[(\alpha + 1)A + \beta B] - f[A] \\
&= (\alpha + 1)f[A] + \beta f[B] - f[A] \\
&= \alpha f[A] + \beta f[B].
\end{aligned}
$$

From this it follows easily that f is linear on $\mathcal{L}$. In future we use the term *affine map* for the unique extended map, $f : \mathcal{L} \to \mathcal{L}$, and we call the restriction $f : \mathcal{V} \to \mathcal{V}$ (which is also linear) its *vector part*.

Example. For every vector W there is a *translation* $t : \mathcal{P} \to \mathcal{P}$ given by

$$t[A] = A + W \quad \text{for every } A \in \mathcal{P}.$$

This is an affine map since for $\alpha + \beta = 1$ we have

$$t[\alpha A + \beta B] = \alpha A + \beta B + W = \alpha(A + W) + \beta(B + W) = \alpha t[A] + \beta t[B]$$

and it leaves every vector $X = B - A$ fixed, since

$$t[X] = t[B] - t[A] = (B + W) - (A + W) = X.$$

In a natural way the translations form a linear space isomorphic to $\mathcal{V}$. □
Let $f : \mathcal{P} \to \mathcal{P}$ be affine. Since

$$f[A + X] = f[A] + f[X],$$

f is determined by its vector part together with its action on a single point A.
Moreover, writing

$$f[A + X] = (A + f[X]) + (f[A] - A)$$
$$= (t \circ g)(A + X),$$

we see that f is the composite with a translation f (by $f[A] - A$) of a map g,
given by

$$g[A + X] = A + f[X] \quad \text{for} \quad X \in \mathcal{V},$$

which is affine, since for $\alpha + \beta = 1$ and $C, D \in \mathcal{P}$, we have

$$g[\alpha C + \beta D] = g[A + \alpha(C - A) + \beta(D - A)]$$
$$= A + f[\alpha(C - A) + \beta(D - A)]$$
$$= A + \alpha f[C - A] + \beta f[D - A]$$
$$= \alpha(A + f[C - A]) + \beta(A + f[D - A])$$
$$= \alpha g[C] + \beta g[D],$$

which has a fixed point, since

$$g[A] = A,$$

and which has the same vector part as f, since for $X \in \mathcal{V}$

$$g[X] = g[A + X] - g[A] = g[A + X] - A = f[X].$$

This effectively reduces the analysis of affine maps to the consideration of linear
maps on vectors.

12   Similitudes

A function $f : \mathcal{P} \to \mathcal{P}$ is a *similitude* if it preserves ratios of distances, or,
equivalently, if there exists $\lambda > 0$ such that

$$(f[B] - f[A])^2 = \lambda^2 (B - A)^2$$

for all points A and B. A similitude for which $\lambda = 1$ is called an *isometry*; thus
isometries preserve distances. Obviously translations are isometries.

Similitudes preserve lines, since a point C lies on the line segment between
A and B if and only if

$$|B - A| = |B - C| + |C - A| \, ;$$

Thus similitudes are affine maps. A similitude f stretches each vector $X = B - A$ by the factor λ :

$$f[X]^2 = (f[B] - f[A])^2 = \lambda^2 (B - A)^2 = \lambda^2 X^2.$$

Example. There is a natural action of angles on vectors: for each angle YZ, where Y and Z are directions, define $f : V \to V$ by

$$f[X] = XYZ$$

for every $X \in V$. Clearly f is linear. The affine map whose vector part is such an f and which has a fixed point A is called the *rotation* about A through the angle YZ. Rotations are isometries since, using the fact that angles commute, we have

$$(XYZ)^2 = (XY)(ZX)(YZ) = (XY)(YZ)(ZX) = X^2.$$

In a natural way the rotations about A form a group which is isomorphic with the group of angles. □

Let f be a similitude, as above. For arbitrary vectors X and Y we have

$$f[X]^2 + f[X]f[Y] + f[Y]f[X] + f[Y]^2 = (f[X] + f[Y])^2$$
$$= \lambda^2 (X + Y)^2$$
$$= \lambda^2 (X^2 + XY + YX + Y^2),$$

and hence

$$\cos \widehat{f[X], f[Y]} = \frac{f[X]f[Y] + f[Y]f[X]}{2\,|f[X]|\,|f[Y]|} = \cos \widehat{X, Y}.$$

It follows that f preserves perpendicularity of vectors. Since, also

$$\sin^2 \widehat{f[X], f[Y]} = 1 - \cos^2 \widehat{f[X], f[Y]} = \sin^2 \widehat{X, Y},$$

f multiplies areas ($\frac{1}{2}|X|\,|Y|\,\left|\sin \widehat{X, Y}\right|$) by the factor λ^2. However, f may reverse orientations.

Let X and Y be directions with $XY = i$, a positive right angle. From the above we know that

$$f[X]f[Y] + f[Y]f[X] = \lambda^2 (XY + YX)$$

and either

$$f[X]f[Y] - f[Y]f[X] = \lambda^2 (XY - YX)$$

or

$$f[X]f[Y] - f[Y]f[X] = \lambda^2 (YX - XY).$$

In the first case we must have

$$f[X]f[Y] = \lambda^2 XY,$$

and hence

$$f[X], f[Y] = \widehat{X, Y},$$

while in the second case we have

$$f[X]f[Y] = \lambda^2 YX$$

and hence

$$f[X], f[Y] = \widehat{Y, X}.$$

In the first case, f preserves all angles and we have what is called a *direct similitude*; for every angle has the form XZ for some direction $Z = \alpha X + \beta Y$, and

$$\begin{aligned} f[X]f[Z] &= f[X](\alpha f[X] + \beta f[Y]) \\ &= \alpha\lambda^2 X^2 + \beta\lambda^2 XY \\ &= \lambda^2 XZ. \end{aligned}$$

Similarly, in the second case f reverses all angles and we have what is called an *indirect similitude*.

Because

$$X^{-1}Y = \frac{|Y|}{|X|}\widehat{X, Y},$$

it must be the case that f carries triangles to similar triangles, preserving orientation if direct and reversing orientation if indirect.

Let A be a point. Define $g : \mathcal{P} \to \mathcal{P}$ by

$$g[B] = A + \frac{1}{\lambda}(f[B] - f[A]),$$

for every point B. Clearly g is an isometry with fixed point A, and

$$\begin{aligned} f[B] &= A + \lambda(g[B] - A) + (f[A] - A) \\ &= (t \circ d \circ g)[B] \end{aligned}$$

for every point B, where t is translation by $f[A] - A$ and $d : \mathcal{P} \to \mathcal{P}$ is the similitude with fixed point A, given by

$$d[B] = A + \lambda(B - A)$$

for every point B, and called a *dilation*. This effectively reduces the problem of finding all similitudes to that of finding all isometries.

13   Isometries

The action of a direct isometry f on vectors is completely determined by its action on a single vector, since the identity

$$f[X]f[Y] = XY$$

entails that f carries each $X \in V$ to

$$f[X] = XY f[Y]^{-1}$$

where Y is some fixed non-zero vector. Indeed

$$Y f[Y]^{-1} = \frac{Y f[Y]}{f[Y]^2} = \frac{Y f[Y]}{|Y||f[Y]|}$$

is an angle, and so the direct isometries are identified as precisely the maps obtained by composing a translation with a rotation.

Example. Consider an indirect isometry f which has a fixed non-zero vector Y, which we take, without loss of generality, to be a direction. Since $f[X]f[Y] = YX$, the action of f on vectors X is given by

$$f[X] = YX f[Y]^{-1} = YXY.$$

If $X = \alpha Y + \beta Y i$ then

$$f[X] = \alpha Y^3 + \beta Y^2 i Y = \alpha Y - \beta Y i.$$

If f has a fixed point A we call f the *reflection* in the line through A with direction Y. Reflections are idempotent, since

$$f^2[X] = Y^2 X Y^2 = X,$$

and the product of the reflections in the lines through A with directions Y and Z is the square of the rotation about A through the angle YZ, since

$$Z(YXY)Z = X(YZ)(YZ).$$

The natural induced action of reflections on angles is conjugation: reflection in a line with direction Y carries the angle YX into

$$(YYY)(YXY) = XY. \quad \square$$

Every indirect isometry is determined by its action on a single direction Y, since, for every vector X,

$$f[X] = YX f[Y]^{-1},$$

and is the composite of a translation, a reflection and a rotation, since

$$f[X] = (YXY)(Y^{-1} f[Y]^{-1}) = (YXY)(Y f[Y]).$$

Every rotation is the composite of two reflections. To see this, let f be a rotation through the angle YZ, where Y and Z are directions. If U is the direction of $\frac{1}{2}(Y + Z)$ then $YU = UZ$, and hence $f[X] = XYZ = (XY)(UY)U = U(YXY)U$.

14 Conclusion

The simplest concrete realisation of our formalism is as the ring generated by
the real numbers and elements O, X and Y, subject to the conditions that real
numbers commute with the other generators, $X^2 = 1$, $Y^2 = 1$ and $YX = -XY$,
but otherwise free. It is straightforward to verify the axioms, with $\mathcal{P} = \{O + \alpha X +
\beta Y : \alpha \in \mathcal{R}, \beta \in \mathcal{R}\}$. Of course many elements have no geometric interpretation,
but our exposition shows that this is not a problem. We have shown how standard
notions of Euclidean plane geometry are realised, and indeed other notions, not
mentioned above, such as line vectors, are also realised.

Note that direct use of such a concrete realisation, rather than our axiomatic
approach, would allow coordinate-based proof of theorems, including those involving points. However, a main point of the paper has been to suggest that it is
necessary — and possible — to move beyond a theorem-proving point of view if
we are to build more autonomous systems that internalize geometric knowledge
in a way that will satisfy the desiderata presented in our first paragraph. The
Euclid ring axiomatization supports direct algebraic reasoning with geometric
concepts.

In a Euclid ring, points, vectors and complex numbers are distinct geometric
entities, with rich interplay between them. Both the identification of points with
vectors (made in expositions of "vector geometry") and with complex numbers
(made in expositions of "complex geometry") are ungeometrical and obscure
those relationships; our treatment shows that these identifications are unnecessary. At the same time, the isomorphisms between $\mathcal{P}$ and $A + \mathcal{V}$ (for any "origin"
A), and between $\mathcal{V}$ and $\mathcal{R} + i\mathcal{R}$, explain why those identifications work, to the
extent that they do.

It is not suggested that the approach to the Euclidean plane presented here
supercedes other approaches. On the contrary, the Euclid ring formalism provides
a context in which other approaches have their places. I conclude with some
remarks on connections with other approaches.

(a) *Complex numbers.* The Argand diagram was of great importance historically because it helped to make complex numbers respectable, and it remains
the way that we picture complex numbers. Nevertheless I want to suggest that
this picture is wrong in roughly the same sense that it is wrong to identify two
structures which are isomorphic but not naturally isomorphic. A drawing board
(physical plane) has no imaginary axis (or real axis, or origin, for that matter).
From a geometric point of view, complex numbers are not points but ratios of
vectors (or positive multiples of angles). Conjugation is not reflection of points
in the real axis but a map which carries each positive multiple of an angle onto
the same positive multiple of the inverse angle. Of course if we choose an origin
O and direction X, every point P is uniquely expressible in the form

$$P = O + \alpha X + \beta Xi,$$

and the map $P \mapsto X^{-1}(P - O) = \alpha + \beta i$ is a bijection between points and
complex numbers, and there is an induced correspondence between reflection in

the real axis and conjugation. Reversing the point of view, one may say that a major advantage of the Euclid ring approach is that it enables one to deal with angles algebraically in the same kind of way that we deal with vectors.

(b) *Coordinates.* People who use geometry, such as engineers and architects, rarely need explicit coordinates. As with (a), application of Occam's razor would excise coordinates from geometry. Nevertheless coordinates are of interest, since to date the most successful geometry theorem provers and systems for the creation and manipulation of geometric objects are based on coordinatisation. The Euclid ring approach evidently supports resort to coordinates for proofs when necessary. At the same time it is reasonable to hope that it may be feasible to build geometric reasoning systems that can produce simple algebraic proofs like those given in this paper. Current work of D. Wang [9] and others, including a group lead by the author, may lead to such systems.

A point to be noted is that whereas the field of complex numbers, by virtue of its algebraic completeness, plays a crucial if somewhat inexplicable role in the coordinate-based proof methods for "real" Euclidean geometry, complex numbers appear naturally as actual geometric objects in the treatment presented in this paper.

(c) *Synthetic geometry.* There is something very appealing about traditional synthetic proofs and they may profitably be mixed with algebraic arguments. For example after the result (proved here) that the angles at the base of an isosceles triangle are equal, one can prove in rapid succession by synthetic arguments a sequence of interesting theorems: that the angle subtended by a chord at the centre of a circle is twice the angle at the circumference, that opposite angles of a cyclic quadrilateral are supplementary, that the product of the distance from a point to the points of intersection with a given circle of a line through the point is constant. One might even choose to carry this approach to the limit by proving as rapidly as possible a set of synthetic axioms (as theorems) and thereafter using only synthetic methods. As we have shown, the Euclid ring also provides a natural setting in which to introduce transformation geometry.

(d) *Inner and outer products.* Of course $X \mapsto |X|$ is a Euclidean norm on $\mathcal{V}$ and by setting

$$X.Y = \Re[XY] = \frac{1}{2}(|X + Y|^2 - |X|^2 - |Y|^2)$$

we obtain the associated inner product. Moreover by setting

$$X \wedge Y = i\, \Im[XY] = \frac{1}{2}(XY - YX)$$

(and $X \wedge Y \wedge Z \equiv 0$) we get an exterior product of vectors. Many of our results may be expressed in terms of these products, but this usually amounts to splitting complex numbers into real and imaginary parts, something which is often best avoided.

This paper is a slightly updated version of [3]. It aspires to bring to bear upon Euclidean geometry with points and numbers as the fundamental objects, the power of ideas which were implicit already in the work of Grassmann, Hamilton and Clifford, and have subsequently been extended by many authors (see, for example, [5]). The modern theory of Clifford algebras was established by Elie Cartan, Hermann Weyl, Claude Chevalley, Marcel Riesz and others, and thorough expositions may be found, for example, in the books [7], [8] and [6]. The dependence of the paper on the work of these authors is acknowledged.

The material presented in this paper is very simple, even trivial. But that is precisely the point: if an automatic reasoning system is to be able to "do" plane Euclidean geometry, and to call upon this ability when needed, it should use a simple formalism that is amenable to computation and that, at the same time, is sufficiently expressive to fully capture the riches of the world that Euclid gave us. That is what the Euclid ring approach attempts to do.

References

1. Fearnley-Sander, D.: Affine Geometry and Exterior Algebra. Houston Math. J. **6** (1980), 53–58
2. Fearnley-Sander, D.: A Royal Road to Geometry. Math. Mag. **53** (1980), 259–268
3. Fearnley-Sander, D.: Plane Euclidean Geometry. Technical Report 161, Department of Mathematics, University of Tasmania, 1981.
4. Fearnley-Sander, D. and Stokes, T.: Area in Grassmann Geometry. In: Wang, D. (ed.): Automated Deduction in Geometry. Lecture Notes in Artificial Intelligence, Vol. 1360. Springer-Verlag, Berlin Heidelberg New York 1998 141–170
5. Forder, H. G.: The Calculus of Extension. Cambridge, 1941
6. Hestenes, D., and Sobczyk, G.: Clifford Algebra to Geometric Calculus. Reidel, 1984
7. Jacobson, N.: Basic Algebra II. Freeman, San Francisco, 1980
8. Porteous, I. R.: Topological Geometry. Cambridge, 1981
9. Wang, D.: Clifford Algebraic Calculus for Geometric Reasoning with Applications to Computer Vision. In: Wang, D. (ed.): Automated Deduction in Geometry. Lecture Notes in Artificial Intelligence, Vol. 1360. Springer-Verlag, Berlin Heidelberg New York 1998 115–140

A Clifford Algebraic Method for Geometric Reasoning

Haiquan Yang, Shugong Zhang, and Guochen Feng

Institute of Mathematics, Jilin University, Changchun 130012, PRC

Abstract. In this paper a method for mechanical theorem proving in geometries is proposed. We first discuss how to describe geometric objects and geometric relations in 2D and/or 3D Euclidean space with Clifford algebraic expression. Then we present some rules to simplify Clifford algebraic polynomials to the so-called final Clifford algebraic polynomials. The key step for proving the theorems is to check if a Clifford algebraic expression can be simplified to zero. With the help of introducing coordinates, we can prove mechanically most of the geometric theorems about lines, conics, planes and so on in plane and/or solid geometry. The proofs produced by machine with our method are readable and geometrically interpretable. Finally, some interesting examples are given.

1 Introduction

It has been a hard work left since the time of Euclid to find a mechanical method to prove difficult geometric theorems to make learning and teaching geometry easier. In history many mathematicians, including Leibniz, Hilbert and others, have tried their best in this field. Modern computer technology and science make it possible to produce proofs of geometric theorems mechanically. Since nineteen fifties, lots of scholars have studied mechanical theorem proving in geometries, but they did not make much progress in proving relatively difficult geometric theorems until nineteen seventies.

In 1978, a breakthrough came with the work of Wu [6], who introduced an algebraic method which, for the first time, can be used to prove hundreds of geometric theorems mechanically. With Wu's method, many difficult theorems whose traditional proofs need an enormous amount of human intelligence can be proved on computer within seconds.

Motivated by Wu's method, Chou, Gao and Zhang developed an area method that can produce short and readable proofs for hundreds of geometric statements in plane and/or solid geometry [1]. This is another great progress in this field.

In 1997, combining Clifford algebraic expression with Wu's method, Li and Cheng proposed a complete method for mechanical theorem proving in plane geometry; this brought people new enlightenment [5]. Soon after this, Fèvre and Wang gave a method for mechanical theorem proving based on rewriting rules for Clifford algebraic expressions [2]. Recently, Boy de la Tour and others have proposed a rewriting system for simplifying Clifford algebraic expressions in

3D and investigated its theoretical properties [3]. Having some similarity to their work, ours presented here emphasizes various applications to reasoning problems in plan and solid geometry.

In the practice of theorem proving, the first task is to represent geometric objects and relations with Clifford algebraic expressions. Then simplifying these expressions to simpler forms with certain rules. To prove a theorem, one needs to simplify the conclusion expression with the condition expressions and the simplifying rules to see if the final expression is zero. With this method we have proved, besides theorems about points, lines, planes and so on in plane and solid geometry, many theorems about conics which were never considered in mechanical theorem proving before.

In addition, by changing the metric, this method can be used to prove theorems in non-Euclidean geometry.

2 Preliminaries

Let V be a vector space over the field R of real numbers and $Q : V \to R$ a quadratic form. The quotient of the tensor algebra $\otimes(V)$ by the two-sided ideal generated by the elements of the form $x \otimes x - Q(x), x \in V$, is called the Clifford algebra of (V, Q), denoted by $\mathcal{G}(V, Q)$ or $\mathcal{G}(V)$. The product in $\mathcal{G}(V)$, denoted by juxtaposition, is called the *geometric product*. Induced by $\otimes(V)$, $\mathcal{G}(V)$ is a graded associative algebra.

Let $0 \neq A \in \mathcal{G}(V)$. A is called a *r-blade* if $A = a_1 a_2 \ldots a_r$, where

$$a_i \in V, \ a_i a_j + a_j a_i = 0, \ \forall i \neq j.$$

The symbols $A_r, B_s, \ldots$, represent a r-blade A, a s-blade B, and so on. All r-blades generate linearly a vector subspace $\mathcal{G}_r(V)$ of $\mathcal{G}(V)$, whose elements are called *r-vectors*. We have $\mathcal{G}(V) = \oplus_{i=0}^{\infty} \mathcal{G}_i(V)$, where $\mathcal{G}_0(V) = R$, $\mathcal{G}_1(V) = V$. So

$$\forall A \in \mathcal{G}(V), \ A = \sum_{i=0}^{\infty} \langle A \rangle_i,$$

where $\langle A \rangle_i \in \mathcal{G}_i(V)$ is the i-vector component.

When $r, s \neq 0$, $\langle A_r B_s \rangle_{|r-s|}$ is called the *inner product* of A_r, B_s, denoted by $A_r \cdot B_s$, $\langle A_r B_s \rangle_{r+s}$ is called the *outer product* of A_r, B_s, denoted by $A_r \wedge B_s$; otherwise we define

$$A_0 \cdot B_s = B_s \cdot A_0 = 0, \ A_0 \wedge B_s = B_s \wedge A_0 = A_0 B_s.$$

The definitions can be extended linearly to $\mathcal{G}(V) \times \mathcal{G}(V)$.

The *commutator product* of $A, B \in \mathcal{G}(V)$ is $A \times B = \frac{1}{2}(AB - BA)$.

The *reverse* of $A \in \mathcal{G}(V)$ is denoted by $A^\dagger$; in which † is a linear operator satisfying $A_r^\dagger = a_r a_{r-1} \ldots a_1$ for any $A_r = a_1 a_2 \ldots a_r$, where $a_i \in V$.

The *scalar product* of $A, B \in \mathcal{G}(V)$ is $A * B = \langle AB \rangle_0$, the *magnitude* of $A \in \mathcal{G}(V)$ is defined as $|A| = \sqrt{|A^\dagger * A|}$.

See [4] for details.

3 Plane Geometry

In this section, we consider the Clifford algebra in the Euclidean plane E^2. The oriented segment from point A to point B, denoted by $\overline{AB}$, is called a *vector*, its magnitude is denoted by $|AB|$. We denote by I_2 the unit 2-vector, and hence $\mathcal{G}(I_2) = \mathcal{G}(E^2)$, $\mathcal{G}_1(I_2) = E^2$. Denote $e^{I_2\theta} = \cos\theta + I_2 \sin\theta$, $\forall\theta \in R$.

Let $line(A, \boldsymbol{a})$ denote the line passing through the point A and parallel to the vector $\boldsymbol{a}$; $circle1(O, P)$ denote the circle with the point O as its center and passing through the point P; $circle2(O, r)$ denote the circle with center O and radius r; $parabola(O, F)$ denote the parabola with vertex O and focus F; $ellipse(O, F, a)$ denote the ellipse with center O, focus F and major semi-axis a; $hyperbola(O, F, a)$ denote the hyperbola with center O, focus F and real semi-axis a;

3.1 Description of the Method

In this section, we restrict ourselves to constructive statements in plane geometry.

By a *geometric object* we mean a point, a scalar, a vector, a multivector, ..., in the plane. A construction means a way to introduce some new geometric objects based on some geometric objects introduced in advance.

Generally, a geometric statement is composed of a sequence of constructions and a conclusion that can be written as a Clifford algebraic expression of some geometric objects.

Definition 1. *A construction is one of the following operations:*

C_0) *(point(X)) Take an arbitrary point X on the plane.*

C_1) *(on_line(X, A, $\boldsymbol{a}$, t)) Take an arbitrary point X on $line(A, \boldsymbol{a})$.*

C_2) *(on_cir1(X, O, P, θ)) Take an arbitrary point X on $circle1(O, P)$.*

C_3) *(inter_line_line(X, A, $\boldsymbol{a}$, B, $\boldsymbol{b}$)) Point X is the intersection of $line(A, \boldsymbol{a})$ and $line(B, \boldsymbol{b})$.*

C_4) *(inter_line_cir1(X, A, $\boldsymbol{a}$, O)) Point X is the intersection of $line(A, \boldsymbol{a})$ and $circle1(O, A)$ other than point A.*

C_4') *(inter_line_cir2(X, A, $\boldsymbol{a}$, O, r, X'), where X' is optional) Points X and X' are the intersections of $line(A, \boldsymbol{a})$ and $circle2(O, r)$.*

C_5) *(inter_cir2_cir2(X, O_1, r_1, O_2, r_2, X'), where X' is optional) Points X and X' are the intersections of $circle2(O_1, r_1)$ and $circle2(O_2, r_2)$.*

C_6) *(on_para(X, O, F, p, t) where $p = |OF|$) Take an arbitrary point X on $parabola(O, F)$.*

C_7) *(para_tan(τ, O, F, p, P), where $p = |OF|$) Vector τ is the tangent vector at point P to $parabola(O, F)$, where point P is on the parabola.*

C_8) *(tan_para(X, O, F, p, σ), where $p = |OF|$) Point X is on $parabola(O, F)$, whose tangent vector is σ.*

C_9) *(inter_para_line(X, O, F, p, B, $\boldsymbol{a}$), where $p = |OF|$) Point X is the intersection of $line(B, \boldsymbol{a})$ and $parabola(O, F)$ other than point B, where point B is on the parabola.*

C_{10}) *(inter_para_linepassfocus(X, O, F, p, a, X'), where $p = |OF|$, X' is optional) Points X and X' are the intersections of line(F, a) and parabola (O, F).*

C_{11}) *(on_elli$(X, O, F, a, b, c, \theta)$, where $c = |OF|$, $b = \sqrt{a^2 - c^2}$) Take an arbitrary point X on ellipse(O, F, a).*

C_{12}) *(elli_tan(τ, O, F, a, b, c, P), where $c = |OF|$, $b = \sqrt{a^2 - c^2}$) Vector τ is the tangent vector at point P to ellipse(O, F, a), where point P is on the ellipse.*

C_{13}) *(tan_elli$(X, O, F, a, b, c, \sigma, X')$, where $c = |OF|$, $b = \sqrt{a^2 - c^2}$, X' is optional) Points X and X' are on ellipse(O, F, a), whose tangent vector is σ.*

C_{14}) *(inter_elli_line$(X, O, F, a, b, c, P, \sigma)$, where $c = |OF|$, $b = \sqrt{a^2 - c^2}$) Point X is the intersection of line(P, σ) and ellipse(O, F, a) other than point P, where point P is on the ellipse.*

C_{15}) *(on_hyper(X, O, F, a, b, c, t), where $c = |OF|$, $b = \sqrt{c^2 - a^2}$) Take an arbitrary point X on hyperbola(O, F, a).*

C_{16}) *(hyper_tan(τ, O, F, a, b, c, P), where $c = |OF|$, $b = \sqrt{c^2 - a^2}$) Vector τ is the tangent vector at point P to hyperbola(O, F, a), where point P is on the hyperbola.*

C_{17}) *(tan_hyper$(X, O, F, a, b, c, \sigma, X')$, where $c = |OF|$, $b = \sqrt{c^2 - a^2}$, X' is optional) Points X and X' are on hyperbola(O, F, a), whose tangent vector is σ.*

C_{18}) *(inter_hyper_line$(X, O, F, a, b, c, P, \sigma)$, where $c = |OF|$, $b = \sqrt{a^2 - c^2}$) Point X is the intersection of line(P, σ) and hyperbola(O, F, a) other than point P, where point P is on the hyperbola.*

The points X, X' or the vector τ in each of the above constructions are said to be introduced by that construction.

Definition 2. *A constructive statement is represented by a list*

$$S = (c_1, c_2, \ldots, c_k, G),$$

where

- *$c_i, i = 1, \ldots, k$, are constructions such that the geometric objects introduced by each c_i must be different from the geometric objects introduced by $c_j, j = 1, \ldots, i-1$, and the other geometric objects occurring in c_i must be introduced before; and*
- *$G = (E_1, E_2)$ where E_1 and E_2 are the Clifford algebraic expressions of some geometric objects introduced by the constructions c_i and $E_1 = E_2$ is the conclusion of S.*

By the propositions in [8], the geometric objects introduced by the constructions can be represented by a Clifford algebraic expression of the known geometric objects. For testing whether a constructive statement is true or not, we simply, from the last construction c_k to the first one c_1, substitute the expressions of the introduced geometric objects into the conclusion's expression, and then simplify the result.

3.2 Algorithm and Implementation of $\mathcal{G}(E^2)$

A Clifford algebraic expression is called *basic* if it is a constant symbol denoting some real number, a constant symbol denoting some vector, or the unit 2-vector I_2. In particular, a *basic vector*, denoted by $a, b, c, \ldots$, is a constant symbol denoting some vector; a *basic scalar*, denoted by $\alpha, \beta, \lambda, \ldots$, is a constant symbol denoting some real number.

Let $A, B, C, \ldots, x, y, z$ denote the Clifford algebraic expressions which are composted of basic scalars, basic vectors and I_2 through Clifford algebraic operations.

Among Clifford algebraic operations, the most useful one is the computation of the grade of a Clifford algebraic element. So we have the following algorithm.

Algorithm 1. *Compute the grade of x if x is a homogeneous multivector, otherwise returns an error message.*

```
Cgrade:=proc(x)
    if x is a basic vector then  RETURN(1)
    elif x = I₂  then  RETURN(2)
    elif x = A · B   then  RETURN(|Cgrade(A)−Cgrade(B)|)
    elif x = A ∧ B   then  RETURN(Cgrade(A)+Cgrade(B))
    elif x = A + B   then
        if Cgrade(A)=Cgrade(B)  then  RETURN(Cgrade(A))
        else  ERROR
        fi
    elif x = λA    then RETURN(Cgrade(A))
    elif x = A⁻¹  then  RETURN(Cgrade(A))
    elif x = AB   then  ERROR
    else RETURN(0)
    fi
end
```

where A, B are not scalars.

We consider the inner and outer products as the basic products and will express other products with them. So we have the following algorithms.

Algorithm 2. *Compute the inverse of x if it exists, otherwise return an error message.*

```
Inverse:=proc(x)
```
$$\text{if } \mathrm{Cgrade}(x) = 0 \quad \text{then} \quad \mathrm{RETURN}(\frac{1}{x})$$
$$\text{elif } \mathrm{Cgrade}(x) = 1 \quad \text{then} \quad \mathrm{RETURN}(\frac{x}{x \cdot x})$$
$$\text{elif } \mathrm{Cgrade}(x) = 2 \quad \text{then} \quad \mathrm{RETURN}(\frac{I_2}{x \cdot I_2})$$
```
    else ERROR
    fi
end.
```

Clearly, the algorithm Inverse cannot deal with all reversible elements, but it is sufficient for our purpose.

Algorithm 3. *Represent a geometric product with inner and outer products.*

```
Eval_g:=proc(x,y)
    if x = ∑ᵢ₌₁ʳ Aᵢ,  y = ∑ⱼ₌₁ˢ Bⱼ and (r > 1 or s > 1)
        then  RETURN (∑ᵢ₌₁ʳ ∑ⱼ₌₁ˢ Eval_g(Aᵢ,Bⱼ))
    elif x = AB  then  RETURN(Eval_g(Eval_g(A,B),y))
    elif y = AB  then  RETURN(Eval_g(x, Eval_g(A,B)))
    elif Cgrade(x) = 0 or Cgrade(y) = 0  then    RETURN(x * y)
    elif Cgrade(x) = 1   then
        if Cgrade(y) = 1  then   RETURN(x · y + x ∧ y)
        else   RETURN(x · y)
        fi
    else   RETURN(x · y)
    fi
end
```

where $x * y$ denotes the scalar multiplication, and A, B are not scalars.

Clearly, if the three conditions at the beginning of the algorithm are not satisfied, then both x and y must be homogeneous multivectors, and the algorithm Cgrade can compute their grades. So the algorithm Eval_g is correct.

With the above algorithms, we can represent various products that appear in our applications with inner and outer products.

A Clifford algebraic expression is called a *Clifford algebraic polynomial* if it does not contain any inverse of non-scalars and any denominators involving non-scalars. What follows will mainly deals with the Clifford algebraic polynomials.

We will use the following five kinds of rules and formulas

Rules 1: Both inner and outer product are bilinear.

Rules 2: For arbitrary scalar $\lambda \in R$ and arbitrary multivector $A \in \mathcal{G}(E^2)$:

$$\lambda \wedge A = A \wedge \lambda = \lambda A, \quad \lambda \cdot A = A \cdot \lambda = 0.$$

Rules 3: For arbitrary r-vector A_r and s-vector B_s:
$$A_r \cdot B_s = (-1)^{r(s-1)} B_s \cdot A_r, \quad \text{if } r \leq s.$$
$$A_r \wedge B_s = (-1)^{rs} B_s \wedge A_r.$$

Rules 4: For arbitrary vectors $a, b \in E^2$:

$$(1) \quad a \wedge (b \cdot I_2) = (a \cdot b)I_2$$
$$(2) \quad a \cdot (b \cdot I_2) = (a \wedge b) \cdot I_2$$
$$(3) \quad (a \cdot I_2) \cdot I_2 = -a$$
$$(4) \quad ((a \wedge b) \cdot I_2)I_2 = -(a \wedge b)$$
$$(5) \quad I_2 \cdot I_2 = -1$$

Rules 5: For any vectors $a, b, c, d \in E^2$:

$$a \cdot (b \wedge c) = (a \cdot b)c - (a \cdot c)b$$
$$(a \wedge b) \cdot (c \wedge d) = (b \cdot c)(a \cdot d) - (b \cdot d)(a \cdot c)$$

Algorithm 4. *Expand a Clifford algebraic polynomial x by Rules 1, Rules 2, and Rules 3.*

```
Eval_i_w1:=proc(x)
   step1: expand x by bilinearity to get a sum, each term of
          the sum does not contain any addition and subtraction.
   step2: process the scalar factors in each term by Rules 2.
   step3: reorder the factors in each term by Rules 3.
end.
```

A Clifford algebraic polynomial is called *expanded* if it has been processed by Eval_i_w1. In the sequel, we will assume that the algorithm Eval_i_w1 is called automatically when it is required.

We now consider the forms of the terms in an expanded Clifford algebraic polynomial.

Class 1: the basic scalars $\alpha, \beta, \ldots$, the basic vectors $a, b, c, d, \ldots$, and the unit 2-vector I_2.

Class 2: the elements which are obtained by taking inner or outer product with factors in Class 1 once. For examples, the scalars $a \cdot b, c \cdot d$, the vectors $a \cdot I_2, b \cdot I_2$, the 2-vectors $a \wedge b, c \wedge d,$.

Class 3: the elements which are obtained by taking inner or outer product with factors in Class 1 or Class 2 once. All of these elements can be simplified to a sum by Rules 4 and Rules 5, the new form occurring in these terms is the scalar $(a \wedge b) \cdot I_2$.

No other new forms will be produced by this processing. So the forms of Clifford algebraic terms are

$$\text{vectors: } a, \ b, \ c, \ d, \ldots\ldots; \ a \cdot I_2, \ b \cdot I_2, \ c \cdot I_2, \ d \cdot I_2, \ldots\ldots$$
$$\text{scalars: } a \cdot b, \ c \cdot d, \ldots; \ (a \wedge b) \cdot I_2, \ (c \wedge d) \cdot I_2, \ldots; \ \alpha, \ \beta, \ldots$$
$$\text{2-vectors: } I_2; \ a \wedge b, \ c \wedge d, \ldots\ldots$$

and an expanded Clifford algebraic polynomial is a sum of products of such elements and the scalars.

From the above, there are five forms of terms in an expanded Clifford algebraic polynomial

$$A, \ AI_2, \ A(a \wedge b), \ Aa, \ A(a \cdot I_2),$$

where

$$A = \alpha_1 \ldots \alpha_i (a_1 \cdot b_1) \ldots (a_j \cdot b_j)((c_1 \wedge d_1) \cdot I_2) \ldots ((c_k \wedge d_k) \cdot I_2).$$

Rules 5′:

$$((a \wedge b) \cdot I_2)((c \wedge d) \cdot I_2) = (b \cdot d)(a \cdot c) - (b \cdot c)(a \cdot d)$$
$$((a \wedge b) \cdot I_2)(c \cdot I_2) = (b \cdot c)a - (a \cdot c)b$$
$$((a \wedge b) \cdot I_2)c = (a \cdot c)(b \cdot I_2) - (b \cdot c)(a \cdot I_2)$$
$$((a \wedge b) \cdot I_2)(c \wedge d) = (b \cdot c)(a \cdot d)I_2 - (b \cdot d)(a \cdot c)I_2$$
$$((a \wedge b) \cdot I_2)I_2 = -(a \wedge b)$$

With the above rules, we can simplify a Clifford algebraic polynomials to a sum of the following six forms

S1: $\alpha_1 \ldots \alpha_i (a_1 \cdot b_1) \ldots (a_j \cdot b_j)$

S2: $\alpha_1 \ldots \alpha_i (a_1 \cdot b_1) \ldots (a_j \cdot b_j)((c \wedge d) \cdot I_2)$

S3: $\alpha_1 \ldots \alpha_i (a_1 \cdot b_1) \ldots (a_j \cdot b_j) I_2$

S4: $\alpha_1 \ldots \alpha_i (a_1 \cdot b_1) \ldots (a_j \cdot b_j)(c \wedge d)$

S5: $\alpha_1 \ldots \alpha_i (a_1 \cdot b_1) \ldots (a_j \cdot b_j) c$

S6: $\alpha_1 \ldots \alpha_i (a_1 \cdot b_1) \ldots (a_j \cdot b_j)(c \cdot I_2)$

where $\alpha_1, \ldots \alpha_i$ are basic scalars, $a_1, \ldots a_j$, $b_1 \ldots b_j$, c, d are basic vectors.

Definition 3. *The forms S1, S2, ..., S6 will be called final forms. A Clifford algebraic polynomial is called a final Clifford algebraic polynomial if its terms are all final forms.*

Theorem 1. *There is an algorithm to simplify a Clifford algebraic polynomial in $\mathcal{G}(I_2)$ to a final Clifford algebraic polynomial.*

Algorithm 5. *Simplify an expanded Clifford algebraic polynomial x by Rules 4 and Rules 5' to a final Clifford algebraic polynomial.*

```
Eval_i_w2:=proc(x)
  step1: Let the terms and/or the factors with grades greater
         than 2 vanish.
  step2: Simplify the 2-vectors in x which are not I2 by formula
         u = -(u · I2)I2, that is in the reverse order of rule (4)
         in Rules 4.
  step3: Simplify x by other rules in Rules 4.
  step4: Simplify x by Rules 5'. x will be a final polynomial.
  end.
```

3.3 Algorithm and Implementation of Prover

We first give Maple procedures for constructions. Each construction will return a list that is composed of the introduced geometric objects and their Clifford algebraic expressions proposed by the propositions in [8].

Examples. Procedure on_para(X, O, F, p, t) returns

$$[\overline{OX}, \frac{\overline{OF}}{p}(\frac{t^2}{4p} + t I_2)].$$

Procedure para_tan(τ, O, F, p, P) returns

$$[\tau, \frac{\overline{OF}}{p}(-\frac{(\overline{OF} \wedge \overline{OP})I_2}{2p^2} + I_2)].$$

Procedure inter_para_linepassfocus(X, O, F, p, a) returns

$$[\overline{OX}, \overline{OF} - \frac{2p^2}{p|a| + \overline{OF} \cdot a} a];$$

Procedure inter_para_linepassfocus(X, O, F, p, a, X') returns

$$[\overline{OX}, \overline{OF} - \frac{2p^2}{p|a| + \overline{OF} \cdot a}a],$$

$$[\overline{OX'}, \overline{OF} + \frac{2p^2}{p|a| - \overline{OF} \cdot a}a].$$

Let L be a list returned by a construction, y a Clifford algebraic expression. The Maple procedure Eliminate(L, y) will substitute $L[2]$ for $L[1]$ in y and return the simplified result, where $L[1]$ and $L[2]$ are the first and the second element of the list L respectively.

Let $S = (c_1, c_2, \ldots, c_k, G)$ be a constructive statement, $(\text{list}_1, \text{list}_2, \ldots, \text{list}_r)$ the sequence of lists returned by the constructions $(c_1, c_2, \ldots, c_k)$, y a Clifford algebraic expression about the points and vectors introduced by the constructions $(c_1, c_2, \ldots, c_k)$. Then y can be simplified by the following algorithm.

Algorithm 6. *Simplify y by $\text{list}_1, \text{list}_2, \ldots, \text{list}_r$.*
```
PreProver:=proc(list₁, list₂, ..., listᵣ, y)
    for i from r by −1 to 1 do
        y := Eliminate(listᵢ, y);
        simplify y by Eval_i_w1 and Eval_i_w2;
    od
    RETURN(y)
end.
```

In the implementation of our program, we use a Maple procedure PreDeal to reduce the computing time and to shorten the proof of geometric theorems.

Algorithm 7. *Do some pretreatment of $\text{list}_1, \text{list}_2, \ldots, \text{list}_r$.*
```
PreDeal:=proc(list₁, list₂,...,listᵣ)
    simplify list₁[2] by Eval_i_w1 and Eval_i_w2;
    for k from 2 to r do
        y := PreProver(list₁,..., listₖ₋₁, listₖ[2]);
        listₖ:=[listₖ[1], y]
    od
end.
```

From the above, we have the following Prover.

Algorithm 8. *Test whether a constructive statement*

$$S = (c_1, c_2, \ldots, c_k, (E, F))$$

is true or not, and print the proof for S. Let $(\text{list}_1, \text{list}_2, \ldots, \text{list}_r)$ be the lists returned by the constructions $(c_1, c_2, \ldots, c_k)$.
```
Prover:=proc(list₁, list₂,..., listᵣ, E,F)
    PreDeal(list₁, list₂,...,listᵣ);
```

```
for  i  to r do print ''list_i[1]=list_i[2])'' od;
E:= PreProver(list_1, list_2, ..., list_r, E);
    and print the steps;
F:= PreProver(list_1, list_2, ..., list_r, F);
    and print the steps;
if  E = F  then  RETURN('True')
else
    y:=Eval_Normal(E - F); and print ''y = E - F''
    if  y = 0  then  RETURN('True')
    else  RETURN('False')
    fi
fi
end.
```

In this algorithm, $E = F$ means that E and F are identical expressions, $y = 0$ means y and 0 are identical expressions too; and the procedure *Eval_normal()* normalizes the final Clifford algebraic expression by introducing coordinates. In fact, Eval_normal() is never used in all examples we met, even those cannot be implemented by our program.

Theorem 2. *Algorithm Prover is correct.*

Proof. Obviously, before the first **if** sentence is called, the expressions E and F will only involve free variables. So the statement S is true if and only if E equals F, or $E - F$ equals 0.

If E and F are not identical expressions in form, we compute y, the normal form of $E - F$ w.r.t. the introduced coordinates. Since normal form is unique, $E - F$ equals 0 if and only if $y = 0$.

3.4 Some Examples

In this subsection, we give some examples to illustrate our general method. See [7] for more examples.

Let O be the origin, we denote vector $\overline{OF}, \overline{OA} \ldots$ by $\overline{F}, \overline{A} \ldots$. respectively in the proofs for the geometric statements. Our program will use the following procedures.

Procedure *para_const*(O, F, p) returns list $[p, |OF|]$.

Procedure *elli_const*(O, F, a, b, c) returns $[c, |OF|], [b, \sqrt{a^2 - c^2}]$.

Procedure *hyper_const*(O, F, a, b, c) returns $[c, |OF|], [b, \sqrt{c^2 - a^2}]$.

Procedure *hyper_sym*$(\tau_1, O, F, a, b, c, \tau_2)$ returns $[\tau_1, a\overline{OF} + b\overline{OF}I_2], [\tau_2, a\overline{OF} - b\overline{OF}I_2]$. They are the asymptotic vectors of the *hyperbola*(O, F, A), and τ_2 is optional.

Example 1. In geometrical optics, the light beams parallel to the main axis of a parabola will focus when they are reflected by the parabola.

This example can be described in the following constructive way.

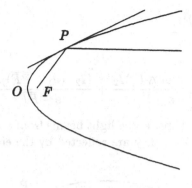

Fig. 1

Constructions:
point(O, F);
para_const(O, F, p);
on_para(P, O, F, p, t);
para_tan(e_1, O, F, p, P).

Conclusion:

$$\frac{\overline{OF} \wedge e_1}{\overline{OF} \cdot e_1} = \frac{e_1 \wedge \overline{FP}}{e_1 \cdot \overline{FP}}.$$

Our program prints the following.

PreDeal:

$$p = \sqrt{\overline{F} \cdot \overline{F}}$$

$$\overline{P} = \frac{t \left(t\overline{F} + 4\,\overline{F} \cdot I_2 \sqrt{\overline{F} \cdot \overline{F}} \right)}{4\,\overline{F} \cdot \overline{F}}$$

$$e_1 = \frac{2\,\overline{F} \cdot I_2 \sqrt{\overline{F} \cdot \overline{F}} + t\overline{F}}{2\,\overline{F} \cdot \overline{F}}$$

Proof:

$$\overset{e_1}{=} \frac{-\dfrac{I_2 \cdot (\overline{F} \wedge e_1) I_2}{\overline{F} \cdot e_1}}{\dfrac{2\sqrt{\overline{F} \cdot \overline{F}} I_2}{t}}$$

$$\overset{s}{=} -\frac{\dfrac{I_2 \cdot (e_1 \wedge \overline{FP}) I_2}{e_1 \cdot \overline{FP}}}{\dfrac{\left(I_2 \cdot (\overline{F} \wedge e_1) + I_2 \cdot (e_1 \wedge \overline{P}) \right) I_2}{-\overline{F} \cdot e_1 + e_1 \cdot \overline{P}}}$$

$$\overset{e_1}{=} \frac{\left(-2\,(\overline{F} \cdot \overline{F})^2 + 2\,\overline{F} \cdot \overline{PF} \cdot \overline{F} + t I_2 \cdot (\overline{F} \wedge \overline{P}) \sqrt{\overline{F} \cdot \overline{F}} \right) I_2}{t\,(\overline{F} \cdot \overline{F})^{3/2} + 2\,I_2 \cdot (\overline{F} \wedge \overline{P}) \overline{F} \cdot \overline{F} - t\overline{F} \cdot \overline{P} \sqrt{\overline{F} \cdot \overline{F}}}$$

$$\stackrel{P}{=} \frac{2\sqrt{\overline{F} \cdot \overline{F}} I_2}{t}$$

So

$$\frac{(I_2 \cdot (e_1 \wedge \overline{F}))I_2}{e_1 \cdot \overline{F}} + \frac{(I_2 \cdot (e_1 \wedge \overline{PF}))I_2}{e_1 \cdot \overline{PF}} = 0$$

Example 2. In geometrical optics, the light beams from a focus of an ellipse pass through the other focus when they are reflected by the ellipse.

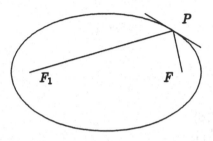

Fig. 2

This example can be described in the following constructive way.
Constructions:
    points(O, F);
    elli_const(O, F, a, b, c);
    on_elli(P, O, F, a, b, c, t_1);
    elli_tan(e_1, O, F, a, b, c, P);
    on_line$(F_1, O, \overline{OF}, -1)$.
Conclusion:

$$\frac{\overline{FP} \wedge e_1}{\overline{FP} \cdot e_1} = \frac{e_1 \wedge \overline{PF_1}}{e_1 \cdot \overline{PF_1}}.$$

Our program prints the following.
PreDeal:

$$\overline{P} = \frac{a\cos(t_1)\overline{F} + b\sin(t_1)(\overline{F} \cdot I_2)}{c}$$

$$e_1 = -\frac{(\overline{F} \cdot \overline{F})\left(-(\overline{F} \cdot I_2)\cos(t_1)b + a\sin(t_1)\overline{F}\right)}{c^3}$$

$$\overline{F_1} = -\overline{F}$$

Proof:

$$\frac{(I_2 \cdot (e_1 \wedge \overline{FP}))I_2}{e_1 \cdot \overline{FP}}$$

$$\overset{s}{=} -\frac{\left(I_2 \cdot (e_1 \wedge \overline{F}) - I_2 \cdot (e_1 \wedge \overline{P})\right) I_2}{-e_1 \cdot \overline{F} + e_1 \cdot \overline{P}}$$

$$\overset{e_1}{=} \frac{\left(-(\overline{F} \cdot \overline{F})\cos(t_1)b + \cos(t_1)b(\overline{P} \cdot \overline{F}) + a\sin(t_1)((\overline{P} \wedge \overline{F}) \cdot I_2)\right) I_2}{(\overline{F} \cdot \overline{F})a\sin(t_1) + \cos(t_1)b((\overline{P} \wedge \overline{F}) \cdot I_2) - a\sin(t_1)(\overline{P} \cdot \overline{F})}$$

$$\overset{P}{=} -\frac{b\left(\cos(t_1)c - \cos(t_1)^2 a - a\sin(t_1)^2\right) I_2}{\sin(t_1)\left(ac + \cos(t_1)b^2 - a^2\cos(t_1)\right)}$$

$$\overset{s}{=} -\frac{bI_2\left(\cos(t_1)c - a\right)}{\sin(t_1)\left(ac + \cos(t_1)b^2 - a^2\cos(t_1)\right)}$$

$$\frac{\left(I_2 \cdot (\overline{F_1 P} \wedge e_1)\right)I_2}{\overline{F_1 P} \cdot e_1}$$

$$\overset{s}{=} \frac{\left(-I_2 \cdot (e_1 \wedge \overline{F_1}) + I_2 \cdot (e_1 \wedge \overline{P})\right) I_2}{e_1 \cdot \overline{F_1} - e_1 \cdot \overline{P}}$$

$$\overset{F_1}{=} -\frac{\left(I_2 \cdot (e_1 \wedge \overline{F}) + I_2 \cdot (e_1 \wedge \overline{P})\right) I_2}{e_1 \cdot \overline{F} + e_1 \cdot \overline{P}}$$

$$\overset{e_1}{=} \frac{\left((\overline{F} \cdot \overline{F})\cos(t_1)b + \cos(t_1)b(\overline{P} \cdot \overline{F}) + a\sin(t_1)((\overline{P} \wedge \overline{F}) \cdot I_2)\right) I_2}{(\overline{F} \cdot \overline{F})a\sin(t_1) - \cos(t_1)b((\overline{P} \wedge \overline{F}) \cdot I_2) + a\sin(t_1)(\overline{P} \cdot \overline{F})}$$

$$\overset{P}{=} \frac{b\left(\cos(t_1)c + \cos(t_1)^2 a + a\sin(t_1)^2\right) I_2}{\sin(t_1)\left(ac - \cos(t_1)b^2 + a^2\cos(t_1)\right)}$$

$$\overset{s}{=} \frac{bI_2\left(\cos(t_1)c + a\right)}{\sin(t_1)\left(ac - \cos(t_1)b^2 + a^2\cos(t_1)\right)}$$

So

$$\frac{\left(I_2 \cdot (e_1 \wedge \overline{FP})\right)I_2}{e_1 \cdot \overline{FP}} - \frac{\left(I_2 \cdot (\overline{F_1 P} \wedge e_1)\right)I_2}{\overline{F_1 P} \cdot e_1}$$
$$= \frac{2\,bI_2\cos(t_1)a\left(c^2 + b^2 - a^2\right)}{\sin(t_1)\left(-a^2c^2 + \cos(t_1)^2 b^4 - 2\cos(t_1)^2 b^2 a^2 + a^4\cos(t_1)^2\right)}$$

Noticed that

$$b = \sqrt{a^2 - c^2}$$

We have

$$\frac{2\,bI_2\cos(t_1)a\left(c^2 + b^2 - a^2\right)}{\sin(t_1)\left(-a^2c^2 + \cos(t_1)^2 b^4 - 2\cos(t_1)^2 b^2 a^2 + a^4\cos(t_1)^2\right)}$$
$$\overset{b}{=} 0$$

4 Solid Geometry

In this section, we consider the Clifford algebra in 3-dimensional Euclidean space E^3. Let I_3 denote the unit 3-vector, $\mathcal{G}(I_3) = \mathcal{G}(E^3)$, $\mathcal{G}_1(I_2) = E^3$ and $\mathcal{G}_2(I_3) = \mathcal{G}_2(E^3)$.

Suppose O is a point in space, $u \in \mathcal{G}_2(I_3)$, the plane with direction u and passing through the point O is defined as

$$plane(O, u) = \{A \in E^3 | \overline{OA} \wedge u = 0\}.$$

the normal vector of this plane is $-uI_3$.

4.1 Description of the Method

Similar to the situation in plane geometry, we consider the constructive statements about solid geometry.

Definition 4. *A construction in E^3 is one of the following operations*
S_0) *(point(X)) Take an arbitrary point X in the space.*
S_1) *(on_plane(X, A, a, t_1, b, t_2)) Take an arbitrary point X on plane($A, a\wedge b$).*
S_2) *(inter_line_line(X, A, a, B, b)) Point X is the intersection of line(A, a) and line(B, b), where $\overline{AB} \wedge a \wedge b = 0$.*
S_3) *(inter_line_plane(X, A, a, B, u)) Point X is the intersection of line(A, a) and plane(B, u).*
S_4) *(inter_plane_plane_plane($X, A, u_1, B, u_2, C, u_3$)) Point X is the intersection of plane(A, u_1), plane(B, u_2) and plane(C, u_3).*
S_5) *(on_line(X, A, a, t)) Point X is on the line(A, a).*
In each case of the above, a point X in the construction is said to be introduced by that construction.

Definition 5. *A constructive statement in solid geometry is represented by a list*

$$S = (s_1, s_2, \ldots, s_k, G),$$

where

- *$s_i, i = 1, \ldots, k$, are constructions in which the geometric objects introduced by each s_i must be different from the geometric objects introduced by $s_j, j = 1, \ldots, i - 1$, and other geometric objects occurring in s_i must be introduced before; and*
- *$G = (E_1, E_2)$ where E_1 and E_2 are the Clifford algebraic expressions of some geometric objects introduced by the constructions s_i and $E_1 = E_2$ is the conclusion of S.*

By the propositions in [8], the geometric objects introduced by the constructions in E^3 can be represented by a Clifford algebraic expression of known geometric objects. To test whether a constructive statement is true or not, we simply, from the last construction s_k to the first s_1, substitute the expressions of the introduced geometric objects into the conclusion's expression, and simplify the result.

4.2 Algorithm and its Implementation in $\mathcal{G}(E^3)$

We only consider the simplification of Clifford algebraic polynomials: Clifford algebraic expressions which contain no inverse of non-scalars and no denominator involving non-scalars.

In solid geometry, we have to consider the commutator product on two 2-vectors. However, the commutator product can also be expressed with the inner and outer products since

$$u_1 \times u_2 = -(u_1 \cdot I_3) \wedge (u_2 \cdot I_3), \quad \forall u_1, u_2 \in \mathcal{G}_2(E^2).$$

It is clear that Rules 1, Rules 2 and Rules 3 in plane geometry and the algorithm Eval_i_w1() are valid in solid geometry too. A Clifford algebraic polynomial is called *expanded* if it has been processed by algorithm Eval_i_w1.

We will use the following rules and formulas.

SRules 1: $(A \wedge B) \wedge C = A \wedge (B \wedge C) = A \wedge B \wedge C$
$\qquad\qquad A_r \cdot (B_s \cdot C_t) = (A_r \wedge B_s) \cdot C_t, \quad$ where $r + s \leq t$ and $r, s > 0$.

SRules 2: For any vectors $a, b, c \in E^3$, 2-vectors $u, u_1, u_2, u_3 \in \mathcal{G}_2(E^3)$, and 3-vector t:

$$
\begin{aligned}
(1) \quad & a \wedge (b \cdot I_3) = (a \cdot b)I_3, \\
(2) \quad & a \wedge (u \cdot I_3) = (a \cdot u) \cdot I_3, \\
(3) \quad & u_1 \wedge (u_2 \cdot I_3) = (u_1 \cdot u_2)I_3, \\
(4) \quad & (a \cdot I_3) \cdot I_3 = -a, \\
(5) \quad & (u \cdot I_3) \cdot I_3 = -u, \\
(6) \quad & (t \cdot I_3)I_3 = -t, \\
(7) \quad & I_3 \cdot I_3 = -1.
\end{aligned}
$$

SRules 3: For any vectors $a, b, c, d \in E^3$:

$$
\begin{aligned}
a \cdot (b \wedge c) &= (a \cdot b)c - (a \cdot c)b \\
(a \cdot b) \wedge (c \cdot d) &= a \cdot (b \cdot (c \wedge d))
\end{aligned}
$$

SRules 4: For any vectors $a, b, c, d, e, f \in E^3$:

$$
\begin{aligned}
a \cdot (b \wedge c \wedge d) &= (a \cdot b)c \wedge d - (a \cdot c)b \wedge d + (a \cdot d)b \wedge c \\
(a \wedge b) \cdot (c \wedge d \wedge e) &= a \cdot (b \cdot (c \wedge d \wedge e)) \\
(a \wedge b \wedge c) \cdot (d \wedge e \wedge f) &= a \cdot (b \cdot (c \cdot (d \wedge e \wedge f)))
\end{aligned}
$$

We consider the forms of the terms in an expanded Clifford algebraic polynomial.

Class 1: the basic scalars $\alpha, \beta, \ldots$, the basic vectors $a, b, c, d, \ldots$, and the unit 3-vector I_3.

Class 2: the elements which are obtained by taking inner or outer product of the elements in Class 1 once. For examples, the scalars $a \cdot b, c \cdot d, \ldots$, the 2-vectors $a \wedge b, c \wedge d, \ldots$ and $a \cdot I_3, b \cdot I_3, \ldots$.

Class 3: the elements which are obtained by taking inner or outer product of the elements in Class 1 or Class 2 once. Some of these elements can be simplified to a sum by SRules 1, SRules 2, SRules 3, and SRules 4, the new forms appearing in their terms are the scalars $(a \wedge b \wedge c) \cdot I_3, \ldots,$ the vectors $(a \wedge b) \cdot I_3, \ldots,$ and the 3-vectors $a \wedge b \wedge c, \ldots$

Class 4: the elements which are obtained by taking inner or outer product of the elements in Class 1, Class 2 or Class 3 once. All of these elements can be simplified to a sum by SRules 1, SRules 2, SRules 3, and SRules 4, and no new form in their terms appears.

Therefore no other new forms can be produced by this processing, and the forms of the Clifford terms are

vectors: $a,\ b,\ c,\ d, \ldots\ldots;\ (a \wedge b) \cdot I_3,\ (c \wedge d) \cdot I_3, \ldots\ldots$
scalars: $a \cdot b,\ c \cdot d, \ldots;\ (a \wedge b \wedge c) \cdot I_3,\ (d \wedge e \wedge f) \cdot I_3, \ldots;\ \alpha,\ \beta, \ldots$
2-vectors: $a \wedge b,\ c \wedge d, \ldots\ldots; a \cdot I_3,\ b \cdot I_3,\ c \cdot I_3, \ldots\ldots$
3-vectors: $I_3;\ a \wedge b \wedge c,\ d \wedge e \wedge f, \ldots\ldots$

and an expanded Clifford algebraic polynomial is a sum of products of such elements and scalars.

From the above, there are seven forms of the terms of an expanded Clifford algebraic polynomial

$$A,\ Aa,\ A((a \wedge b) \cdot I_3),\ A(a \wedge b),\ A(a \cdot I_3),\ AI_3,\ A(a \wedge b \wedge c),$$

where

$$A = \alpha_1 \ldots \alpha_i (a_1 \cdot b_1) \ldots (a_j \cdot b_j)((c_1 \wedge d_1 \wedge e_1) \cdot I_3) \ldots ((c_k \wedge d_k \wedge e_k) \cdot I_3).$$

SRules 5:

$$((a \wedge b \wedge c) \cdot I_3)((d \wedge e \wedge f) \cdot I_3) = -(a \wedge b \wedge c) \cdot (d \wedge e \wedge f)$$
$$((a \wedge b \wedge c) \cdot I_3)d = (d \cdot (a \wedge b \wedge c)) \cdot I_3$$
$$((a \wedge b \wedge c) \cdot I_3)((d \wedge e) \cdot I_3) = -(a \wedge b \wedge c) \cdot (d \wedge e)$$
$$((a \wedge b \wedge c) \cdot I_3)(d \wedge e) = ((a \wedge b \wedge c) \cdot (d \wedge e) \cdot I_3$$
$$((a \wedge b \wedge c) \cdot I_3)(d \cdot I_3) = -(a \wedge b \wedge c) \cdot d$$
$$((a \wedge b \wedge c) \cdot I_3)I_3 = -(a \wedge b \wedge c)$$
$$((a \wedge b \wedge c) \cdot I_3)(d \wedge e \wedge f) = ((a \wedge b \wedge c) \cdot (d \wedge e \wedge f))I_3.$$

With the above rules and SRules 4, we can simplify a Clifford algebraic polynomial to a sum of the following eight forms

S1: $\alpha_1 \ldots \alpha_i (a_1 \cdot b_1) \ldots (a_j \cdot b_j)$
S2: $\alpha_1 \ldots \alpha_i (a_1 \cdot b_1) \ldots (a_j \cdot b_j)((c \wedge d \wedge e) \cdot I_3)$
S3: $\alpha_1 \ldots \alpha_i (a_1 \cdot b_1) \ldots (a_j \cdot b_j)c$
S4: $\alpha_1 \ldots \alpha_i (a_1 \cdot b_1) \ldots (a_j \cdot b_j)((c \wedge d) \cdot I_3)$
S5: $\alpha_1 \ldots \alpha_i (a_1 \cdot b_1) \ldots (a_j \cdot b_j)(c \wedge d)$
S6: $\alpha_1 \ldots \alpha_i (a_1 \cdot b_1) \ldots (a_j \cdot b_j)(c \cdot I_3)$
S7: $\alpha_1 \ldots \alpha_i (a_1 \cdot b_1) \ldots (a_j \cdot b_j)I_3$
S8: $\alpha_1 \ldots \alpha_i (a_1 \cdot b_1) \ldots (a_j \cdot b_j)(c \wedge d \wedge e)$

where $\alpha_1, \ldots \alpha_i$ are basic scalars, $a_1, \ldots a_j,\ b_1 \ldots b_j,\ c, d, e$ are basic vectors.

Definition 6. *The forms S1, S2, ..., S8 will be called final forms in $\mathcal{G}(E^3)$. A Clifford algebraic polynomial in $\mathcal{G}(E^3)$ is called a final Clifford algebraic polynomial if its terms are all final forms.*

Theorem 3. *There is an algorithm to simplify a Clifford algebraic polynomial in $\mathcal{G}(I_3)$ to a final Clifford algebraic polynomial.*

Algorithm 9. *Simplify an expanded Clifford algebraic polynomial x by SRules 1, SRules 2, ..., SRules 5 to a final Clifford algebraic polynomial.*

```
Eval_i_w2:=proc(x)
    step1: Let the terms or the factors with grades greater than 3
           vanish.
    step2: Simplify the 3-vectors in x which is not I3 by t = -(t ·
           I3)I3, that is in the reverse order of rule (6) in
           SRules 2.
    step3: Simplify x by other rules in SRules 2, SRules 1 and
           SRules 3.
    step4: Simplify x by SRules 5.
    step5: Simplify x by SRules 4. x will be a final polynomial.
    end.
```

4.3 Algorithm and Implementation of Prover

This is very similar to the case of plane geometry, so we omitted the details.

4.4 Some Examples

Example 3 (Monge). The six planes passing through the midpoint of the edges of a tetrahedron and perpendicular to the edges respectively opposite have a point in common. This point is called the Monge point of the tetrahedron.

This example can be described in the following constructive way.

Constructions:
    point(A, B, C, D);
    on_line$(H, A, \overline{AB}, 1/2)$;
    on_line$(J, A, \overline{AC}, 1/2)$;
    on_line$(K, A, \overline{AD}, 1/2)$;
    on_line$(L, B, \overline{BC}, 1/2)$;
    on_line$(M, B, \overline{BD}, 1/2)$;
    on_line$(N, C, \overline{CD}, 1/2)$;
    inter_plane_plane_plane$(P, H, \overline{CD} \cdot I_3, J, \overline{BD} \cdot I_3, L, \overline{AD} \cdot I_3)$.
Conclusions:

$$\overline{KP} \cdot \overline{BC} = 0; \quad \overline{MP} \cdot \overline{AC} = 0; \quad \overline{NP} \cdot \overline{AB} = 0.$$

Let point A be the origin. Then our program prints the following.

PreDeal:

$$\overline{H} = \frac{\overline{B}}{2}$$

$$\overline{J} = \frac{\overline{C}}{2}$$

$$\overline{K} = \frac{\overline{D}}{2}$$

$$\overline{LB} = \frac{\overline{B}}{2} - \frac{\overline{C}}{2}$$

$$\overline{MB} = -\frac{\overline{D}}{2} + \frac{\overline{B}}{2}$$

$$\overline{NC} = -\frac{\overline{D}}{2} + \frac{\overline{C}}{2}$$

$$\overline{P} = -\frac{((\overline{B} \wedge \overline{D}) \cdot I_3)(\overline{B} \cdot \overline{C}) - ((\overline{C} \wedge \overline{D}) \cdot I_3)(\overline{B} \cdot \overline{C}) - ((\overline{C} \wedge \overline{D}) \cdot I_3)(\overline{B} \cdot \overline{D})}{2 (I_3 \cdot (\overline{B} \wedge \overline{C} \wedge \overline{D}))}$$
$$- \frac{((\overline{B} \wedge \overline{D}) \cdot I_3)(\overline{C} \cdot \overline{D}) - ((\overline{B} \wedge \overline{C}) \cdot I_3)(\overline{B} \cdot \overline{D}) - ((\overline{B} \wedge \overline{C}) \cdot I_3)(\overline{C} \cdot \overline{D})}{2 (I_3 \cdot (\overline{B} \wedge \overline{C} \wedge \overline{D}))}$$

Proof:

$$\overline{BC} \cdot \overline{KP}$$
$$\overset{s}{=} \overline{B} \cdot \overline{K} - \overline{C} \cdot \overline{K} - \overline{B} \cdot \overline{P} + \overline{C} \cdot \overline{P}$$
$$\overset{P}{=} \overline{B} \cdot \overline{K} - \overline{C} \cdot \overline{K} - \frac{\overline{B} \cdot \overline{D}}{2} + \frac{\overline{C} \cdot \overline{D}}{2}$$
$$\overset{K}{=} 0$$
$$\overline{C} \cdot \overline{MP}$$
$$\overset{s}{=} -\overline{C} \cdot \overline{M} + \overline{C} \cdot \overline{P}$$
$$\overset{P}{=} -\overline{C} \cdot \overline{M} + \frac{\overline{B} \cdot \overline{C}}{2} + \frac{\overline{C} \cdot \overline{D}}{2}$$
$$\overset{M}{=} 0$$
$$\overline{B} \cdot \overline{NP}$$
$$\overset{s}{=} -\overline{B} \cdot \overline{N} + \overline{B} \cdot \overline{P}$$
$$\overset{P}{=} -\overline{B} \cdot \overline{N} + \frac{\overline{B} \cdot \overline{C}}{2} + \frac{\overline{B} \cdot \overline{D}}{2}$$
$$\overset{N}{=} 0$$

This example is Example 4.92 in [1]; the proof produced by [1] is too long to print.

Example 4. Let $ABCD$ be a tetrahedron and G a point in space. The lines passing through points A, B and C and parallel to line DG meet their opposite face at P, Q, and R respectively. Try to find the condition for $V_{GPQR} = 3V_{ABCD}$.

This example can be described in the following constructive way.

Constructions:

points(A, B, C, D, G);

inter_line_plane$(P, A, \overline{DG}, D, \overline{DB} \wedge \overline{DC})$;

inter_line_plane$(Q, B, \overline{DG}, D, \overline{DA} \wedge \overline{DC})$;

inter_line_plane$(R, C, \overline{DG}, D, \overline{DA} \wedge \overline{DB})$.

Conclusion: Compute $\overline{GP} \wedge \overline{GQ} \wedge \overline{GR}$.

Let point A be the origin. Then our program prints the following.

PreDeal:

$$\overline{P} = \frac{(\overline{D} - \overline{G}) \, I_3 \cdot (\overline{D} \wedge \overline{C} \wedge \overline{B})}{-I_3 \cdot (\overline{D} \wedge \overline{G} \wedge \overline{B}) - I_3 \cdot (\overline{D} \wedge \overline{C} \wedge \overline{G}) + I_3 \cdot (\overline{D} \wedge \overline{C} \wedge \overline{B}) + I_3 \cdot (\overline{C} \wedge \overline{G} \wedge \overline{B})}$$

$$\overline{Q} = -\frac{-I_3 \cdot (\overline{D} \wedge \overline{C} \wedge \overline{B})\overline{D} + I_3 \cdot (\overline{D} \wedge \overline{C} \wedge \overline{B})\overline{G} - \overline{B}I_3 \cdot (\overline{D} \wedge \overline{C} \wedge \overline{G})}{I_3 \cdot (\overline{D} \wedge \overline{C} \wedge \overline{G})}$$

$$\overline{R} = \frac{I_3 \cdot (\overline{D} \wedge \overline{C} \wedge \overline{B})\overline{D} - I_3 \cdot (\overline{D} \wedge \overline{C} \wedge \overline{B})\overline{G} + \overline{C}I_3 \cdot (\overline{D} \wedge \overline{G} \wedge \overline{B})}{I_3 \cdot (\overline{D} \wedge \overline{G} \wedge \overline{B})}$$

Computation:

$$-I_3 \cdot (\overline{GP} \wedge \overline{GQ} \wedge \overline{GR})I_3$$
$$= -\left(I_3 \cdot (\overline{P} \wedge \overline{G} \wedge \overline{Q}) + I_3 \cdot (\overline{P} \wedge \overline{R} \wedge \overline{G}) - I_3 \cdot (\overline{R} \wedge \overline{G} \wedge \overline{Q}) - I_3 \cdot (\overline{P} \wedge \overline{R} \wedge \overline{Q})\right) I_3$$
$$= I_3 \left(I_3 \cdot (\overline{C} \wedge \overline{G} \wedge \overline{B}) + 3 \, I_3 \cdot (\overline{D} \wedge \overline{C} \wedge \overline{B})\right)$$

So $V_{GPQR} = 3V_{ABCD}$ if and only if the points A, B, C, G are on the same plane.

This is an extended version of a problem from the 1964 International Mathematical Olympiad.

References

1. Chou, S.-C., Gao, X.-S., Zhang, J.-Z.: Machine proofs in geometry. World Scientific, Singapore (1995).
2. Fèvre, S., Wang, D.: Proving geometric theorems using Clifford algebra and rewrite rules. In: Proc. CADE-15 (Lindau, Germany, July 5–10, 1998), LNAI 1421, pp. 17–32.
3. Boy de la Tour, T., Fèvre, S., Wang, D.: Clifford term rewriting for geometric reasoning in 3D. In this volume.
4. Hestenes, D., Sobczyk, G.: Clifford algebra to geometric calculus. D. Reidel, Dordrecht, Boston (1984).
5. Li, H.-B., Cheng, M.-t.: Proving theorems in elementary geometry with Clifford algebraic method. Chinese Math. Progress 26: 357–371 (1997).
6. Wu, W.-t.: Mechanical theorem proving in geometries: Basic principles. Springer, Wien New York (1994).
7. Yang, H.-Q.: Clifford algebra and mechanical theorem proving in geometries. Ph.D thesis, Jilin University, Changchun (1998).
8. Yang, H.-Q., Zhang, S.-G., Feng, G.-C.: Clifford algebra and mechanical geometry theorem proving. In: Proc. 3rd ASCM (Lanzhou, China, August 6–8, 1998), pp. 49–63.

Clifford Term Rewriting for Geometric Reasoning in 3D

Thierry Boy de la Tour, Stéphane Fèvre, and Dongming Wang

LEIBNIZ–IMAG, 46, avenue Félix Viallet, 38031 Grenoble Cedex, France

Abstract. Clifford algebra formalism has been used recently in combination with standard term-rewriting techniques for proving many nontrivial geometric theorems. The key issue in this approach consists in verifying whether two Clifford expressions are equal. This paper is concerned with the generalization of the work to 3D geometric problems. A rewriting system is proposed and its theoretical properties are investigated. Some examples and potential applications are also presented.

1 Introduction

Proving geometric theorems has become an important activity with respect to many domains such as geometry education, computer-aided design and computer vision. Several methods have been very successful in proving many difficult theorems [3, 10, 12], and researchers are motivated to investigate the use of general techniques of automated deduction for finding machine proofs.

Recently simple methods for proving geometric theorems have been developed based on Clifford algebra formalism [5, 7, 8, 11, 13]. Their main purpose is to prove geometric theorems in a synthetic way, that is without using coordinates. Traditionally synthetic geometry is only used at an elementary level for performing simple computations. Analytic geometry has been much more developed under the influence of physicists and engineers for which geometric problems are reduced to algebraic ones. Actually the Clifford algebra formalism provides an elegant way for solving geometric problems in a synthetic style though it is possible to keep the advantages of the analytic point of view.

The expressive power of Clifford algebra allows to reduce a lot of problems to the proof that a Clifford expression is equal to 0, or to the reduction of an expression to a normal form. This makes very easy understanding the methodology for solving problems. However defining a normal form for Clifford expressions and computing such a normal form is a non-trivial task. This has already been achieved for the Euclidean plane [5, 6] but the case of the Euclidean space appears more complex. As many simplification rules may be applied to Clifford expressions, we have found convenient to use rewriting systems for simplifying such expressions.

This paper investigates a rewriting system for systematically simplifying Clifford algebraic expressions associated with problems in three-dimensional (3D) Euclidean space. We also study its theoretical properties (completeness, termination) and give some insights on the normalization problem explaining why performing this task efficiently is so complex. Last, significant examples and applications of our simplification method are presented.

2 A Synthetic Presentation

This section provides some background on the Clifford algebra used in this paper and presents Clifford algebra from a synthetic point of view.

First, we define what is a Clifford algebra. Let $\mathbb{K}$ be a field of characteristic not equal to 2, $\mathcal{V}$ a vector space of dimension n over the base field $\mathbb{K}$ and q a quadratic form over $\mathcal{V}$. Let

$$\mathcal{V}^+ = \mathbb{K} \oplus \left(\bigoplus_{i=1}^{n} \left(\bigotimes_{j=1}^{i} \mathcal{V} \right) \right),$$

where $\oplus$ is the direct sum and $\otimes$ is the tensor product. $\mathcal{V}^+$ is the tensor algebra over $\mathcal{V}$ and its dimension is 2^n. Let $\mathcal{I}(q)$ be the two-sided ideal of $\mathcal{V}^+$ generated by the elements $\mathbf{v} \otimes \mathbf{v} - q(\mathbf{v})$ for every $\mathbf{v} \in \mathcal{V}$. Let $\mathcal{C}$ be the quotient of $\mathcal{V}^+$ by $\mathcal{I}(q)$. $\mathcal{C}$ is called the *Clifford algebra* associated to the quadratic form q.

This definition of Clifford algebra leads to introducing coordinates over a basis of $\mathcal{V}$ and extending computations over $\mathcal{V}$ to $\mathcal{C}$. This approach has been developed for instance in [1].

Instead of using it, we propose to consider an axiomatization of Clifford algebras specialized to some fixed geometry. For instance, an equational presentation of such an axiomatization for dimension two has been given in [5]. The basic idea is to consider an internal law · representing the unique bilinear symmetric form associated with q. The outer product is represented by a binary associative operator $\wedge$ and a unary dual operator $\sim$ is also introduced to strengthen the expressive power. Last, the sum of two elements A and B of $\mathcal{C}$ is denoted by $A + B$. There is also a well-defined product in Clifford algebra (the geometric product) but we do not use it in this paper except for multiplying elements of the basic field $\mathbb{K}$ with elements of $\mathcal{C}$. In this case, it may be considered as the extension to $\mathcal{C}$ of the external law of $\mathcal{V}$.

More formally, we consider the Clifford algebra $\mathcal{C}$ of dimension 8 associated with the 3D Euclidean space $\mathcal{E}$. Points in $\mathcal{E}$ are represented by vectors in $\mathcal{C}$. To every element in $\mathcal{C}$ is associated a *grade* (intuitively, its *"dimension"*). Scalars are elements of grade 0 and vectors are elements of grade 1. The outer product of k vectors is called a *k-vector* and an element of $\mathcal{C}$ a Clifford number. A bivector is the outer product of two vectors, and a trivector is the outer product of three vectors or of a bivector and a vector. In our case, there is no quadrivector as we are restricted to dimension 3. The rules for computing the grade of an arbitrary expression are given in the next section. In what follows, an expression in $\mathcal{C}$ is called a *Clifford expression*.

Let $\mathbf{a}$ be a vector. The dual of $\mathbf{a}$, denoted by $\mathbf{a}^\sim$, is a bivector and may be geometrically interpreted as an equivalence class of parallelograms in a plane orthogonal to $\mathbf{a}$. Reciprocally, any bivector may be interpreted as the equivalence class of parallelograms of given direction, orientation and area. The dual of a bivector is a vector. Last, any trivector is the dual of a scalar and reciprocally. As the space of trivectors is a line, one can take $I\!\!I$ as a basis for trivectors and define it such as $I\!\!I^{\sim} = 1$; $I\!\!I$ is called the *unit pseudo-scalar*.

Computation rules in C are derived from the definition of C and the operations: sum $+$, inner product $\cdot$, outer product $\wedge$ and (restricted) geometric product $\circ$ (this symbol will often be omitted in expressions). These rules will be given in detail in the next section (see system R). The basic step of applying Clifford algebra to geometric problems is to express geometric constructions using Clifford expressions. Such constructions are used in Section 8.

The next section considers the problem of simplifying Clifford expressions represented by ground terms and provides a set of simplification rules. Although this set is not sufficient, its significance will appear when pseudo-coordinates are introduced in Section 6.

3   Basic Simplification Rules

Let $\mathbb{V}$ be a finite set of elements of $\mathcal{V}$. We consider Clifford expressions as first-order terms built on constant symbols denoting some elements of $\mathbb{K}$, constant symbols from $\mathbb{V}$, and the pseudo-scalar $I\!\!I$ by means of function symbols with one argument (the dual operator) or two arguments (the sum, the geometric, inner and outer products). In the sequel we will need to define rewriting systems by means of conditional rules, and most conditions apply to the grades of the terms involved. We will obviously provide a way to compute these grades, but efficiency requires that the results of such computations be memorized, and hence to work not exactly with first-order terms but with labeled trees, where each vertex is labeled by a function symbol and a grade. However, in order to keep notations simple (and standard), this labeling with a grade will be left implicit. This structure will be called a Clifford term. From now on, $t = t'$ means that t and t' are identical terms, while $t =_c t'$ means that t and t' are equal up to the axioms of Clifford algebra (the values $\lceil t \rceil$ and $\lceil t' \rceil$ of t and t' in C are equal for every interpretation $\lceil a \rceil$ of the elements $\mathbf{a} \in \mathbb{V}$).

Of course it is not always possible to attribute a grade to a Clifford expression, but we will consider ungraded expressions as expressions of grade $\perp$. It is also essential that the grades keep unchanged during the computation. But our aim is to rewrite expressions of grade possibly non-zero to 0, which is a scalar. This is the only case of non-preservation of the grade, and for this reason, we create a new grade especially for 0; this is ∞. More precisely, ∞ is the grade of expressions that are trivially equal to 0 due to absorption properties of 0 and a few others as shown below.

It is important to remind here the basic restriction we impose on Clifford expressions: the geometric product is only allowed with at least one scalar as argument, that is a term of grade 0 or ∞. The rules for computing the grades, denoted as $\mathrm{gr}(\cdot)$, are the following:

$-\ \mathrm{gr}(0) = \infty;\ \forall x \in \mathbb{K}^*, \mathrm{gr}(x) = 0;\ \forall v \in \mathbb{V}, \mathrm{gr}(v) = 1$ and $\mathrm{gr}(I\!\!I)=3$.

- If $g_1 = \mathrm{gr}(t_1)$ and $g_2 = \mathrm{gr}(t_2)$ then:
  - $\mathrm{gr}(t_1{}^\sim) \quad = 3 \dot{-} g_1;$

  - $\mathrm{gr}(t_1 \wedge t_2) = \begin{cases} \infty & \text{if } \perp \neq g_1 \dot{+} g_2 > 3, \\ g_1 \dot{+} g_2 & \text{otherwise}; \end{cases}$

  - $\mathrm{gr}(t_1 \cdot t_2) = \begin{cases} \infty & \text{if } g_1 = 0 \text{ or } g_2 = 0, \\ |g_1 \dot{-} g_2| & \text{otherwise}; \end{cases}$

  - $\mathrm{gr}(t_1 t_2) \quad = \max(g_1, g_2)$ (we have g_1 or g_2 in $\{0, \infty\}$);

  - $\mathrm{gr}(t_1 + t_2) = \begin{cases} \min(g_1, g_2) & \text{if } g_1 = \infty \text{ or } g_2 = \infty, \\ g_1 & \text{if } g_1 = g_2, \\ \perp & \text{otherwise}; \end{cases}$

where $\mathbb{K}^*$ is an arbitrary extension field of $\mathbb{K}$ and
- $\dot{+}$ is the extension of $+$ with $x \dot{+} \infty = \infty \dot{+} x = \infty$, and $x \dot{+} \perp = \perp \dot{+} x = \perp$ when $x \neq \infty$,
- $\dot{-}$ is the extension of $-$ with $x \dot{-} \infty = \infty \dot{-} x = \infty$, and $x \dot{-} \perp = \perp \dot{-} x = \perp$ when $x \neq \infty$,
- $|\infty| = \infty$ and $|\perp| = \perp$,
- the order on $\mathbb{N}$ (the set of natural numbers) is extended with $\forall x \in \mathbb{N}, x < \perp < \infty$.

In the sequel we will define rewriting systems as sets of conditional rules, i.e. expressions of the form $l \to r$ if C, where l and r are terms built with the same symbols as above and variables, and C is some condition on these variables. Such a set R defines a relation between Clifford terms defined by: $t \to_R t'$ if and only if there is a subterm s of t and a rule $l \to r$ if C in R and a substitution σ such that $\sigma(l) = s$ and $\sigma(C)$ is true and t' is the term t after replacement of s by $\sigma(r)$ (and $\mathrm{gr}(\sigma(r))$, and possibly a recomputation of some grades). A term t is a R-normal form if there is no t' such that $t \to_R t'$. R is terminating if for any t there is no infinite sequence $t \to_R t_1 \to_R t_2 \to \cdots$. For any binary relation B we note by B^- its reflexive closure, B^+ its transitive closure and B^* its reflexive and transitive closure.

In order to state conditions about grades implicitly we use the following convention: the variables A, B, C will match with any Clifford term, the variables h, k will match with terms of grade 0 or ∞ (i.e. h always comes with the condition $\mathrm{gr}(h) \in \{0, \infty\}$), the variables X, Y, Z will match with terms of grade different from 0 or ∞, and the variables x, y, z, u will match with terms of grade 1.

We now consider the rewriting system E containing the following conditional rules:

$$0A \to 0, \quad 0 \wedge A \to 0, \quad h \cdot A \to 0, \quad 0^\sim \to 0, \quad 0 + 0 \to 0,$$
$$A0 \to 0, \quad A \wedge 0 \to 0, \quad A \cdot h \to 0, \quad A \wedge B \to 0 \text{ if } \mathrm{gr}(A) + \mathrm{gr}(B) > 3.$$

It is rather easy to prove that grades are preserved by this rewriting system:

Theorem 1. *If* $t \to_E^* t'$ *then* $\mathrm{gr}(t) = \mathrm{gr}(t')$.

Proof. We prove by structural induction on t that if $t \to_{\bar{E}} t'$ then $\mathrm{gr}(t) = \mathrm{gr}(t')$. In the base cases this is obvious since $t = t'$. If t is of the form $t_1{}^\sim$ (resp. $t_1 \wedge t_2$, $t_1 \cdot t_2$, $t_1 t_2$, $t_1 + t_2$), and $t' \neq 0$ then t' is of the form $t_1'{}^\sim$ (resp. $t_1' \wedge t_2'$, $t_1' \cdot t_2'$, $t_1' t_2'$, $t_1' + t_2'$) and $\exists i$ such that $t_i \to_{\bar{E}} t_i'$ (and $t_j = t_j'$ for $j \neq i$); hence by induction hypothesis $\mathrm{gr}(t_i) = \mathrm{gr}(t_i')$, and thus $\mathrm{gr}(t) = \mathrm{gr}(t')$. If $t' = 0$ then $\mathrm{gr}(t') = \infty$, and either t is $0^\sim$, or t is $t_1 \wedge t_2$ or $t_1 t_2$ and $\exists i$ such that $t_i = 0$, or t is $t_1 + t_2$ and $t_1 = t_2 = 0$, or t is $t_1 \cdot t_2$ and $\exists i$ such that $\mathrm{gr}(t_i) \in \{0, \infty\}$, or $t = t_1 \wedge t_2$ and $\mathrm{gr}(t_1) + \mathrm{gr}(t_2) > 3$, and in all these cases we have $\mathrm{gr}(t) = \infty$.

From Theorem 1 we can easily prove:

Theorem 2. $\mathrm{gr}(t) = \infty$ *if and only if* $t \to_{\bar{E}}^* 0$.

Proof. If $t \to_{\bar{E}}^* 0$ then by Theorem 1 we have $\mathrm{gr}(t) = \mathrm{gr}(0) = \infty$. We prove the converse by induction on t. In the base case $\mathrm{gr}(t) = \infty$ implies $t = 0$; hence $t \to_{\bar{E}}^* 0$. If $t = t_1{}^\sim$ and $\mathrm{gr}(t) = \infty$ then $\mathrm{gr}(t_1) = \infty$; hence $t_1 \to_{\bar{E}}^* 0$ by induction hypothesis, and thus $t \to_{\bar{E}}^+ 0$. If $t = t_1 t_2$ then $\exists i$ such that $\mathrm{gr}(t_i) = \infty$; hence $t_i \to_{\bar{E}}^* 0$ and $t \to_{\bar{E}}^+ 0$. If $t = t_1 + t_2$ then $\mathrm{gr}(t_1) = \mathrm{gr}(t_2) = \infty$; hence $t \to_{\bar{E}}^+ 0$. If $t = t_1 \cdot t_2$ then $\exists i$ such that $\mathrm{gr}(t_i) \in \{0, \infty\}$; hence $t \to_{\bar{E}} 0$. Finally, if $t = t_1 \wedge t_2$, then either $\exists i$ such that $\mathrm{gr}(t_i) = \infty$, and $t \to_{\bar{E}}^+ 0$, or $\mathrm{gr}(t_1) + \mathrm{gr}(t_2) > 3$ and $t \to_{\bar{E}} 0$.

It is therefore clear that once the grades are computed, we do not really need to apply the rewriting system E. Instead, we apply the system E' defined as:

$$A \to 0 \text{ if } \mathrm{gr}(A) = \infty, \qquad A + 0 \to A, \qquad 0 + A \to A.$$

By Theorem 2 it is clear that $t \to_{E'}^* t'$ implies $t =_C t'$, since the same obviously holds of E. It is easy to prove by induction on t that if t is an E'-normal form, then either $t' = 0$ or t' does not contain any occurrence of 0; and hence no subterm of grade ∞. Moreover, the termination of E' is obvious, and hence an E'-normal form is easily obtained. However, if $t \neq 0$, we may still have $\mathrm{gr}(t) = \perp$, which can be the case only if t contains a subterm $t_1 + t_2$ with $\mathrm{gr}(t_1) \neq \mathrm{gr}(t_2)$. We consider the following system H of distributivity rules applied only to sums of grade $\perp$:

if $\mathrm{gr}(A) \neq \mathrm{gr}(B)$ then

$$h(A + B) \to hA + hB, \qquad (A + B)h \to hA + hB,$$
$$X \cdot (A + B) \to X \cdot A + X \cdot B, \qquad (A + B) \cdot X \to A \cdot X + B \cdot X,$$
$$C \wedge (A + B) \to C \wedge A + C \wedge B, \qquad (A + B) \wedge C \to A \wedge C + B \wedge C,$$
$$(A + B)^\sim \to A^\sim + B^\sim.$$

Once again the termination of H is obvious (the depth of sums decreases), and an H-normal form is either 0 or a sum $t_1 + \cdots + t_n$ of terms t_i of grades non $\perp$ (with any possible parentheses). However, H may produce subterms $t_1 \wedge t_2$ with $\mathrm{gr}(t_1) + \mathrm{gr}(t_2) > 3$ (from $t_1 \wedge (t_2 + t_3)$ with $\mathrm{gr}(t_2) \neq \mathrm{gr}(t_3)$); hence it is necessary to recompute the grades and apply the system E' once more. After these steps, we therefore obtain an H and E' normal form, that is either 0 or a sum of terms t_i's whose subterms are non-zero and of grade between 0 and 3.

In the sequel, the rules of E' will be used again, since some 0's may be produced, while the rules from H will not, since only sums of definite grades may be produced by the forthcoming rules.

4 Reduction to Normal Form

The exact structure of the terms obtained after applying the basic simplification rules can be made explicit by looking at the expressions for computing the grades, from which we can read the following grammar for terms of grades 0 to 3:

$$L_0 ::= \mathbb{K}^* \mid L_0 + L_0 \mid L_0 L_0 \mid L_1 \cdot L_1 \mid L_2 \cdot L_2 \mid L_3 \cdot L_3 \mid L_0 \wedge L_0 \mid L_3{}^\sim,$$

$$L_1 ::= \mathbb{V} \mid L_1 + L_1 \mid L_0 L_1 \mid L_1 L_0 \mid L_1 \cdot L_2 \mid L_2 \cdot L_3 \mid L_3 \cdot L_2 \mid L_2 \cdot L_1 \mid L_0 \wedge L_1 \mid L_1 \wedge L_0 \mid L_2{}^\sim,$$

$$L_2 ::= L_2 + L_2 \mid L_0 L_2 \mid L_2 L_0 \mid L_1 \cdot L_3 \mid L_3 \cdot L_1 \mid L_0 \wedge L_2 \mid L_1 \wedge L_1 \mid L_2 \wedge L_0 \mid L_1{}^\sim,$$

$$L_3 ::= \mathit{I\!I} \mid L_3 + L_3 \mid L_0 L_3 \mid L_3 L_0 \mid L_0 \wedge L_3 \mid L_1 \wedge L_2 \mid L_2 \wedge L_1 \mid L_3 \wedge L_0 \mid L_0{}^\sim,$$

where L_i is the language of terms of grade i. Hence the structure of these terms is still quite complex, and they can certainly be reduced by using computation rules (equations) from Clifford algebra. For instance, computations in $\mathbb{K}$ can be performed by standard means of computer algebra. Similarly, we can consider terms of grade 0, such as $(1 + \mathbf{a} \cdot \mathbf{a})\mathbf{a} \cdot \mathbf{b} + (-1)\mathbf{b} \cdot \mathbf{a}$, as polynomials over indeterminates $\mathbf{a} \cdot \mathbf{a}$, $\mathbf{a} \cdot \mathbf{b}$, $\mathbf{b} \cdot \mathbf{a}$, and perform computations using a computer algebra system. But this is clearly not sufficient, and we need Clifford algebra to tell us that $\mathbf{b} \cdot \mathbf{a} = \mathbf{a} \cdot \mathbf{b}$, and to reduce complex products.

But we must first make clear what we shall consider as irreducible terms, i.e. sublanguages of L_i of terms which cannot be reduced further simply by applying rules of Clifford algebra. More precisely, we aim at defining a syntactic notion of irreducibility, one that does not depend on particular relations which may exist among the basic vectors (elements of $\mathbb{V}$) appearing in a Clifford term; these relations will then be easier to cope with.

It is quite obvious that the expressions $\mathbf{a}, \mathbf{a}^\sim, \mathbf{a} \cdot \mathbf{b}, \mathbf{a} \wedge \mathbf{b}, \mathit{I\!I}$ cannot be reduced further, except by eventually swapping arguments. This is also the case of $\mathbf{a} \cdot \mathbf{b}^\sim$ and $\mathbf{a} \cdot (\mathbf{b} \cdot \mathbf{c}^\sim)$. This gives a (tentative) grammar for irreducible terms:

$$L_0^\circ ::= \mathbb{K}^* \mid \mathbb{V} \cdot \mathbb{V} \mid \mathbb{V} \cdot (\mathbb{V} \cdot \mathbb{V}^\sim) \mid L_0^\circ + L_0^\circ \mid L_0^\circ L_0^\circ,$$

$$L_1^\circ ::= \mathbb{V} \mid \mathbb{V} \cdot \mathbb{V}^\sim \mid L_1^\circ + L_1^\circ \mid L_1^\circ L_1^\circ,$$

$$L_2^\circ ::= \mathbb{V}^\sim \mid \mathbb{V} \wedge \mathbb{V} \mid L_2^\circ + L_2^\circ \mid L_2^\circ L_2^\circ,$$

$$L_3^\circ ::= \mathit{I\!I} \mid L_3^\circ + L_3^\circ \mid L_3^\circ L_3^\circ,$$

and it is difficult to see how it could be made simpler (except if an expression is null, such as $\mathbf{a} \wedge \mathbf{b} + \mathbf{b} \wedge \mathbf{a}$, but it is convenient to keep a simple syntactic definition of irreducibility). We are going to prove that these irreducible terms can always be reached as normal forms of a rewriting system R. The rules of R are divided into the following 5 groups.

(1). Distributivity rules:

$$h(X + Y) \rightarrow hX + hY, \qquad (X + Y)^{\sim} \rightarrow X^{\sim} + Y^{\sim},$$
$$Z \cdot (X + Y) \rightarrow Z \cdot X + Z \cdot Y, \qquad (X + Y) \cdot Z \rightarrow X \cdot Z + Y \cdot Z,$$
$$Z \wedge (X + Y) \rightarrow Z \wedge X + Z \wedge Y, \qquad (X + Y) \wedge Z \rightarrow X \wedge Z + Y \wedge Z.$$

(2). Linearity rules:

$$(hX) \cdot Y \rightarrow h(X \cdot Y), \qquad X \cdot (hY) \rightarrow h(X \cdot Y),$$
$$(hX) \wedge Y \rightarrow h(X \wedge Y), \qquad X \wedge (hY) \rightarrow h(X \wedge Y),$$
$$h \wedge A \rightarrow hA, \qquad X \wedge h \rightarrow hX,$$
$$Xh \rightarrow hX, \qquad h(kX) \rightarrow (hk)X,$$
$$(hX)^{\sim} \rightarrow hX^{\sim}.$$

(3). Duality rules:

$$h^{\sim} \rightarrow -h\mathbb{I}, \quad \mathbb{I}^{\sim} \rightarrow 1, \quad X \cdot \mathbb{I} \rightarrow -X^{\sim}, \quad (X^{\sim})^{\sim} \rightarrow -X,$$

where $-t$ stands for $(-1)t$.

(4). Ordering rules:

$$X \wedge Y \rightarrow (-1)^{ab} Y \wedge X \text{ if } Y \prec X,$$
$$X \cdot Y \rightarrow (-1)^{|a-b|-\min(a,b)} Y \cdot X \text{ if } Y \prec X,$$

where $a = \mathrm{gr}(X)$ and $b = \mathrm{gr}(Y)$, and $\prec$ is an order on the set of ground terms.

The following definition of $\prec$ is convenient. We assume an order $<$ on $\mathbb{V}$, and also that for any constants c, c' such that $\mathrm{gr}(c) < \mathrm{gr}(c')$ we have $c < c'$. This order is extended to the set of function symbols in the following way: $\cdot < \wedge < \sim < \circ < +$. This symbol precedence is used to define an order on ground terms. The definition is given by structural induction and extends the order $<$ on constants. Let t and t' be two terms. Then $t \prec t'$ if and only if one of the following conditions is satisfied:

- $\mathrm{gr}(t) < \mathrm{gr}(t')$;
- $\mathrm{gr}(t) = \mathrm{gr}(t')$, $\circ \notin \{\mathrm{hd}(t), \mathrm{hd}(t')\}$ and $\mathrm{hd}(t) < \mathrm{hd}(t')$, where $\mathrm{hd}(\cdot)$ denotes the head symbol (i.e. the operator);
- $\mathrm{gr}(t) = \mathrm{gr}(t')$ and $t = hA$ and $t' = kB$, where either $A \prec B$ or $A = B$ and $h \prec k$;
- $\mathrm{gr}(t) = \mathrm{gr}(t')$ and $t = A$ and $t' = kB$, where either $A \prec B$ or $A = B$;
- $\mathrm{gr}(t) = \mathrm{gr}(t')$ and $t = hA$ and $t' = B$, where $A \prec B$;
- $\mathrm{gr}(t) = \mathrm{gr}(t')$ and $\mathrm{hd}(t) = \mathrm{hd}(t') = f \neq \circ$, where $t = f(t_1, \ldots, t_m)$ and $t' = f(t'_1, \ldots, t'_m)$, and there is some $i \in \{1, \ldots, m\}$ such that $t_i \prec t'_i$ and for every $j \in \{1, \ldots, i\}$ $t_j = t'_j$.

In these rules, the scalar $(-1)^k$ has to be computed each time a rule is applied. Alternatively, we can consider these rules as a shortcut for the set of rules obtained from these by fixing the values of a and b (in $\{1, 2, 3\}$). The reason for these two ordering rules is to reduce the number of rules in R. Remark that $\prec$ allows to perform fast comparisons between terms.

(5). Reduction rules:

$$(5.1). \ (x \wedge y)^{\sim} \to x \cdot y^{\sim},$$
$$(5.2). \ (x \cdot y^{\sim})^{\sim} \to -x \wedge y,$$
$$(5.3). \ x \wedge (y \cdot z^{\sim}) \to (x \cdot y)z^{\sim} - (x \cdot z)y^{\sim},$$
$$(5.4). \ x \wedge y^{\sim} \to -(x \cdot y)I\!\!I,$$
$$(5.5). \ x \wedge (y \wedge z) \to x \cdot (y \cdot z^{\sim})I\!\!I,$$
$$(5.6). \ x \cdot (y \wedge z) \to (x \cdot y)z - (x \cdot z)y,$$
$$(5.7). \ (x \cdot y^{\sim}) \cdot z^{\sim} \to (x \cdot z)y - (y \cdot z)x,$$
$$(5.8). \ (x \cdot y^{\sim}) \cdot (z \cdot u^{\sim}) \to (x \cdot z)(y \cdot u) - (x \cdot u)(y \cdot z),$$
$$(5.9). \ x^{\sim} \cdot y^{\sim} \to -x \cdot y,$$
$$(5.10). \ (x \wedge y) \cdot z^{\sim} \to x \cdot (y \cdot z^{\sim}),$$
$$(5.11). \ (x \wedge y) \cdot (z \wedge u) \to (x \cdot u)(y \cdot z) - (x \cdot z)(y \cdot u).$$

We can now prove that the R-normal forms are exactly the irreducible terms. Here, we consider that the order $<$ on $\mathbb{V}$ is $\emptyset$, which results in the smallest possible $\prec$.

Theorem 3. $\forall t \in L_i, t$ is a R-normal form if and only if $t \in L_i^{\circ}$.

Proof. The "if" part is easy: we prove by induction on $t \in L_i^{\circ}$ that t is a R-normal form. The base cases are $\mathbf{a} \cdot \mathbf{b}$ and $\mathbf{a} \cdot (\mathbf{b} \cdot \mathbf{c}^{\sim})$ for $i = 0$, $\mathbf{a}$ and $\mathbf{a} \cdot \mathbf{b}^{\sim}$ for $i = 1$, $\mathbf{a}^{\sim}$ and $\mathbf{a} \wedge \mathbf{b}$ for $i = 2$, and $I\!\!I$ for $i = 3$, and these are obviously R-normal forms. If $t = t_1 + t_2 \in L_i^{\circ}$, then $t_1, t_2 \in L_i^{\circ}$; hence by induction hypothesis they are R-normal forms, and so is t. If $t = t_1 t_2 \in L_i^{\circ}$, then $t_1 \in L_0^{\circ}$ and $t_2 \in L_i^{\circ}$ are R-normal forms, and so is t.

We now prove by induction on $t \in L_i$ that if t is a R-normal form then $t \in L_i^{\circ}$. The base cases are $t \in \mathbb{K}^*$ for $i = 0$, $t \in \mathbb{V}$ for $i = 1$, and $t = I\!\!I$ for $i = 3$, which are R-normal forms and are in L_i°. We now consider $t \in L_i$ a R-normal form; then every strict subterm t' of t is a R-normal form as well, and hence $t' \in L_{\mathrm{gr}(t')}^{\circ}$ by induction hypothesis. If $t = t_1 + t_2$ then $t_1, t_2 \in L_i^{\circ}$; hence $t \in L_i^{\circ}$. If $t = t_1 t_2$, then according to a linearity rule $t_1 \in L_0^{\circ}$ and $t_2 \in L_i^{\circ}$; hence $t \in L_i^{\circ}$. We now consider the case where t is not a sum or a geometric product: then t does not contain any sum or geometric product; otherwise some distributivity or scalar rule could be applied to t. Hence any strict subterm t' (with grade j) of t, being in L_j°, is of the form $\mathbf{a} \cdot \mathbf{b}$ or $\mathbf{a} \cdot (\mathbf{b} \cdot \mathbf{c}^{\sim})$ if $j = 0$, $\mathbf{a}$ or $\mathbf{a} \cdot \mathbf{b}^{\sim}$ if $j = 1$, $\mathbf{a}^{\sim}$ or $\mathbf{a} \wedge \mathbf{b}$ if $i = 2$, and $I\!\!I$ if $j = 3$. We consider all possible cases:

If $t = t_1^{\sim}$ with $t_1 \in L_j^{\circ}$, then $j \neq 0$ and $j \neq 3$ by duality rules. If $j = 1$ ($i = 2$) then t_1 cannot be of the form $\mathbf{a} \cdot \mathbf{b}^{\sim}$ because of rule (5.2); hence t_1 is of the form $\mathbf{a}$, and $t \in L_i^{\circ}$. If $j = 2$ ($i = 1$) then t_1 cannot be of the form $\mathbf{a} \wedge \mathbf{b}$ because of rule (5.1), and $t_1 \neq \mathbf{a}^{\sim}$ because of duality; hence $j = 2$ is impossible.

If $t = t_1 \wedge t_2$ with $t_1 \in L_j^{\circ}, t_2 \in L_k^{\circ}$, then $j \neq 0, k \neq 0$ due to scalar rules, $j \leq k$ due to ordering rules, and of course $j + k \leq 3$. We have $t \in L_i^{\circ}$ when $j = k = 1$ ($i = 2$) and $t_1, t_2 \in \mathbb{V}$. In all other cases $t \notin L_i^{\circ}$, and we prove that t is not a R-normal form. In case $j = k = 1$, if $t_2 = \mathbf{a} \cdot \mathbf{b}^{\sim}$ then rule (5.3) applies, and if $t_1 = \mathbf{a} \cdot \mathbf{b}^{\sim}$ and $t_2 \in \mathbb{V}$ then an ordering rule applies. If $j = 1, k = 2$ ($i = 3$), then $t_2 \neq \mathbf{a}^{\sim}$ because of rule (5.4) and $t_2 \neq \mathbf{a} \wedge \mathbf{b}$ because of rule (5.5).

If $t = t_1 \cdot t_2$ with $t_1 \in L_j^o, t_2 \in L_k^o$, then $j \neq 0, k \neq 0$ since $t \in L_i$ and $j \leq k$ by ordering. If $k = 3$ then $t_2 = I\!\!I$, which is impossible by duality rules. Hence we have 3 possible cases for j and k. In case $j = k = 1$ '($i = 0$), if $t_1 \in V$ we have $t \in L_i^o$; otherwise $t_1 = \mathbf{a} \cdot \mathbf{b}^\sim$, and $t_2 \in V$ is impossible by ordering, while $t_2 \in V \cdot V^\sim$ is impossible due to rule (5.8). If $j = 1, k = 2$ ($i = 1$), then $t \in L_i^o$ when $t_1 \in V, t_2 \in V^\sim$, and the other cases are impossible: $t_2 = \mathbf{b} \wedge \mathbf{c}$ because of rule (5.6), and $t_1 = \mathbf{a} \cdot \mathbf{b}^\sim, t_2 = \mathbf{c}^\sim$ because of rule (5.7). The last case $j = k = 2$ is impossible: when $t_1 = \mathbf{a}^\sim$, $t_2 = \mathbf{b} \wedge \mathbf{c}$ is impossible by ordering, and $t_2 = \mathbf{b}^\sim$ because of rule (5.9), and when $t_1 = \mathbf{a} \wedge \mathbf{b}$, $t_2 = \mathbf{c}^\sim$ is impossible because of rule (5.10), and $t_2 = \mathbf{c} \wedge \mathbf{d}$ because of rule (5.11).

It is clear that the above result can be generalized to any order $<$, by restricting the L_i^o's to the smallest elements w.r.t. the corresponding $\prec$. It is however not necessary to give further details of this, though in the sequel we consider only *linear* orders $<$ on V, so that $\prec$ is also a linear order.

The previous theorem does not assert that the system is terminating but only gives syntactic criteria for irreducible terms. Termination is the focus of the next theorem.

Theorem 4. $R \cup E'$ *is a ground terminating rewriting system.*

Proof. We proceed in four steps. First we prove that the system $(1) + (2)$ is terminating. Then we prove that $(3) + (5)$ also has this property. In the third step we prove that $(1) + (2) + (3) + (5)$ is terminating by introducing a function from the algebra of terms to $\mathbb{N}$. We will conclude by considering the applicability of rules in the group (4) together with the system E' and the computation of grades.

Let $<$ be an order on function symbols: $+ < \circ < \cdot < \wedge < \sim$. Let $<_m$ be the multiset path ordering extension of $<$ (see [4]); then $<_m$ is a simplification ordering and for every rule $l \to r$ in (1) or (2) it is clear that $r <_m l$. This proves the termination of $(1) + (2)$.

We note by $t[p \leftarrow u]$ the result of the substitution by the term u at position p in the term t. Let μ be the following map, defined by structural induction, from the algebra of terms to $(\mathbb{N}, <)$; when x is a functional symbol of arity n, $\mu(x)$ is the map from $\mathbb{N}^n$ to $\mathbb{N}$ associated to x:

$$\mu(\wedge)(x, y) = 7xy,$$
$$\mu(\sim)(x) = 4x,$$
$$\mu(\cdot)(x, y) = 2xy,$$
$$\mu(\circ)(x, y) = xy,$$
$$\mu(+)(x, y) = x + y,$$
$$\mu(I\!\!I) = 3,$$
$$\mu(c) = 1 \text{ for every constant } c \neq I\!\!I.$$

It is easy to check that for every rule $l \to r$ in $(3) + (5)$ we have $\mu(r) < \mu(l)$. Moreover, if t is a ground term, p a position in t, u and v two terms such that $\mu(v) < \mu(u)$ then $\mu(t[p \leftarrow u]) < \mu(t[p \leftarrow v])$, i.e. μ is strictly monotone. As $(\mathbb{N}, <)$ is well-founded, $(3) + (5)$ is ground terminating.

The map μ also has the following property: for every rule $l \to r$ in $(1) + (2)$ $\mu(l) = \mu(r)$. Let us assume that $(1) + (2) + (3) + (5)$ is not terminating. As $(1) + (2)$ is, there would be an infinite sequence of terms with strictly decreasing measure μ, which is impossible.

A similar argument may be used for proving the termination of R. Let (t_i) be an infinite sequence of terms successively rewritten by rules in R. As for every rule $l \to r$ in (4) $\mu(l) = \mu(r)$, there exists an i such that for every natural $j \geq i$, t_j is irreducible by any rule in $(3) + (5)$. We can assume without restriction that $j = 0$. Let us prove by induction on n the number of occurrences of $+$ in t_0 that such an infinite sequence cannot exist.

If $n = 0$, only the rules in (2) and (4) may be used. Let ν be the map defined as follows:

$$\nu(x \wedge y) = \begin{cases} 2(\nu(x) + \nu(y)) & \text{if } x \prec y, \\ 2(\nu(x) + \nu(y)) + 1 & \text{otherwise}, \end{cases}$$
$$\nu(x \cdot y) = \nu(x \wedge y),$$
$$\nu(x^\sim) = \nu(x),$$
$$\nu(xy) = \nu(x) + \nu(y),$$
$$\nu(x + y) = \nu(x) + \nu(y),$$
$$\nu(c) = \begin{cases} 0 & \text{for every constant } c \in \mathbb{K}, \\ 1 & \text{otherwise}. \end{cases}$$

For every rule $l \to r$ in (2), $\nu(r) \leq \nu(l)$. But for every rule $l \to r$ in (4), $\nu(r) < \nu(l)$. As ν is strictly monotone and $(\mathbb{N}, <)$ well-founded, there exists an i such that the only rules applied to the t_j $(j \geq i)$ are those in (2) and as (2) is terminating, (t_i) cannot be infinite.

If $n > 0$, it is easy to see that there exists an i such that the root symbol of t_i is $+$. As the number of occurrences of the symbol $+$ in a term cannot increase, this means that there exist u and v such that $t_i = u + v$ and the number of symbol $+$ in u and v is less than n. Moreover, no rule can be applied to the root of t_i. By induction hypothesis, the sequence is finite.

We see that for every rule $l \to r$ in E' we have $\mu(r) < \mu(l)$. So even when the rules in the system E' are taken into account, it is necessarily terminating. Note also that according to the previous argument, we can add any rewrite rules $l \to r$ with $\mu(r) < \mu(l)$ while preserving the termination property.

5 Further Simplifications

We have therefore proved that an irreducible form of a Clifford expression can indeed be reached in finite time. However, we have seen that the rewriting system R may not be able to rewrite a term to 0, even if $t =_c 0$; this is due to the fact that what we have called irreducibility is only a syntactic property, as is clear from the fact that all rules in R are left-linear (there is at most one occurrence of variables in their left-hand side), and hence that basic relations among constants are not exploited. We do this by means of the following geometric rules:

$$x \wedge x \to 0, \quad x \cdot x^\sim \to 0, \quad x \cdot (x \cdot y^\sim) \to 0.$$

Clearly these rules are still not enough to rewrite $\mathbf{a} - \mathbf{a}$ to 0; but for this only standard algebraic computations are required, and it is convenient to sum up the corresponding rules into a single one:

$$X \to Y \text{ if } X \text{ computes to } Y \text{ and } Y \neq X.$$

This computation is to be performed by some computer algebra system, by considering the subterms of X which are not elements of $\mathbb{K}$ as indeterminates. Of course, we have to trust the system that $\mu(Y) \leq \mu(X)$, but this is a very reasonable assumption.

In order to minimize the number of indeterminates, we have to care for equalities such as $\mathbf{a} \cdot \mathbf{b}^\sim = -\mathbf{b} \cdot \mathbf{a}^\sim$. In order to avoid non-termination we use the order $\prec$ previously defined. Then

$$x \cdot y^\sim \to -y \cdot x^\sim \text{ if } y \prec x, \quad x \cdot (y \cdot z^\sim) \to -y \cdot (x \cdot z^\sim) \text{ if } y \prec x.$$

Note that the measure μ is unchanged by application of these basic ordering rules, and hence the argument used in Theorem 4 can be applied here.

The set of these rules (geometric, algebraic computations, basic orderings) will be noted Z. Applying the rewriting system Z (obviously terminating) to a term may replace subterms by 0; hence it is necessary to include the elimination rules from E' (and to recompute the grades each time a 0 is produced). Hence the rewriting system to be applied to a term in L_i in order to reduce it to 0 is $E' \cup R \cup Z$, which is terminating as shown by the above remark.

The rules from R are all needed but sufficient for obtaining the reduced form defined above. Unfortunately, the system $E' \cup R \cup Z$ is not complete with respect to the reduction to 0, due to the fact that R is not confluent. For instance, take $t = ((\mathbf{a} \cdot \mathbf{b}^\sim) \cdot \mathbf{c}^\sim) \cdot \mathbf{d}^\sim$, which is not a normal form: rule (5.7) applies, but in two different ways, yielding either

$$t_1 = ((\mathbf{a} \cdot \mathbf{b}^\sim) \cdot \mathbf{d})\mathbf{c} - (\mathbf{c} \cdot \mathbf{d})(\mathbf{a} \cdot \mathbf{b}^\sim) \quad \text{or} \quad t_2 = (\mathbf{a} \cdot \mathbf{c})(\mathbf{b} \cdot \mathbf{d}^\sim) - (\mathbf{b} \cdot \mathbf{c})(\mathbf{a} \cdot \mathbf{d}^\sim),$$

which are both R-normal forms. Hence $t \to_R^* t_1$ or $t \to_R^* t_2$ (t_1, t_2 form a critical pair of R) according to the strategy used to apply the rules; hence either $t - t_2 \to_R^* t_1 - t_2$ or $t - t_1 \to_R^* t_2 - t_1$, and $t_1 - t_2$ is a $E' \cup R \cup Z$-normal form, despite the fact that $t_1 - t_2 =_c 0$. The system cannot be complete if it is not confluent.

Of course, the standard way to make a rewriting system confluent is to add rules derived from the critical pairs, either $t_1 \to t_2$ or $t_2 \to t_1$, or some other rules able to rewrite t_1 and t_2 to a common term. The many critical pairs of R can actually be solved by the following three rules:

$(a).\ (x \cdot y)(z \cdot (u \cdot v^\sim)) - (x \cdot z)(y \cdot (u \cdot v^\sim)) \to$
$\qquad\qquad (x \cdot u)(v \cdot (z \cdot y^\sim)) - (x \cdot v)(u \cdot (z \cdot y^\sim)),$

$(b).\ (x \cdot (y \cdot z^\sim))u \to (x \cdot u)(y \cdot z^\sim) + (y \cdot u)(z \cdot x^\sim) + (z \cdot u)(x \cdot y^\sim),$

$(c).\ (x \cdot (y \cdot z^\sim))u^\sim \to (x \cdot u)(z \wedge y) + (y \cdot u)(x \wedge z) + (z \cdot u)(y \wedge x).$

However, these rules are non-terminating, and hence they do not guarantee confluence. From a practical point of view, rule (b) may be difficult to control (or we may restrict it to the case where u is not of the form $t \cdot t'^{\sim}$), and rule (a) needs a special order to ensure termination. Moreover, they are here written up to the associative-commutative properties of the geometric product and the sum, and rule (a) is not left linear, and this implies complex and expensive matchings. They are not right-linear either, and hence also imply term duplications. For all these reasons we postpone the study and use of these rules.

It should be noted that confluence does not imply completeness, and does not solve either the problem of the strategy for applying the rules. Even if all sequences of rewriting steps converge to the same result, some may converge much faster than others. The complexity of the application of the rules (higher if non-linear) should also be addressed.

Since some rules are not right-linear (the distributivity rules and rules (5.3), (5.6)–(5.8) and (5.11)), we should be careful to apply such rules so as to avoid the duplication of unnormalized terms, and hence of rewriting steps. This implies an innermost strategy: if two rules can be applied, one to t and the other to t' and t' is a subterm of t, then the rule on t' is applied first. We also give priority to linear over non-linear rules.

We only have three rules which are not left-linear: the geometric rules. The matching of the left-hand side of these rules therefore involves syntactic equality testing between terms. However, in the context of the innermost strategy, we can restrict the application of these rules to the case where the variables are matched with elements of $\mathbb{V}$, so that testing the equality becomes trivial. It is indeed clear that with this strategy, the geometric rules can only be applied to terms $t \wedge t$, $t \cdot t^{\sim}$ and $t \cdot (t \cdot t'^{\sim})$ where t and t' are R-normal forms. By noting Z' the system Z with the above-mentioned restriction on the geometric rules, and $R' = E' \cup R \cup Z'$, we have:

Theorem 5. *If* $t, t' \in L_1^\circ$ *then*

$$t \wedge t \rightarrow_{R'}^* 0,$$
$$t \cdot t^{\sim} \rightarrow_{R'}^* 0,$$
$$t \cdot (t \cdot t'^{\sim}) \rightarrow_{R'}^* 0.$$

This remains true with an innermost strategy.

Proof. We prove the first two, by induction on t. If $t \in \mathbb{V}$, the result is trivial by applying a restricted geometric rule. If $t \in \mathbb{V} \cdot \mathbb{V}^{\sim}$, say $t = \mathbf{a} \cdot \mathbf{b}^{\sim}$, then $t \wedge t$ rewrites to $((\mathbf{a} \cdot \mathbf{b}^{\sim}) \cdot \mathbf{a})\mathbf{b}^{\sim} - ((\mathbf{a} \cdot \mathbf{b}^{\sim}) \cdot \mathbf{b})\mathbf{a}^{\sim}$ by rule (5.3), and thus to 0 by restricted geometric rules (and algebraic computations). Similarly, $t \cdot t^{\sim}$ rewrites to $(\mathbf{a} \cdot \mathbf{b}^{\sim}) \cdot (-\mathbf{a} \cdot \mathbf{b})$ by rule (5.2), then to $-[((\mathbf{a} \cdot \mathbf{b}^{\sim}) \cdot \mathbf{a})\mathbf{b} - ((\mathbf{a} \cdot \mathbf{b}^{\sim}) \cdot \mathbf{b})\mathbf{a}]$ by rule (5.6) (and linearity), and finally to 0 by geometric rules and algebraic computations.

If $t = t_1 + t_2$, then $t \wedge t$ rewrites to $t_1 \wedge t_1 + t_1 \wedge t_2 + t_2 \wedge t_1 + t_2 \wedge t_2$ by distributivity, and to 0 by induction hypothesis, anticommutativity of $\wedge$ and algebraic computations, and similarly for $t \cdot t^{\sim}$. If $t = t_0 t_1$ with $t_0 \in L_0^\circ$, then similarly $t \wedge t$ and $t \cdot t^{\sim}$ rewrite to 0 by linearity rules and induction hypothesis.

We prove the last line by induction on t, then on t'. If $t \in \mathbb{V}$, say $t = \mathbf{a}$, we prove that $\forall t', t \cdot (t \cdot t'^{\sim}) \rightarrow_{R'}^* 0$. For $t' \in \mathbb{V}$ this is trivial. If $t' = \mathbf{a} \cdot \mathbf{b}^{\sim}$, then $t \cdot (t \cdot t'^{\sim})$ rewrites to $-\mathbf{a} \cdot (\mathbf{a} \cdot \mathbf{b} \wedge \mathbf{c})$ by rule (5.2) and linearity, then to $-[(\mathbf{a} \cdot \mathbf{b})(\mathbf{a} \cdot \mathbf{c}) - (\mathbf{a} \cdot \mathbf{c})(\mathbf{a} \cdot \mathbf{b})]$ by rule (5.6), linearity and distributivity, and finally to 0 by algebraic computation. The inductive cases $t' = t'_1 + t'_2$ and $t' = t'_0 t'_1$ are proved as above. The case $t = \mathbf{a} \cdot \mathbf{b}^{\sim}$ is similar: $t' \in \mathbb{V}$ is solved by rule (5.7), and $t' \in \mathbb{V} \cdot \mathbb{V}^{\sim}$ by rules (5.2) and (5.6). The inductive cases for t' are solved as usual. The induction on t can then proceed: if $t = t_1 + t_2$, the base case $t' \in \mathbb{V}$ is trivial, and $t' = \mathbf{a} \cdot \mathbf{b}^{\sim}$ is solved by rules (5.2) and (5.6), and the induction on t' is as above. If $t = t_0 t_1$, then the base case $t' \in \mathbb{V}$ is trivial, and the case $t' = \mathbf{a} \cdot \mathbf{b}^{\sim}$ is solved by rules (5.2) and (5.6), and the induction on t' is as above. All these rewriting steps follow an innermost strategy.

It is therefore clear that the rewriting system R' is complete over $E' \cup R \cup Z$ for rewriting terms to 0, as far as an innermost strategy is used for applying the rules. Although it is not complete over equality in Clifford algebra, it provides simple normal forms for Clifford terms. Intuitively, the incompleteness of this rewriting system is due to the lack of a general mechanism to use the linear dependence of any four vectors.

6 Pseudo-coordinates

A possibility for computing normal forms of expressions in Clifford algebra and proving equality consists in using both the rewriting system presented in the previous sections and pseudo-coordinates. Although it is not possible to get a complete rewriting system, it is possible to simplify a null expression to 0 by using the previous rules on transformed expressions. The transformation consists in replacing any vector by a linear combination of three independent vectors. In some sense, this transformation is equivalent to the introduction of coordinates for any vector. That is why it is called *pseudo-coordinates expansion*.

Pseudo-coordinates are a practical means of using coordinates in our synthetic framework. They also ensure completeness of our approach. The main idea is the following: let $\mathbf{a}, \mathbf{b}$ be two non-zero non-collinear vectors, with $\mathbf{a} \prec \mathbf{b}$. Then $\mathbf{a}, \mathbf{a} \cdot \mathbf{b}^{\sim}$ are two orthogonal vectors and $\mathbf{a}, \mathbf{a} \cdot \mathbf{b}^{\sim}, (\mathbf{a} \wedge (\mathbf{a} \cdot \mathbf{b}^{\sim}))^{\sim}$ form an orthogonal basis. Here $\mathbf{a}$ and $\mathbf{b}$ are called the *reference vectors*. It follows that any vector $\mathbf{c}$ may be expressed as a linear combination of the three latter vectors with Clifford coefficients. If $\mathbf{c} =_c x\mathbf{a} + y(\mathbf{a} \cdot \mathbf{b}^{\sim}) + z(\mathbf{a} \wedge (\mathbf{a} \cdot \mathbf{b}^{\sim}))^{\sim}$, then $\mathbf{c} \cdot \mathbf{a} =_c x(\mathbf{a} \cdot \mathbf{a})$; hence

$$x =_c \frac{\mathbf{c} \cdot \mathbf{a}}{\mathbf{a} \cdot \mathbf{a}}$$

and similarly:

$$y =_c \frac{\mathbf{c} \cdot (\mathbf{a} \cdot \mathbf{b}^{\sim})}{(\mathbf{a} \cdot \mathbf{b}^{\sim}) \cdot (\mathbf{a} \cdot \mathbf{b}^{\sim})}, \qquad z =_c \frac{\mathbf{c} \cdot (\mathbf{a} \wedge (\mathbf{a} \cdot \mathbf{b}^{\sim}))^{\sim}}{(\mathbf{a} \wedge (\mathbf{a} \cdot \mathbf{b}^{\sim}))^{\sim} \cdot (\mathbf{a} \wedge (\mathbf{a} \cdot \mathbf{b}^{\sim}))^{\sim}}.$$

Using the reduction rules given in the previous section, one gets

$$(\mathbf{a} \wedge (\mathbf{a} \cdot \mathbf{b}^{\sim}))^{\sim} =_c (\mathbf{a} \cdot \mathbf{b})\mathbf{a} - (\mathbf{a} \cdot \mathbf{a})\mathbf{b}.$$

Let
$$\kappa = (\mathbf{a} \cdot \mathbf{a})(\mathbf{b} \cdot \mathbf{b}) - (\mathbf{a} \cdot \mathbf{b})^2;$$
we have
$$\kappa =_C (\mathbf{a} \wedge (\mathbf{a} \cdot \mathbf{b}^\sim))^\sim \cdot (\mathbf{a} \wedge (\mathbf{a} \cdot \mathbf{b}^\sim))^\sim =_C (\mathbf{a} \cdot \mathbf{b}^\sim) \cdot (\mathbf{a} \cdot \mathbf{b}^\sim),$$

and hence $\lceil \kappa \rceil = 0$ if and only if $\lceil \mathbf{a} \cdot \mathbf{b}^\sim \rceil = 0$ if and only if $\lceil \mathbf{a} \rceil$ and $\lceil \mathbf{b} \rceil$ are collinear. By replacing these expressions in x, y, z and factorizing $\mathbf{a}, \mathbf{b}$ and $\mathbf{a} \cdot \mathbf{b}^\sim$, we get the *pseudo-coordinate expression* of $\mathbf{c}$ relative to $\mathbf{a}$ and $\mathbf{b}$:

$$\mathbf{c} =_C \mathrm{pce}(\mathbf{c}, \mathbf{a}, \mathbf{b}) = \alpha \mathbf{a} + \beta \mathbf{b} + \gamma(\mathbf{a} \cdot \mathbf{b}^\sim)$$

with
$$\alpha = \frac{(\mathbf{c} \cdot \mathbf{a})(\mathbf{b} \cdot \mathbf{b}) - (\mathbf{c} \cdot \mathbf{b})(\mathbf{a} \cdot \mathbf{b})}{\kappa},$$
$$\beta = \frac{(\mathbf{c} \cdot \mathbf{b})(\mathbf{a} \cdot \mathbf{a}) - (\mathbf{c} \cdot \mathbf{a})(\mathbf{a} \cdot \mathbf{b})}{\kappa},$$
$$\gamma = \frac{\mathbf{c} \cdot (\mathbf{a} \cdot \mathbf{b}^\sim)}{\kappa}.$$

Pseudo-coordinates may be introduced in a Clifford term t by replacing $\mathbf{c}$ with $\mathrm{pce}(\mathbf{c}, \mathbf{a}, \mathbf{b})$. After all vectors in $\mathbb{V}$ (except $\mathbf{a}$ and $\mathbf{b}$) are expanded this way, we obtain an expression $g =_C t$ containing fractions: g is the *pseudo-coordinate expansion* of t with respect to the reference vectors $\mathbf{a}$ and $\mathbf{b}$. In order to eliminate the denominators, we first multiply the whole expression by a suitable power of κ, and compute the normal form of the resulting term; we obtain the *reduced term* ε associated to the expression g. Hence there is an n such that $\kappa^n t =_C \kappa^n g =_C \varepsilon$. It is clear that $\forall \mathbf{c} \in \mathbb{V} \setminus \{\mathbf{a}, \mathbf{b}\}$, $\kappa \mathrm{pce}(\mathbf{c}, \mathbf{a}, \mathbf{b})$ is a linear combination of $\mathbf{a}, \mathbf{b}$ and $\mathbf{a} \cdot \mathbf{b}^\sim$, where the coefficients are polynomials over $\mathbb{K}$ in the indeterminates $\mathbf{a} \cdot \mathbf{a}$, $\mathbf{a} \cdot \mathbf{b}, \mathbf{b} \cdot \mathbf{b}, \mathbf{c} \cdot \mathbf{a}, \mathbf{c} \cdot \mathbf{b}$ and $\mathbf{c} \cdot (\mathbf{a} \cdot \mathbf{b}^\sim)$.

Theorem 6. ε *is a linear combination of* $1, \mathbf{a}, \mathbf{b}, \mathbf{a} \cdot \mathbf{b}^\sim, \mathbf{a}^\sim, \mathbf{b}^\sim, \mathbf{a} \wedge \mathbf{b}$ *and* $I\!\!I$ *with coefficients being polynomials over* $\mathbb{K}$ *in the indeterminates* $\mathbf{a} \cdot \mathbf{a}, \mathbf{a} \cdot \mathbf{b}, \mathbf{b} \cdot \mathbf{b},$ *and all* $\mathbf{c} \cdot \mathbf{a}$ *(or* $\mathbf{a} \cdot \mathbf{c}$ *if* $\mathbf{a} \prec \mathbf{c}$), $\mathbf{c} \cdot \mathbf{b}$ *(or* $\mathbf{b} \cdot \mathbf{c}$), $\mathbf{c} \cdot (\mathbf{a} \cdot \mathbf{b}^\sim)$ *(or* $\mathbf{a} \cdot (\mathbf{c} \cdot \mathbf{b}^\sim)$ *or* $\mathbf{a} \cdot (\mathbf{b} \cdot \mathbf{c}^\sim)$) *for every* $\mathbf{c} \in \mathbb{V} \setminus \{\mathbf{a}, \mathbf{b}\}$.

Proof. By induction on t: In the base cases $t = I\!\!I, t = \mathbf{a}, t = \mathbf{b}, t \in \mathbb{K}$, this is obvious. If $t = \mathbf{c}$, this is true of $\kappa \mathrm{pce}(\mathbf{c}, \mathbf{a}, \mathbf{b})$, and hence of ε which is identical to $\kappa \mathrm{pce}(\mathbf{c}, \mathbf{a}, \mathbf{b})$ up to basic orderings. Suppose now that it is true of t_1 and t_2; then it is obviously true of $t_1 + t_2$ and $t_1 t_2$ (if $\mathrm{gr}(t_1) \in \{0, \infty\}$). To see that it is also true of $t_1{}^\sim, t_1 \cdot t_2$ and $t_1 \wedge t_2$, it is sufficient to check that it is true of the normal forms of:

- the duals of $1, \mathbf{a}, \mathbf{b}, \mathbf{a} \cdot \mathbf{b}^\sim, \mathbf{a}^\sim, \mathbf{b}^\sim, \mathbf{a} \wedge \mathbf{b}$ and $I\!\!I$,
- the inner and outer products of any two terms in $\{1, \mathbf{a}, \mathbf{b}, \mathbf{a} \cdot \mathbf{b}^\sim, \mathbf{a}^\sim, \mathbf{b}^\sim, \mathbf{a} \wedge \mathbf{b}, I\!\!I\}$,

and to apply linearity and distributivity rules.

The following theorem is crucial in our soundness proof.

Theorem 7 (Extension of generic Clifford algebraic 3D identities). *Let h be a Clifford term of grade 0 and t be any Clifford term. If $h \neq_C 0$ and for every value $\lceil h \rceil \neq 0 \Rightarrow \lceil t \rceil = 0$ then $t =_C 0$.*

Proof. It is clear that $ht =_C 0$. Let e_1, e_2, e_3 be three constants not in $\mathbb{V}$ denoting vectors. Let t' be the term t in which all $c \in \mathbb{V}$ are replaced by $X_1^c e_1 + X_2^c e_2 + X_3^c e_3$ where X_i^c are constants of grade 0. We note $\mathbf{X}$ the set of these constants. By applying the same transformation to h we get the term h'.

We consider an orthonormal basis of $\mathcal{V}$ as a fixed interpretation of (e_1, e_2, e_3), such that $\lceil e_3 \rceil^\sim = \lceil e_1 \rceil \wedge \lceil e_2 \rceil$. Every interpretation of $\mathbf{X}$ can be extended this way: the result is an *orthonormal* interpretation of $\mathbf{X}$. We note $=_C^n$ the equality modulo orthonormal interpretations. It is clear that there is a bijective correspondence between $\mathbb{V}$-interpretations and orthonormal $\mathbf{X}$-interpretations compatible with the replacement above: hence $t =_C 0$ if and only if $t' =_C^n 0$, and $ht =_C 0$ if and only if $h't' =_C^n 0$. It is easy to check that:

$$e_1 \cdot e_1 =_C^n e_2 \cdot e_2 =_C^n e_3 \cdot e_3 =_C^n 1,$$
$$e_1 \cdot e_2 =_C^n e_2 \cdot e_3 =_C^n e_1 \cdot e_3 =_C^n 0,$$
$$e_1 \wedge e_2 =_C^n e_3{}^\sim, \quad e_2 \wedge e_3 =_C^n e_1{}^\sim, \quad e_1 \wedge e_3 =_C^n -e_2{}^\sim,$$
$$e_1 \cdot e_2^\sim =_C^n -e_3, \quad e_2 \cdot e_3^\sim =_C^n -e_1, \quad e_1 \cdot e_3^\sim =_C^n e_2.$$

Following the previous theorem, and using the rewriting system R' as well as the equations above, $h't'$ can be reduced to a linear combination of $1, e_1, e_2, e_3, e_1{}^\sim$, $e_2{}^\sim$, $e_3{}^\sim$ and $I\!\!I$, with coefficients being polynomials in $\mathbb{K}[\mathbf{X}]$. Since $h't' =_C^n 0$, its scalar, vector, bivector and trivector parts are null. The scalar (resp. trivector) part is the polynomial coefficient of 1 (resp. $I\!\!I$); hence this polynomial is null. The vector part is a linear combination of e_1, e_2, e_3, and these three vectors are independent; hence their polynomial coefficients are null. The bivector part is a linear combination of $e_1{}^\sim, e_2{}^\sim, e_3{}^\sim$, and these three bivectors are independent; hence their polynomial coefficients are null. But these coefficients are the product by h' of the corresponding coefficients of t', which are therefore null since $h' \neq_C^n 0$. Hence $t' =_C^n 0$, and $t =_C 0$.

Completeness and soundness are direct consequences of the two previous theorems:

Theorem 8. $t =_C 0$ *if and only if $\varepsilon = 0$.*

Proof. If $t =_C 0$, then $\kappa^n t =_C \varepsilon =_C 0$; since ε is a linear combination of $1, \mathbf{a}, \mathbf{b}, \mathbf{a} \cdot \mathbf{b}^\sim$ (which are independent vectors), $\mathbf{a}^\sim, \mathbf{b}^\sim, \mathbf{a} \wedge \mathbf{b}$ (which are independent bivectors) and $I\!\!I$, as in the previous proof their polynomial coefficients are null, and can be rewritten to 0 by standard algebraic computations. Since ε is in normal form, we have $\varepsilon = 0$.

If $\varepsilon = 0$, $\kappa^n t =_C 0$ and as $\kappa^n \neq_C 0$, it follows from the previous theorem that $t =_C 0$.

Pseudo-coordinate expansion can be applied to any Clifford expressions, but it is important to compute the normal form of t first, because even if it does not yield 0, it may considerably reduce t and hence ease the process of expansion.

Remark that applying pseudo-coordinate expansion only involves basic properties of the equality and rules from R', which proves that the equational theory obtained from R' is complete for $=_C$.

Experimental considerations

The time necessary for simplifying an expression according to the previous method highly depends on the choice of the basic vectors. In the worst case, this is equivalent to the introduction of scalar coordinates. However it appears that in many cases a good choice of the reference vectors reduces significantly the number of simplification steps to be performed. The whole method and a good strategy for introducing pseudo-coordinates are presented below.

The main problem in pseudo-coordinate expansion consists in selecting the two reference vectors. In geometric constructions, some vectors are taken to be free. As we deal with space problems, there are usually two or more free vectors. It is thus possible to consider two of them as the reference vectors. Following this strategy, any vector c different from the reference vectors, say a and b, is replaced by its expansion pce(c, a, b) using pseudo-coordinates. We call this *global expansion*.

However a more efficient strategy may be used. Instead of choosing the reference vectors once for all, one may decide to choose them according to the neighborhood of the constant in a term. For instance, let us consider the following (sub-)expression

$$(c \cdot d)(c \cdot e).$$

If we decide to expand it using a, b as the reference vectors, one gets a much larger expression than using d, e. In the latter case, the expression is already expanded. The key idea consists in minimizing the number of monomials generated by expansion. To achieve this goal, we propose the following two heuristics.

- *Order-based local expansion*. First, the set of constant vectors is ordered. As innermost terms are reduced first, we compute the two maximal vectors according to this order in each subterm and take them as the reference vectors for the expansion in this subterm.
- *Count-based local expansion*. Occurrences of vectors in a term are counted, and the two vectors with the two largest numbers of occurrences are taken as the reference vectors.

As pseudo-coordinate expansion introduces a sum, this may lead to a combinatorial explosion of the number of terms. That is why in practice this expansion should be applied only after a first normalization. It appears that the proof of many geometric theorems does not require pseudo-coordinates (see Example 1 in Section 8).

Of course this depends on the formulation of the theorems. In some sense we can assert that any constructive geometric theorem may be formulated in a way such that it does not require to introduce pseudo-coordinates: simply imagine that constructive expressions are built directly with pseudo-coordinates. However, this would hardly qualify as a coordinate-free method! In practice, we want to formulate theorems in the simplest possible way, and postpone the use of pseudo-coordinates as far as possible.

Another important point concerns the use of computer algebra algorithms for simplifying scalar expressions. This point is crucial for efficiency and the reader may refer to [6] for further details. Last we stress the fact that the considerations on the priority of rules (presented in the previous sections) are of prime importance for the efficiency of rewriting.

7 Geometric Reasoning

The rewriting techniques and system for proving Clifford identities presented in the previous sections have an interesting application to geometric reasoning. This application is based on the observation that many geometric concepts and relations can be represented by means of Clifford algebraic expressions. Therefore, reasoning about geometric problems may be reduced to manipulating such expressions and to investigating their relationships.

Consider for example the case of proving geometric theorems. From [5, 7, 11], one sees that many theorems in plane and solid Euclidean geometry can be formulated by using Clifford algebra, where the hypothesis-relations of a theorem in question are expressed as a system of Clifford algebraic equations. By means of vectorial equations solving as presented in [8], the hypothesis-relations may often be transformed into the following triangular form

$$h_1(\mu, \chi_1) = 0,$$
$$h_2(\mu, \chi_1, \chi_2) = 0, \qquad \qquad \text{(H)}$$
$$\cdots \cdots$$
$$h_r(\mu, \chi_1, \chi_2, \ldots, \chi_r) = 0,$$

where each h_i is composed using Clifford operators, $\mu = (\mu_1, \ldots, \mu_d)$ are free *geometric entities* (called *parameters*), and $\chi_1, \ldots, \chi_r$ are constrained by the geometric hypotheses (called *dependents*). Examples of geometric entities are points, areas of triangles and other geometric invariants.

To make the triangularization possible and easy, it is usually required that the theorem under consideration is stated constructively. To reduce the triangularizing cost, the last two authors [6, 11] have considered a restricted class of geometric theorems for which each dependent χ_i can be solved in terms of μ and $\chi_1, \ldots, \chi_{i-1}$ using Clifford operators; namely,

$$\chi_i = f_i(\mu, \chi_1, \ldots, \chi_{i-1}), \quad 1 \leq i \leq r. \qquad \text{(H*)}$$

A number of popular constructions with the corresponding expressions of f_i in plane Euclidean geometry are provided in [6, 11]. In fact, by solving vectorial equations as in [7] similar constructions and their corresponding Clifford representations may be established for 3D Euclidean geometry, where the vector space $\mathcal{V}$ is of dimension 3. Here are two examples.

- Construct the vertex X such that $P_1 P_2 P_3 X$ is a parallelogram:

$$X = \mathsf{parg}(P_1, P_2, P_3) = P_1 + P_3 - P_2.$$

This representation holds for plane Euclidean geometry as well.
- Construct the intersection point X of line $P_1 P_2$ and plane $P_3 P_4 P_5$:

$$X = \mathsf{intp}(P_1, P_2, P_3, P_4, P_5) = \lambda P_1 + (1 - \lambda) P_2,$$

where

$$\lambda = \frac{[P_3 \wedge P_4 \wedge P_5 - P_2 \wedge (P_4 \wedge P_5 + P_3 \wedge P_4 - P_3 \wedge P_5)]^\sim}{[(P_1 - P_2) \wedge (P_4 \wedge P_5 + P_3 \wedge P_4 - P_3 \wedge P_5)]^\sim}.$$

The non-degeneracy condition is: the denominator is non-zero, i.e., $P_1 P_2 \nparallel P_3 P_4 P_5$.

Such explicit Clifford representations may help make the proving method more effective. In deriving them, some (non-automated yet elementary) geometric reasoning is often needed. We do not get into the details in this paper.

Meanwhile, constructive geometric theorems of this type have also be studied by Yang, Zhang and Feng [13], using Clifford algebra formalism and similar proving techniques. A set of constructions with geometric objects including points, lines, circles, and ellipses in both 2D and 3D with the corresponding Clifford expressions have been given by them.

Suppose that the conclusion of the theorem to be proved is given as another Clifford expression

$$g = g(\mu, \chi_1, \ldots, \chi_r) = 0. \tag{C}$$

Let d_i be the denominator of f_i in (H*) for $1 \le i \le r$. Proving the theorem amounts to verifying whether (C) follows from (H*), under the subsidiary (non-degeneracy) conditions $d_i \ne 0$.

The class under discussion (referred to as class $\mathbb{C}$), though restricted, is large enough to cover many interesting theorems in elementary geometry. The method we use as proposed in [5, 6, 11] for confirming theorems in $\mathbb{C}$ consists of the following three steps.

Step A. Let $f_1^* = f_1$. Do the following substitution for $i = 2, \ldots, r$:

$$f_i^* = f_i^*(\mu) = f_i|_{\chi_1 = f_1^*, \ldots, \chi_{i-1} = f_{i-1}^*} = f_i(\mu, f_1^*, \ldots, f_{i-1}^*).$$

Step B. Substitute all $\chi_i = f_i^*$ into g:

$$g^* = g^*(\mu) = g|_{\chi_1 = f_1^*, \ldots, \chi_r = f_r^*} = g(\mu, f_1^*, \ldots, f_r^*).$$

Let h be the numerator of g^*.

Step C. Apply a term-rewriting system with suitably chosen rewrite rules to h. If h is rewritten to 0, then the theorem is true under the non-degeneracy conditions $d_i \neq 0$ for $1 \leq i \leq r$. If h does not evaluate to 0 when the (pseudo-)coordinate rules are applied, then the theorem is false under the conditions $d_i \neq 0$.

It is clear that in the above method both Clifford algebraic computing and term-rewriting are required. The former can be done in a computer algebra system, and the latter has been discussed in detail in the previous sections.

The entire method has been implemented in **Maple V** for algebraic computing and in **Objective Caml** for term-rewriting with an interface between the two systems. Here we do not enter into the implementation details (see [6] and a forthcoming paper by us).

8 Application Examples

In this section we present some examples to illustrate the application of our rewriting techniques and system to 3D geometric problems.

For any two vectors **a** and **b**, their *cross product* is defined as

$$\mathbf{a} \times \mathbf{b} = (\mathbf{a} \wedge \mathbf{b})^{\sim}.$$

Our first example shows how the rewriting procedure works by proving the well-known *Jacobi identity*

$$\mathbf{a} \times (\mathbf{b} \times \mathbf{c}) + \mathbf{b} \times (\mathbf{c} \times \mathbf{a}) + \mathbf{c} \times (\mathbf{a} \times \mathbf{b}) = 0$$

for arbitrary 3D vectors $\mathbf{a}, \mathbf{b}, \mathbf{c}$.

Example 1. Prove that

$$\mathbf{a} \wedge (\mathbf{b} \wedge \mathbf{c})^{\sim} + \mathbf{b} \wedge (\mathbf{c} \wedge \mathbf{a})^{\sim} + \mathbf{c} \wedge (\mathbf{a} \wedge \mathbf{b})^{\sim} = 0$$

holds for any 3D vectors $\mathbf{a}, \mathbf{b}, \mathbf{c}$.

Our system is able to produce a trace of the rewriting steps. We give below some of them; the others similar or unessential for understanding the proof are omitted.

```
rewriting (b^c)~
        -> b.c~  done.

rewriting a^(b.c~)
        -> (a.b)*c~+(-1)*((a.c)*b~)        done.

rewriting (c^a)~
        -> c.a~  done.

rewriting b^(c.a~)
        -> (b.c)*a~+(-1)*((b.a)*c~)        done.
```

```
rewriting b.a
        -> (1)*(a.b)      done.

(...)

rewriting (a.b)*c~+(-1)*(a.b)*c~
        -> 0*c~  done.

(...)

0

(S+U) Rewriting Time = 0.04s
(S+U) Simplification Time = 0.00s
(S+U) Communication Time = 0.01s
```

The next two examples are taken from [7]. The proofs produced here are somewhat simpler than those given by Li therein.

Example 2. Let a plane intersect the sides AB, AC, CD, DB of an arbitrary tetrahedron $ABCD$ at four points M, N, E, F respectively such that $MNEF$ is a parallelogram. Then the center of $\square MNEF$ lies on the line connecting the midpoints of AD and BC.

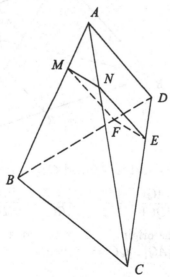

Let D, E, F, M be free points. Without loss of generality, we can take M as the origin: $M = 0$. The other dependent points may be constructed and represented as follows

$$N = \text{parg}(0, F, E) = E - F,$$
$$C = \text{on\_line}(D, E) = uD + (1 - u)E,$$

$$A = \text{intp}(C, N, 0, E, D) = D + \frac{1-u}{u}F,$$

$$B = \text{int}(A, 0, D, F) = uD + (1-u)F,$$

where a scalar parameter u is introduced, and the expressions of the constructed points have already been substituted into the expression of the point being constructed (see step A of the method in the preceding section). The exact meanings of on_line, int and other predicates, which are not really needed for understanding the examples in this section, can be found in [6, 11]. The conclusion of the theorem to be proved is

$$g = (E-A-D) \wedge (B+C-A-D) = 0. \ \% \ \text{col}(\text{midp}(M,E), \text{midp}(A,D), \text{midp}(B,C))$$

Substituting the expressions of the dependents into g, one finds that the resulting expression may be easily rewritten to 0. Therefore, the theorem is proved to be true under the non-degeneracy condition $u \neq 0$ (i.e., C does not coincide with E).

Example 3. Let $ABCD$ be an arbitrary tetrahedron, and a plane which is parallel to both AB and CD cut the tetrahedron into two parts, intersecting the sides AD, AC, BC, BD at points P, Q, R, S respectively. Let the ratio of the distances from the plane to AB and CD respectively be $r > 0$. Determine the ratio s of the signed volume of the two parts.

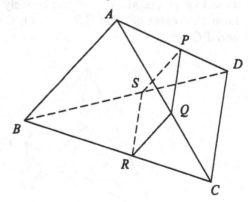

Since the plane is parallel to AB and CD, it is easy to see that

$$\frac{|AQ|}{|QC|} = \frac{|AP|}{|PD|} = \frac{|BR|}{|RC|} = \frac{|BS|}{|SD|} = r.$$

Let Q be located at the origin, i.e. $Q = 0$, and A, B, P be free points. Take C on line AQ such that $|AQ| : |QC| = r$; then

$$C = -\frac{A}{r}.$$

Similarly, take R on BC such that $|BR| : |RC| = r$, D on AP such that $|AP| : |PD| = r$, and S on BD such that $|BS| : |SD| = r$. We have

$$R = \frac{B+rC}{r+1}, \quad D = \frac{(r+1)P - A}{r}, \quad S = \frac{B+rD}{r+1}.$$

The points P, Q, R, S constructed above lie necessarily in the same plane. The non-degeneracy conditions are $r \neq 0$ and $r \neq -1$, which are already satisfied by the hypothesis. Let

$$\mathsf{vol}(P_1, P_2, P_3, P_4) = (P_2 - P_1) \wedge (P_3 - P_1) \wedge (P_4 - P_1),$$

whose magnitude is six times the volume of the tetrahedron $P_1 P_2 P_3 P_4$. Then the ratio we wanted to find is

$$s = -\frac{\mathsf{vol}(A, P, Q, R) + \mathsf{vol}(A, P, R, S) + \mathsf{vol}(A, R, B, S)}{\mathsf{vol}(D, P, Q, R) + \mathsf{vol}(D, P, R, S) + \mathsf{vol}(D, Q, C, R)}.$$

Simple substitution of the expressions of the dependent points with rewriting yields

$$s = \frac{r^2(r + 3)}{3r + 1}.$$

The following example is the one that takes the largest amount of proving time among 20 examples given in [2].

Example 4. Let M and N be two opposite vertices of an arbitrary parallelepiped. Then the diagonals of the hexagon formed by the mid-points of the edges that do not pass through M or N are concurrent.

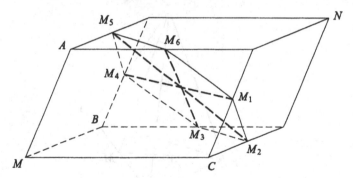

Let M be located at the origin (i.e., $M = 0$) and the vertices A, B, C be free points, and denote the six mid-points of the edges by $M_1, \ldots, M_6$ as in the above figure. Obviously, we have

$$M_1 = \frac{A}{2} + C, \quad M_2 = \frac{B}{2} + C,$$

$$M_3 = B + \frac{C}{2}, \quad M_4 = \frac{A}{2} + B,$$

$$M_5 = A + \frac{B}{2}, \quad M_6 = A + \frac{C}{2}.$$

It is easy to verify that

$$(M_1 - M_4) \wedge (M_2 - M_5) \wedge (M_3 - M_6) = 0,$$

so the three diagonals M_1M_4, M_2M_5, M_3M_6 are coplanar. Let M_2M_5 and M_3M_6 meet at point H; then

$$H = \text{intp}(M_2, M_5, M_3, M_6, 0) = \frac{A + B + C}{2}.$$

The conclusion we need to prove is

$$g = (M_1 - H) \wedge (M_1 - M_4) = 0.$$

This may be done easily by substituting the above expressions of M_1, M_4, H into g (and rewriting the result to 0). As a side effect, we can see from the above proof that the intersection point of the diagonals is actually the centroid of the parallelepiped.

Term-rewriting is not heavily used for the above examples, while proving the following theorem requires quite extensive rewriting to simplify the conclusion-expression to 0.

Example 5. Let $ABCD$ be an arbitrary tetrahedron and M be the intersection point of the line CD and the bisector of the dihedral angle formed by the two planes ABC and ABD. Denote the altitudes of $\triangle ABC$ and $\triangle ABD$ on the base AB by h_C and h_D respectively. Then $h_C : h_D = |CM| : |DM|$.

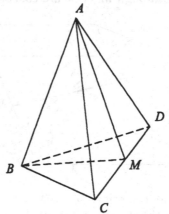

Without loss of generality, we take A as the origin; i.e., $A = 0$. Let B, C, M be free points. Note that the normal vectors of the planes ABC and ABM are $(B \wedge C)^\sim$ and $(B \wedge M)^\sim$, respectively. Let the normal vector of the plane ABD be denoted by $\mathbf{n}$. Then $\mathbf{n}$ is the symmetrical vector of $(B \wedge C)^\sim$ with respect to $(B \wedge M)^\sim$. It is thus easy to represent $\mathbf{n}$ in terms of B, C, M as follows

$$\mathbf{n} = 2(B \wedge C)^\sim \cdot (B \wedge M)^\sim (B \wedge M)^\sim - (B \wedge M)^\sim \cdot (B \wedge M)^\sim (B \wedge C)^\sim.$$

The point D is the intersection of the line CM and the plane passing through 0 with normal vector $\mathbf{n}$, so

$$D = \frac{\mathbf{n} \cdot (M \wedge C)}{\mathbf{n} \cdot (M - C)}.$$

The altitudes h_C and h_D can be easily represented:

$$h_C = C \cdot C - \frac{(B \cdot C)^2}{B \cdot B},$$

$$h_D = D \cdot D - \frac{(B \cdot D)^2}{B \cdot B}.$$

The conclusion of the theorem is

$$g = h_D(M - C) \cdot (M - C) - h_C(M - D) \cdot (M - D) = 0.$$

It is proved by substituting the expressions of $\mathbf{n}, D, h_C, h_D$ into g and rewriting the resulting expression to 0.

As the ratio of the areas of $\triangle ABC$ and $\triangle ABD$ is the same as $h_C : h_D$, the theorem stated as Example 4 in [13] may be considered as a corollary of the above theorem. With the following conclusion-equation

$$g^* = (B \wedge D) \cdot (B \wedge D)(M - C) \cdot (M - C) - (B \wedge C) \cdot (B \wedge C))(M - D) \cdot (M - D) = 0,$$

we have proved the corollary directly with less difficulty.

An *essential* matrix is a 3×3 matrix that can be expressed as a skew-symmetrical matrix postmultiplied by a rotation matrix. It plays a central role in two-view estimation as shown in the computer vision literature [9]. A rank two matrix E is an essential matrix if and only if the equality (1) below holds. There are several other properties about the essential matrix; we shall prove one of them in the following example. Let

$$E = (\mathbf{e}_1 \ \mathbf{e}_2 \ \mathbf{e}_3)$$

be any 3×3 matrix, where $\mathbf{e}_i$ is the ith column vector for each i. Assume that $\mathbf{e}_i \cdot \mathbf{e}_j \neq 0$ for all i, j. For any matrix M, let M^τ and $\mathrm{tr}(M)$ denote the *transposed* matrix and *trace* of M respectively.

Example 6. If

$$EE^\tau E = \frac{1}{2}\mathrm{tr}(EE^\tau)E, \tag{1}$$

then

$$g = \frac{\mathbf{e}_3}{\mathbf{e}_1 \cdot \mathbf{e}_2} + \frac{\mathbf{e}_1}{\mathbf{e}_2 \cdot \mathbf{e}_3} + \frac{\mathbf{e}_2}{\mathbf{e}_3 \cdot \mathbf{e}_1} = 0. \tag{2}$$

Let $t = \mathrm{tr}(EE^\tau)$; then

$$t = \mathbf{e}_1 \cdot \mathbf{e}_1 + \mathbf{e}_2 \cdot \mathbf{e}_2 + \mathbf{e}_3 \cdot \mathbf{e}_3.$$

The equality (1) is thus equivalent to the following set of conditions

$$h_1 = 2(\mathbf{e}_1 \cdot \mathbf{e}_1\,\mathbf{e}_1 + \mathbf{e}_2 \cdot \mathbf{e}_1\,\mathbf{e}_2 + \mathbf{e}_3 \cdot \mathbf{e}_1\,\mathbf{e}_3) - t\mathbf{e}_1 = 0,$$

$$h_2 = 2(\mathbf{e}_1 \cdot \mathbf{e}_2\,\mathbf{e}_1 + \mathbf{e}_2 \cdot \mathbf{e}_2\,\mathbf{e}_2 + \mathbf{e}_3 \cdot \mathbf{e}_2\,\mathbf{e}_3) - t\mathbf{e}_2 = 0,$$

$$h_3 = 2(\mathbf{e}_1 \cdot \mathbf{e}_3\,\mathbf{e}_1 + \mathbf{e}_2 \cdot \mathbf{e}_3\,\mathbf{e}_2 + \mathbf{e}_3 \cdot \mathbf{e}_3\,\mathbf{e}_3) - t\mathbf{e}_3 = 0.$$

Therefore, proving that (1) implies (2) is reduced to proving that

$$[h_1 = 0, h_2 = 0, h_3 = 0] \implies g = 0. \tag{3}$$

However, this cannot be done with a straightforward application of our method. For none of the e_i can be represented in terms of the other vectors, nor can the hypothesis-expressions be easily triangularized. Even using coordinates, it is also not easy to prove (3). We have tried Wu's method without success, yet a machine proof can be obtained by using Gröbner bases only with respect to the total degree term ordering.

Fortunately, we are able to give a simple proof of (3) with two rewriting steps: Let g^* be the numerator of g. The first step eliminates the term $e_1 \cdot e_3 \, e_2 \cdot e_3 \, e_3$ from g^* using $h_1 = 0$, and the second step of *generalized* rewriting reduces the resulting expression to 0 using $h_2 \cdot e_3 = 0$. More precisely, we have

$$g^* \longrightarrow g^* - e_2 \cdot e_3 \, h_1 \longrightarrow g^* - e_2 \cdot e_3 \, h_1 - h_2 \cdot e_3 \, e_1 \longrightarrow 0,$$

so (3) is proved. Moreover, we have seen that the hypothesis $h_3 = 0$ is not needed for proving the conclusion (2).

Although the above proof was produced on computer not automatically, we would be able to find such a proof automatically when the generalized rewriting technique is appropriately incorporated into our program. A systematic investigation on generalized rewriting techniques and their mechanization, presenting machine-generated proofs, and interpreting Clifford algebraic non-degeneracy conditions geometrically all remain for future research.

Acknowledgments

This work has been supported by CEC under Reactive LTR Project 21914 (CU-MULI) and by CNRS/CAS under a cooperation project between LEIBNIZ and CICA.

References

1. Ablamowicz, R., Lounesto, P., Parra, J. M.: Clifford algebras with numeric and symbolic computations. Birkhäuser, Boston (1996).
2. Chou, S.-C., Gao, X.-S., Zhang, J.-Z.: Automated production of traditional proofs for theorems in Euclidian geometry - II. The volume method. Tech. Rep. WSUCS-92-5, Department of Computer Science, Wichita State University, USA (1992).
3. Chou, S.-C., Gao, X.-S., Zhang, J.-Z.: Machine proofs in geometry. World Scientific, Singapore (1995).
4. Dershowitz, N., Jouannaud, J.-P.: Rewrite systems. In: Handbook of theoretical computer science, vol. B (J. van Leeuwen, ed.), Elsevier, Amsterdam, pp. 243–320 (1990).
5. Fèvre, S., Wang, D.: Proving geometric theorems using Clifford algebra and rewrite rules. In: Proc. CADE-15 (Lindau, Germany, July 5–10, 1998), LNAI **1421**, Springer, Berlin Heidelberg, pp. 17–32.

6. Fèvre, S., Wang, D.: Combining algebraic computing and term-rewriting for geometry theorem proving. In: Proc. AISC '98 (Plattsburgh, USA, September 16–18, 1998), LNAI **1476**, Springer, Berlin Heidelberg, pp. 145–156.
7. Li, H.: Vectorial equations-solving for mechanical geometry theorem proving. J. Automat. Reason. (to appear).
8. Li, H., Cheng, M.-t.: Proving theorems in elementary geometry with Clifford algebraic method. Adv. Math. (Beijing) **26**: 357–371 (1997).
9. Maybank, S. J.: Applications of algebraic geometry to computer vision. In: Computational algebraic geometry (F. Eyssette and A. Galligo, eds.), Birkhäuser, Boston, pp. 185–194 (1993).
10. Wang, D.: Geometry machines: From AI to SMC. In: Proc. AISMC-3 (Steyr, Austria, September 23–25, 1996), LNCS **1138**, Springer, Berlin Heidelberg, pp. 213–239.
11. Wang, D.: Clifford algebraic calculus for geometric reasoning with application to computer vision. In: Automated deduction in geometry (D. Wang, ed.), LNAI **1360**, Springer, Berlin Heidelberg, pp. 115–140 (1997).
12. Wu, W.-t.: Mechanical theorem proving in geometries: Basic principles. Springer, Wien New York (1994).
13. Yang, H., Zhang, S., Feng, G.: Clifford algebra and mechanical geometry theorem proving. In: Proc. 3rd ASCM (Lanzhou, China, August 6–8, 1998), Lanzhou Univ. Press, Lanzhou, pp. 49–63.

Some Applications of Clifford Algebra to Geometries

Hongbo Li

Department of Physics and Astronomy
Arizona State University
Tempe, AZ 85287-1504, USA

Abstract. This paper focuses on a Clifford algebra model for geometric computations in 2D and 3D geometries. The model integrates symbolic representation of geometric entities, such as points, lines, planes, circles and spheres, with that of geometric constraints such as angles and distances, and is appropriate for both symbolic and numeric computations. Details on how to apply this model are provided and examples are given to illustrate the application.

1 Background and Notations

Applications of Clifford algebra to geometries were originated in the work of Grassmann and Hamilton in the 19th century [23]. For nearly 150 years, there have been various applications of Clifford algebra to geometries, in the names of Grassmann algebra, Grassmann-Cayley algebra, quaternions, dual quaternions, Dirac algebra, Pauli algebra, spinor representation, Clifford algebra, dual vector algebra, geometric algebra, etc. In the language of geometric algebra formulated by Hestenes [16], for nD Euclidean geometry, there are four typical Clifford algebra models:

(1) The Clifford model $\mathcal{G}_n$: Points and directions of the nD Euclidean space are both represented by vectors in $\mathcal{R}^n$, and these vectors generate the Clifford algebra $\mathcal{G}_n$ [7, 8, 16, 17].

The Clifford model has it origin in Grassmann's work on geometric computing [19, 34], and is completed by Clifford who combined Grassmann's work with Hamilton's quaternions [23]. As a result, vector algebra and quaternions are naturally included in this model. The model is very convenient for geometric computations involving Euclidean transformations of points and orthogonal transformations of directions.

(2) The Grassmann model $\mathcal{G}_{n+1}$: The nD affine Euclidean space is embedded as a hyperplane of $\mathcal{R}^{n+1}$, with unit distance from the origin. Points are represented by vectors in $\mathcal{R}^{n+1}$ while directions are represented by vectors in $\mathcal{R}^n$. The space $\mathcal{R}^{n+1}$ generates the Clifford algebra $\mathcal{G}_{n+1}$ [18, 19, 34].

The Grassmann model was first proposed by Grassmann [34]. In coordinate form, this model gives the homogeneous coordinates representation of points in projective space. This model is very efficient in geometric computations involving projective transformations, and has been extensively used in the study of computer vision from the geometric viewpoint [1–4, 9, 10, 25, 36].

(3) The degenerate model $\mathcal{G}_{n,0,1}$: The Clifford algebra $\mathcal{G}_{n,0,1}$ is generated by vectors in $\mathcal{R}^n \oplus \langle e_0 \rangle$, where vector e_0 is null and is perpendicular to $\mathcal{R}^n$; the nD Euclidean space $\mathcal{R}^n$ as a subspace of $\mathcal{R}^n \oplus \langle e_0 \rangle$ is imbedded in the Clifford algebra $\mathcal{G}_{n,0,1}$ by the mapping $x \mapsto 1 + e_0 \wedge x$, for $x \in \mathcal{R}^n$ [5, 6, 20, 21, 24, 37, 38]. This model usually occurs in the form of dual quaternions and dual vector algebra.

The degenerate model was first proposed by Clifford in the form of dual quaternions, and further developed by Study [37], Blaschke [5] and others. Yang and Freudenstein [38] applied another version of this model, the dual vector algebra, in mechanism analysis. This model is an important computing tool for Euclidean transformations of straight lines and planes.

(4) The Wachter model $\mathcal{G}_{n+1,1}$: The nD Euclidean space $\mathcal{R}^n$ as a subspace of the Minkowski space $\mathcal{R}^{n+1,1}$, is imbedded in the null cone of the Minkowski space through the mapping $x \mapsto f_0 + x + (x^2/2)f$, where f, f_0 are two null vectors orthogonal to $\mathcal{R}^n$. $\mathcal{G}_{n+1,1}$ is the Clifford algebra generated by the Minkowski space.

The Wachter model was first found in the work of Wachter (1792–1817) [35], a student of Gauss. It is closely connected with distance geometry [12, 13, 35], conformal geometry [14] and the geometry of spheres [31, 32]. Various applications in engineering, computer vision and molecular structures can be found in the literature [12–14, 31, 32].

In a recent series of papers [22, 27–30], Li, Hestenes and Rockwood established a universal Clifford algebra model for conformal geometries of Euclidean, spherical and double-hyperbolic spaces. It is shown that the three geometric spaces can be obtained from the null cone of the same Minkowski space through three kinds of projective splits [18], and all the important analytical models for hyperbolic geometry can be obtained by simply changing the viewpoint of a projective split. The Wachter model is part of the universal model in the case of Euclidean geometry.

In this paper, the unified theory of the Wachter model will be applied to 2D and 3D Euclidean geometries, which are the most commonly used ones in application. Details of algebraic representations and manipulations of geometric entities and constraints will be presented in various lists and examples. The present paper intends to show that, the Wachter model is capable of preserving the advantages of the other three models while making the representations and computations of distances, lines, planes, circles and spheres more easily.

This paper is organized as follows: an introduction of the Wachter model in 2D case is given in Section 2. A discussion of the connections of the Wachter model with other models is given in Section 3. The Wachter model in 3D geometry is presented in Section 4. Section 5 presents some applications of the Wachter model in geometries.

The definition and basic operators of Clifford algebra can be found in various literature, for example [16, 17, 22, 26]. To facilitate the readers, we present below a brief explanation of the definition and basic operators of the algebra.

A real *Clifford algebra* generated by an n-dimensional real inner product space V^n of signature[1] (p, q, r), is the quentient algebra of the tensor algebra $\otimes(V^n)$ by the two-sided ideal generated by elements of the form $x \otimes x - x \cdot x$, where $x \in V^n$ and the dot is the inner product in V^n. The Clifford algebra is denoted by $\mathcal{G}(V^n)$, or $\mathcal{G}_{p,q,r}$ when we stress the signature of the space. An element of a Clifford algebra is called a *multivector*, and the multiplication of the algebra induced from the tensor product is called the *geometric product*.

The *inner product* of the space V^n can be extended to the whole of $\mathcal{G}(V^n)$. The geometric product also induces a product, called the *outer product*, in the algebra. In fact $\mathcal{G}(V^n)$ taken as a vector space, together with the outer product, form a *Grassmann algebra*. The outer product is denoted by "$\wedge$". There is another important product in Clifford algebra, called the *cross product*. The cross product of two multivectors x, y is $x \times y = (xy - yx)/2$, where the juxtaposition of multivectors represents the geometric product. We follow the notation $x^2 = xx$ for a multivector x, and call it the *square* of x.

For two multivectors x, y, if $xy = yx = 1$, we say x, y are both *invertible* and write $x = y^{-1}$, $y = x^{-1}$. Not all multivectors are invertible. For two multivectors x, y, if $xy^{-1} = y^{-1}x$, we write $xy^{-1} = y^{-1}x = x/y$.

A fundamental concept in Clifford algebra is r-*blade* (blade of *grade* r). Let $a_1, \ldots, a_r$ be vectors of V^n, then $a_1 \wedge \cdots \wedge a_r$ is called an r-blade if it is nonzero. The square of a blade is a scalar, which equals the inner product of the blade with itself. The *magnitude* of a blade x is defined as $|x| = \sqrt{|x^2|}$. The *reverse* of an r-blade x is defined as $x^\dagger = (-1)^{r(r-1)/2}x$. The *main anti-automorphism* of an r-blade x is $x^* = (-1)^r x$. The latter two concepts can be extended to any multivector by linearity.

Let x be an invertible blade. The *projection* of a multivector y onto x is defined by $P_x(y) = (y \cdot x)x^{-1}$. The *dual* of a multivector y with respect to x is defined by $yx^\dagger$. In particular, when x is an n-blade, called a *pseudoscalar*, the dual is denoted as $y^\sim$. The *meet* of two multivectors y, z is defined as $y \vee z = y^\sim \cdot z$.

The set of all r-blades where r is even, together with $\mathcal{R}$, generates a vector subspace $\mathcal{G}^+(V^n)$ of $\mathcal{G}(V^n)$. It is closed under the geometric product, and therefore is a subalgebra of $\mathcal{G}(V^n)$, called the *even subalgebra*.

Let $a_1, \ldots, a_r$ be invertible vectors of V^n. Then $a_1 \cdots a_r$ is called a *versor*. When r is even, $a_1 \cdots a_r$ is called a *spinor*. The *adjugation* of a multivector y by a versor x is defined as $Ad_x^*(y) = x^{*-1}yx$. When x is a spinor, since $Ad_x^*(y) = x^{-1}yx$, we can use a different symbol $Ad_x(y)$ to represent $Ad_x^*(y)$.

---

[1] The *signature* of a real inner product space is a triplet of nonnegative integers whose sum equals the dimension of the space; the first (second, third) integer equals the dimension of a maximal subspace where every vector x satisfies $x \cdot x > 0 \, (< 0, = 0)$. A real inner product space with signature (p, q, r) is often denoted by $\mathcal{R}^{p,q,r}$; when $r = 0$ it is denoted by $\mathcal{R}^{p,q}$; when $q = r = 0$ it is denoted by $\mathcal{R}^p$; when $p = r = 0$ it is denoted by $\mathcal{R}^{-q}$.

2 The Wachter Model for the Plane

2.1 Basic representations

Consider the Minkowski space $\mathcal{R}^{3,1}$. It contains the plane $\mathcal{R}^2$ as a subspace. The orthogonal complement of $\mathcal{R}^2$ in the Minkowski space is denoted as $\mathcal{R}^{1,1}$, which is a Minkowski plane. Let (f, f_0) be a hyperbolic pair of the Minkowski plane, i.e.,

$$f^2 = f_0^2 = 0, \quad f \cdot f_0 = -1. \tag{1}$$

The following mapping

$$A \mapsto \acute{A} = f_0 + A + \frac{A^2}{2}f, \tag{2}$$

maps a vector $A \in \mathcal{R}^2$ to a null vector $\acute{A}$ in $\mathcal{R}^{3,1}$, i.e., $\acute{A}^2 = 0$. A fundamental property of this mapping is that the mapping, being one-to-one and onto the set

$$\{x \in \mathcal{R}^{3,1} | x^2 = 0, \ x \cdot f = -1\}, \tag{3}$$

satisfies

$$\acute{A} \cdot \acute{B} = -\frac{|A - B|^2}{2}, \tag{4}$$

for $A, B \in \mathcal{R}^2$. The set (3) together with its connection with the plane (2), is called the *Wachter model* of the plane. (4) says that in the Wachter model, the square of the distance between two points on the plane can be represented by the inner product of their null-vector representations. This property decides the importance of the Wachter model in the study of distance geometry [12,35].

Another main application of the Wachter model is in the conformal geometry of the plane, i.e., in the study of the transformations that always map a line or circle to a line or circle. In the Wachter model, a line passing through two points A, B has the representation $f \wedge \acute{A} \wedge \acute{B}$; a circle passing through three points A, B, C has the representation $\acute{A} \wedge \acute{B} \wedge \acute{C}$. The representation is unique up to a nonzero scale.

To derive the above representations of lines and circles, we need the *Ptolemy's Theorem* in plane geometry, which says that four points A_i, $i = 1, \ldots, 4$, are on the same line or circle if and only if $\det(|A_i - A_j|^2)_{4\times 4} = 0$. Since

$$(\acute{A}_1 \wedge \acute{A}_2 \wedge \acute{A}_3 \wedge \acute{A}_4)^2 = \det(\acute{A}_i \cdot \acute{A}_j)_{4\times 4} = \frac{1}{16}\det(|A_i - A_j|^2)_{4\times 4}, \tag{5}$$

the four points A_i, $i = 1, \ldots, 4$, are on the same line or circle if and only if $\acute{A}_1 \wedge \acute{A}_2 \wedge \acute{A}_3 \wedge \acute{A}_4 = 0$. Therefore we can use $\acute{A} \wedge \acute{B} \wedge \acute{C}$ to represent the line passing through A, B, C if they are collinear, or the circle passing through them otherwise. A point D is on the line or circle if and only if $\acute{A} \wedge \acute{B} \wedge \acute{C} \wedge \acute{D} = 0$.

Next we show that the vector f is always on the blade $\acute{A} \wedge \acute{B} \wedge \acute{C}$ if points A, B, C are collinear. Assume that $A \neq B$, then $C = \lambda A + (1 - \lambda)B$ for some $\lambda \in \mathcal{R}$. As a result,

$$f \wedge \acute{A} \wedge \acute{B} \wedge \acute{C} = f \wedge (f_0 + A) \wedge (f_0 + B) \wedge (f_0 + \lambda A + (1 - \lambda)B) = 0. \tag{6}$$

This conclusion guarantees that we can use $f \wedge \acute{A} \wedge \acute{B}$ to represent line AB, as it differs from $\acute{A} \wedge \acute{B} \wedge \acute{C}$ only by a nonzero scale.

Indeed, the null vector f represents the point at infinity of the plane (not the concept in projective geometry), and is on every line of the plane. So line $f \wedge \acute{A} \wedge \acute{B}$ can also be explained as the circle passing through the point at infinity and points A, B. The null vector f_0 represents the origin of the plane, as it is the representation of the zero vector of the plane.

For a circle on the plane, the usual way to describe it is through its center and radius; for a line, the usually way is through its normal direction and distance from the origin. In the Wachter model, a circle with center O and radius ρ is represented by $(\acute{O} - (\rho^2/2)f)^\sim$; a line with unit normal n and n-distance δ (the signed distance from the origin to the line in the direction of n) is represented by $(n + \delta f)^\sim$.

These representations can be obtained as follows: let A be a point on the plane. First, since

$$\acute{A} \wedge (\acute{O} - \frac{\rho^2}{2} f)^\sim = (\acute{A} \cdot (\acute{O} - \frac{\rho^2}{2} f))^\sim = \frac{(\rho^2 - |A - O|^2)^\sim}{2}, \qquad (7)$$

A is on the circle with center O and radius ρ, i.e., $|A - O| = \rho$, if and only if $\acute{A} \wedge (\acute{O} - (\rho^2/2)f)^\sim = 0$, therefore the circle can be represented by $(\acute{O} - (\rho^2/2)f)^\sim$. Second, since

$$\acute{A} \wedge (n + \delta f)^\sim = (\acute{A} \cdot (n + \delta f))^\sim = (A \cdot n - \delta)^\sim, \qquad (8)$$

A is on the line with unit normal n and n-distance δ, i.e., $A \cdot n = \delta$, if and only if $\acute{A} \wedge (n + \delta f)^\sim = 0$, therefore the line can be represented by $(n + \delta f)^\sim$.

The representations of circles and lines through their geometric data, together with the previous representations through their points, enable us to derive the center and radius of a circle from its three points, and the normal and distance to the origin of a line from its two points. The formulas are provided in the following table, which is on representing basic geometric entities and constraints in the 2D geometry within the Wachter model.

Geometric entities and constraints	Representations
The origin	f_0
The point at infinity	f
The square of the distance between points A, B	$-2\acute{A} \cdot \acute{B}$

Line AB	$f \wedge \acute{A} \wedge \acute{B}$
Line (A, l), where l is the direction	$f \wedge \acute{A} \wedge l$
The line with unit normal n and signed distance δ away from the origin	$(n + \delta f)\tilde{\ }$
The ratio of collinear line segments AB and CD	$\dfrac{f \wedge \acute{A} \wedge \acute{B}}{f \wedge \acute{C} \wedge \acute{D}}$
The signed distance from line (A, l) to point B	$(f \wedge \acute{A} \wedge l \wedge \acute{B})\tilde{\ }$
The signed area of triangle ABC	$(f \wedge \acute{A} \wedge \acute{B} \wedge \acute{C})\tilde{\ }$
The inner product of vectors $A - B$ and $C - D$	$(f \wedge \acute{A} \wedge \acute{B}) \cdot (f \wedge \acute{C} \wedge \acute{D})$
The intersection of lines AB and CD	$\dfrac{(f \wedge \acute{A} \wedge \acute{B} \wedge \acute{C})\tilde{\ } D}{(f \wedge \acute{A} \wedge \acute{B} \wedge (\acute{C} - \acute{D}))\tilde{\ }}$ $- \dfrac{(f \wedge \acute{A} \wedge \acute{B} \wedge \acute{D})\tilde{\ } C}{(f \wedge \acute{A} \wedge \acute{B} \wedge (\acute{C} - \acute{D}))\tilde{\ }}$
The circle passing through points A, B, C	$\acute{A} \wedge \acute{B} \wedge \acute{C}$
The circle with center O and radius ρ	$(\acute{O} - \dfrac{\rho^2}{2} f)\tilde{\ }$
The center of circle ABC	$(\acute{A} \wedge \acute{B} \wedge \acute{C})^2 f$ $-2(\acute{A} \wedge \acute{B} \wedge \acute{C}) \cdot (f \wedge \acute{A} \wedge \acute{B} \wedge \acute{C})$
The square of the radius of circle ABC	$-\dfrac{(\acute{A} \wedge \acute{B} \wedge \acute{C})^2}{(f \wedge \acute{A} \wedge \acute{B} \wedge \acute{C})^2}$
The square of the distance between point D and the center of circle ABC	$\rho^2 - 2\dfrac{\acute{A} \wedge \acute{B} \wedge \acute{C} \wedge \acute{D}}{f \wedge \acute{A} \wedge \acute{B} \wedge \acute{C}}$
The signed distance from line $(n + \delta f)\tilde{\ }$ to the center of the circle ABC	$-\delta - \dfrac{n \wedge \acute{A} \wedge \acute{B} \wedge \acute{C}}{f \wedge \acute{A} \wedge \acute{B} \wedge \acute{C}}$
The intersection of circles ABC and DEF	$(\acute{A} \wedge \acute{B} \wedge \acute{C}) \vee (\acute{D} \wedge \acute{E} \wedge \acute{F})$

The intersection of line AB and circle CDE	$(f \wedge \acute{A} \wedge \acute{B}) \vee (\acute{C} \wedge \acute{D} \wedge \acute{E})$
The point of tangency of line AB and circle CDE	$((\acute{C} \wedge \acute{D} \wedge \acute{E}) \vee (f \wedge \acute{A} \wedge \acute{B})) \cdot (f \wedge \acute{A} \wedge \acute{B})$
The point of tangency of circles ABC and DEF	$((\acute{D} \wedge \acute{E} \wedge \acute{F}) \vee (\acute{A} \wedge \acute{B} \wedge \acute{C})) \cdot (\acute{A} \wedge \acute{B} \wedge \acute{C})$
The tangent line of circle ABC at point D	$f \wedge (\acute{D} \cdot (\acute{A} \wedge \acute{B} \wedge \acute{C}))$
The constraint that points A, B, C are collinear	$f \wedge \acute{A} \wedge \acute{B} \wedge \acute{C} = 0$
The constraint that points A, B, C, D are on the same circle	$\acute{A} \wedge \acute{B} \wedge \acute{C} \wedge \acute{D} = 0$
The constraint that lines AB, CD are parallel	$f \wedge \acute{A} \wedge \acute{B} \wedge \acute{C} = f \wedge \acute{A} \wedge \acute{B} \wedge \acute{D}$
The constraint that lines AB, CD are perpendicular	$(f \wedge \acute{A} \wedge \acute{B}) \cdot (f \wedge \acute{C} \wedge \acute{D}) = 0$
The constraint that lines AB, CD, EF are concurrent	$(f \wedge \acute{A} \wedge \acute{B} \wedge \acute{C}) \cdot (f \wedge \acute{D} \wedge \acute{E} \wedge \acute{F})$ $= (f \wedge \acute{A} \wedge \acute{B} \wedge \acute{D}) \cdot (f \wedge \acute{C} \wedge \acute{E} \wedge \acute{F})$

2.2 Conformal transformations

A basic theorem in 2D conformal geometry with the Wachter model is that any conformal transformation can be realized by the adjugation of a versor in $\mathcal{G}_{3,1}$, and any versor realizes a conformal transformation through its adjugation. Two versors induce the same conformal transformation if and only if they are the same up to a nonzero scalar or pseudoscalar factor [30]. In this section we provide Clifford algebra representations for several typical conformal transformations.

Let M be an algebraic entity in the Wachter model representing a geometric object on the plane, i.e., a point, line or circle.

(1) The *reflection* of M with respect to line AB is represented by $Ad^*_{f \wedge \acute{A} \wedge \acute{B}}(M)$, as $f \wedge \acute{A} \wedge \acute{B}$ represents the line.

(2) The *inversion* of M with respect to circle ABC is $Ad^*_{\acute{A} \wedge \acute{B} \wedge \acute{C}}(M)$.

(3) The *rotation* of M, centered at the origin and with signed angle θ relative to the orientation of the plane, is represented by $Ad_{e^{\theta I_2/2}}(M)$, where I_2 is the unit pseudoscalar representing the oriented plane [16].

(4) The *translation* of M along vector l is represented by $Ad_{1-f \wedge l/2}(M)$.

(5) The *dilation* of M, centered at the origin and with positive ratio e^λ, is represented by $Ad_{e^{-\lambda f \wedge f_0/2}}(M)$. For example, when $M = \acute{A}$ where A is a point, then $Ad_{e^{-\lambda f \wedge f_0/2}}(M)$ represents point $e^\lambda A$.

(6) The *dilation* of M, centered at the origin and with negative ratio $-e^\lambda$, is represented by $Ad_{(f \wedge f_0)e^{-\lambda f \wedge f_0/2}}(M)$. For example, when $M = \acute{A}$ where A is a point, then $Ad_{(f \wedge f_0)e^{-\lambda f \wedge f_0/2}}(M)$ represents point $-e^\lambda A$.

(7) The *special conformal transformation* of a point X along vector l is defined to be point $1/(1/X - 1/l)$. For a geometric entity M, its image under this transformation is represented by $Ad_{1+f_0 \wedge l/2}(M)$.

2.3 Geometric relations between two entities

The geometric entities are assumed to be points, lines and circles. The following results are direct applications of the theories proposed in [27].

1). Point D is inside circle ABC if and only if $\dfrac{\acute{A} \wedge \acute{B} \wedge \acute{C} \wedge \acute{D}}{f \wedge \acute{A} \wedge \acute{B} \wedge \acute{C}} > 0$. This can be obtained by comparing the radius of the circle with the distance between D and the center of the circle.

2). For circles ABC, DEF, they intersect, are tangent, or do not intersect if and only if

$$((\acute{A} \wedge \acute{B} \wedge \acute{C}) \vee (\acute{D} \wedge \acute{E} \wedge \acute{F}))^2 > 0, = 0, < 0,$$

respectively.

3). When circles ABC, DEF do not intersect, there is a pair of points that are inversive to each other with respect to both circles. The two points, called *Poncelet points*, correspond to the two null vectors

$$|\acute{A} \wedge \acute{B} \wedge \acute{C}||A_2|(\acute{A} \wedge \acute{B} \wedge \acute{C})^\sim \pm |\acute{D} \wedge \acute{E} \wedge \acute{F}|A_2 \cdot (\acute{A} \wedge \acute{B} \wedge \acute{C}),$$

where $A_2 = (\acute{A} \wedge \acute{B} \wedge \acute{C}) \vee (\acute{D} \wedge \acute{E} \wedge \acute{F})$.

4). For line AB and circle CDE, they intersect, are tangent or do not intersect if and only if

$$((f \wedge \acute{A} \wedge \acute{B}) \vee (\acute{C} \wedge \acute{D} \wedge \acute{E}))^2 > 0, = 0, < 0,$$

respectively.

5). For two intersecting circles ABC and DEF, where A, B, C and D, E, F are both ordered anticlockwise, the inner product of the two outward normal vectors of radius length of the circles at any point of the intersection equals

$$t = \frac{(\acute{A} \wedge \acute{B} \wedge \acute{C}) \cdot (\acute{D} \wedge \acute{E} \wedge \acute{F})}{(f \wedge \acute{A} \wedge \acute{B} \wedge \acute{C}) \cdot (f \wedge \acute{D} \wedge \acute{E} \wedge \acute{F})}.$$

Therefore t represents a geometric invariant. We say the two circles are *near*, *orthogonal*, or *far*, if $t > 0, = 0, < 0$, respectively.

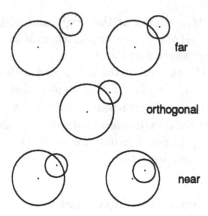

Fig. 1. The geometric relations between two circles.

2.4 Geometric relations among three entities

There is an algebraic expression characterizing most of the relations among three geometric entities. Let a, b, c be vectors in $\mathcal{R}^{3,1}$, and let

$$s(a, b, c) = (a \cdot b)(b \cdot c)(c \cdot a). \tag{9}$$

When a, b, c are null or of positive square, the sign of the scalar $s(a, b, c)$ is independent of the length and sign of each vector, and therefore is a geometric invariant. Below we discuss some inequality constraints of three geometric entities using this invariant. The proofs are omitted because they are direct applications of the tables in Section 2.1.

1. For points A, B, they are on different sides of line CD if and only if $s(\acute{A}, \acute{B}, (f \wedge \acute{C} \wedge \acute{D})^{\sim}) > 0$, i.e.,

$$(f \wedge \acute{A} \wedge \acute{C} \wedge \acute{D}) \cdot (f \wedge \acute{B} \wedge \acute{C} \wedge \acute{D}) < 0.$$

2. For points A, B, they are on different sides of circle CDE if and only if $s(\acute{A}, \acute{B}, (\acute{C} \wedge \acute{D} \wedge \acute{E})^{\sim}) > 0$, i.e.,

$$(\acute{A} \wedge \acute{C} \wedge \acute{D} \wedge \acute{E}) \cdot (\acute{B} \wedge \acute{C} \wedge \acute{D} \wedge \acute{E}) < 0.$$

3. For point A and parallel lines BC, DE, A is between the two lines if and only if

$$s(\acute{A}, (f \wedge \acute{B} \wedge \acute{C})^{\sim}, (f \wedge \acute{D} \wedge \acute{E})^{\sim}) < 0.$$

4. For point A and intersecting lines BC, DE, A is in one of the two acute-angled regions of the plane divided by the lines if and only if

$$s(\acute{A}, (f \wedge \acute{B} \wedge \acute{C})^{\sim}, (f \wedge \acute{D} \wedge \acute{E})^{\sim}) < 0.$$

5. For line AB, circle CDE whose center does not lie on AB, and point F inside the circle, F and the center of the circle are on the same side of the line if and only if

$$s(\acute{F}, (f \wedge \acute{A} \wedge \acute{B})^\sim, (\acute{C} \wedge \acute{D} \wedge \acute{E})^\sim) > 0.$$

6. For line AB, circle CDE whose center does not lie on AB, and point F outside the circle, F and the center of the circle are on the same side of the line if and only if

$$s(\acute{F}, (f \wedge \acute{A} \wedge \acute{B})^\sim, (\acute{C} \wedge \acute{D} \wedge \acute{E})^\sim) < 0.$$

7. For far circles ABC, DEF, point X is inside or outside both of them if and only if

$$s(\acute{X}, (\acute{A} \wedge \acute{B} \wedge \acute{C})^\sim, (\acute{D} \wedge \acute{E} \wedge \acute{F})^\sim) < 0.$$

8. For near circles ABC, DEF, point X is inside or outside both of them if and only if

$$s(\acute{X}, (\acute{A} \wedge \acute{B} \wedge \acute{C})^\sim, (\acute{D} \wedge \acute{E} \wedge \acute{F})^\sim) > 0.$$

9. For intersecting lines AB, CD, the center of circle EFG is in one of the two acute-angled regions of the plane divided by the lines if and only if

$$s((\acute{E} \wedge \acute{F} \wedge \acute{G})^\sim, (f \wedge \acute{A} \wedge \acute{B})^\sim, (f \wedge \acute{C} \wedge \acute{D})^\sim) < 0.$$

10. For parallel lines AB, CD, the center of circle EFG is between the lines if and only if

$$s((\acute{E} \wedge \acute{F} \wedge \acute{G})^\sim, (f \wedge \acute{A} \wedge \acute{B})^\sim, (f \wedge \acute{C} \wedge \acute{D})^\sim) < 0.$$

11. For far circles ABC, DEF, their centers are on the same side of line GH if and only if

$$s((f \wedge \acute{G} \wedge \acute{H})^\sim, (\acute{A} \wedge \acute{B} \wedge \acute{C})^\sim, (\acute{D} \wedge \acute{E} \wedge \acute{F})^\sim) < 0.$$

12. For near circles ABC, DEF, their centers are on the same side of line GH if and only if

$$s((f \wedge \acute{G} \wedge \acute{H})^\sim, (\acute{A} \wedge \acute{B} \wedge \acute{C})^\sim, (\acute{D} \wedge \acute{E} \wedge \acute{F})^\sim) > 0.$$

13. For three circles ABC, DEF, GHI, they are all far from each other, or two are far and both near to the third, if and only if

$$s((\acute{A} \wedge \acute{B} \wedge \acute{C})^\sim, (\acute{D} \wedge \acute{E} \wedge \acute{F})^\sim, (\acute{G} \wedge \acute{H} \wedge \acute{I})^\sim) < 0.$$

3 Connections with Other Clifford Algebra Models

Having obtained some feeling about the Wachter model, we now discuss the connections and differences of this model with other Clifford algebra models: the Clifford, Grassmann and degenerate models.

In what follows, within the framework of 2D geometry, we present in each model representations of typical geometric entities and constraints. Some typical geometric entities are: points, directions, angles, areas, distances, ratios, line segments, triangles, circles. Some typical geometric constraints are: collinearity of three points; concurrence of three lines; parallelism (perpendicularity) of two lines; co-circularity of four points.

In the Clifford model, a point A is represented by the vector from the origin, which is on the plane, to the point. A direction is represented by a vector. Also,

- an angle as a scalar, is represented by the inner product of two unit vectors along the sides of the angle;
- the signed area of triangle ABC is $((A - B) \wedge (A - C))^{\sim}/2$;
- the square of the distance between points A, B is $(A - B)^2$;
- the signed ratio of collinear line segments AB with CD is $(A - B)/(C - D)$;
- a line passing through point A and with unit direction l is represented by the pair (A, l), i.e., a point B is on the line if and only if $(B - A) \wedge l = 0$;
- the signed distance from line (A, l) to point C is $((C - A) \wedge l)^{\sim}$;
- the intersection of two lines $(A, l), (B, m)$ is $\dfrac{B \wedge m}{l \wedge m}l - \dfrac{A \wedge l}{l \wedge m}m$;
- a circle with center O and radius r is represented by the pair (O, r^2), i.e., a point A is on (inside) the circle if and only if $(A - O)^2 - r^2 = 0 \ (< 0)$;
- the projection of a point C onto a line (A, l) is $A + P_l(C - A) = A + ((C - A) \cdot l)l$;
- the reflection of point B with respect to line (A, l) is $A + Ad_l^*(B - A)$;
- the rotation of point A by angle θ is $Ad_{e^{\theta/2I_2}}(A)$;
- the translation of point A along vector l is $A + l$;
- the constraint that three points A, B, C are collinear is represented by $(A - B) \wedge (A - C) = 0$;
- the constraint that two lines AB, CD are parallel is represented by $(A - B) \wedge (C - D) = 0$;
- the constraint that two lines AB, CD are perpendicular is represented by $(A - B) \cdot (C - D) = 0$.

Since $\mathcal{R}^2$ is a subspace of $\mathcal{R}^{3,1}$, the Clifford model $\mathcal{G}_2$ is a subalgebra of the Clifford algebra $\mathcal{G}_{3,1}$ realizing the Wachter model. So this model is naturally included in the Wachter model. Another realization of the Clifford model is to use the fact that a point A in the Clifford model is represented by the vector from the origin to the point, which corresponds to the directed line segment $f \wedge f_0 \wedge \acute{A}$. The set $\{f \wedge f_0 \wedge \acute{A} | A \in \mathcal{R}^2\}$, together with the inner product in $\mathcal{G}_{3,1}$ and the following outer product g:

$$g(f \wedge f_0 \wedge \acute{A}, f \wedge f_0 \wedge \acute{B}) = f \wedge f_0 \wedge \acute{A} \wedge \acute{B},$$

for $A, B \in \mathcal{R}^2$, realize the Clifford model within the Wachter model. This realization is called the *theorem of conformal split* [19] of $\mathcal{R}^2$.

We see that in the Clifford model, circles and lines do not have algebraic representations that can join algebraic manipulations as single terms. Also the representations of affine properties, for example the signed area, the collinear constraint, are not as simple as in the Wachter model.

To describe affine properties, it is convenient to embed the plane into a 3D vector space, which is denoted by $\mathcal{R}^2 \oplus \langle e_0 \rangle$, where vector e_0 is orthogonal to $\mathcal{R}^2$. By doing so we can represent an affine transformation with a 3×3 matrix.

When we choose $e_0^2 = 1$, we obtain the Grassmann model, where the plane is the one in $\mathcal{R}^3 = \mathcal{R}^2 \oplus \langle e_0 \rangle$ passing through the end of vector e_0. A point in the plane corresponds to the vector from the origin to the point. A direction is represented by a vector in $\mathcal{R}^2$. Moreover,

- line AB is represented by $A \wedge B$;
- the signed area of triangle ABC is $(A \wedge B \wedge C)^\sim / 2$;
- the signed ratio of collinear line segments AB with CD is $A \wedge B / C \wedge D$;
- the constraint that three points A, B, C are collinear is represented by $A \wedge B \wedge C = 0$;
- the constraint that two lines AB, CD are parallel is represented by $A \wedge C \wedge D = B \wedge C \wedge D$;
- the constraint that three lines AB, CD, EF are concurrent is represented by $(A \wedge B \wedge C)^\sim (D \wedge E \wedge F)^\sim = (A \wedge B \wedge D)^\sim (C \wedge E \wedge F)^\sim$;
- vector a represents a point if and only if $\partial(a) = e_0 \cdot a = 1$; it represents a direction if and only if $\partial(a) = 0$;
- the direction of line AB is represented by $\partial(A \wedge B) = e_0 \cdot (A \wedge B) = B - A$;
- the square of the distance between points A, B is $(\partial(A \wedge B))^2$;
- the distance between line AB and point C is $|A \wedge B \wedge C| / |\partial(A \wedge B)|$;
- a circle with center O and radius r is represented by the pair (O, r^2);
- the intersection of two lines AB and CD is $(A \wedge B) \vee (C \wedge D) / \partial((A \wedge B) \vee (C \wedge D))$;
- the constraint that two lines AB, CD are perpendicular is represented by $\partial(A \wedge B) \cdot \partial(C \wedge D) = 0$.

Compared with the Clifford model, the simplicity in representing affine properties here is obvious. One drawback of the Grassmann model is that the inner product has geometric meaning only when restricted to vectors in the plane. Another drawback is the representation of a circle. This model can be realized within the Wachter model, as will be shown in Example 3 of the last section.

When we choose $e_0^2 = 0$, we get the degenerate model. Historically, this model occurred in the form of dual quaternions and dual vector algebra, and has important applications in 3D geometry and mechanics. In this model, a point A is represented by multivector $1 + e_0 \wedge A$, where A is the vector in $\mathcal{R}^2$ from the origin to the point. The origin is represented by scalar 1. Moreover,

- the signed area of triangle ABC is $((A - B) \wedge (A - C))^\sim / 2$;
- the square of the distance between points A, B is $(A - B)^2$;

- the signed ratio of collinear line segments AB with CD is $(A-B)/(C-D)$;
- a circle with center O and radius r is represented by the pair (O, r^2);
- a line passing through point A and with direction l is represented by $l + e_0 \wedge A \wedge l$;
- the reflection of point C with respect to line (A, l) is $(l + e_0 \wedge A \wedge l)(1 - e_0 \wedge C)(l + e_0 \wedge A \wedge l) = 1 + e_0 \wedge (2A - C - 2((A - C) \cdot l)l)$;
- the rotation of a point C induced by a spinor U is $Ad_U(1 + e_0 \wedge C) = 1 + e_0 \wedge Ad_U(C)$;
- the translation of a point C along a vector l is $(1 + e_0 \wedge l/2)(1 + e_0 \wedge C)(1 + e_0 \wedge l/2) = 1 + e_0 \wedge (C + l)$;
- the constraint that three points A, B, C are collinear is represented by $(A - B) \wedge (A - C) = 0$;
- the constraint that two lines AB, CD are parallel is represented by $(A - B) \wedge (C - D) = 0$;
- the constraint that two lines AB, CD are perpendicular is represented by $(A - B) \cdot (C - D) = 0$.

We see that in many representations in this model, we have to use the representations of points and lines in the Clifford model, instead of the ones in the model itself. We also have to represent a circle with a pair of algebraic entities. Another drawback is that the Euclidean transformations of the plane do not have a uniform representation, as can be seen from the representations of reflections, rotations and translations in the previous list.

One advantage of this model is that a line has an algebraic representation that can be used directly in algebraic manipulations, although not of a single term. The degenerate model is more suitable for computations on Euclidean transformations of lines. This model can also be realized in the Wachter model, as will be shown in Example 4 in the last section.

The Clifford, Grassmann and degenerate models can all be realized within the Wachter model. The Wachter model is capable of preserving the advantages of the other three models while making the representations and computations of distances, lines, planes, circles and spheres more easily. We shall given more examples in later sections.

4 The Wachter Model for the Space

4.1 Points, lines, planes, circles and spheres

Similar to the 2D case, now consider the Minkowski space $\mathcal{R}^{4,1}$, which contains the space $\mathcal{R}^3$ as a subspace. Let (f, f_0) be a hyperbolic pair of the Minkowski plane which is the orthogonal complement of $\mathcal{R}^3$. The set $\{x \in \mathcal{R}^{3,1} | x^2 = 0, \ x \cdot f = -1\}$, together with the mapping $A \mapsto \acute{A} = f_0 + A + (A^2/2)f$, forms a model for the Euclidean space, called the *Wachter model* of the space.

In this model, the representations of a line and a circle by their points are the same as in the 2D case. A plane passing through three points A, B, C has the representation $f \wedge \acute{A} \wedge \acute{B} \wedge \acute{C}$; a sphere passing through four points A, B, C, D has the representation $\acute{A} \wedge \acute{B} \wedge \acute{C} \wedge \acute{D}$. Using geometric data, a sphere with center O and radius ρ is represented by $(\acute{O} - (\rho^2/2)f)^\sim$; a plane with unit normal n and n-distance δ is represented by $(n + \delta f)^\sim$.

A line with direction l and n-distance δ, where n is a unit normal of the line and is in the plane decided by the line and the origin of the space, can be represented by $f \wedge f_0 \wedge l + \delta f \wedge n \wedge l$. This is because points δn and $\delta n + l$ are both on the line. A circle on plane ABC, with center O and radius ρ, is the intersection of the plane with the sphere whose center is O and whose radius is ρ, therefore can be represented by $(\acute{O} - (\rho^2/2)f)^\sim \vee (f \wedge \acute{A} \wedge \acute{B} \wedge \acute{C})$.

The following table gives more representations of typical geometric entities in the 3D geometry.

Geometric entities and constraints	Representations
Point A	$\acute{A}$
Line AB	$f \wedge \acute{A} \wedge \acute{B}$
Circle ABC	$\acute{A} \wedge \acute{B} \wedge \acute{C}$
Plane ABC	$f \wedge \acute{A} \wedge \acute{B} \wedge \acute{C}$
The plane normal to unit vector n and with n-distance δ	$(n + \delta f)^\sim$
The plane passing through point A and normal to vector n	$(n + (A \cdot n)f)^\sim$
The plane passing through point A and directions l, m	$f \wedge \acute{A} \wedge l \wedge m$
The normal vector of plane ABC	$(f \wedge \acute{A} \wedge \acute{B} \wedge \acute{C})^\sim - (f \wedge f_0 \wedge \acute{A} \wedge \acute{B} \wedge \acute{C})^\sim f$
The signed distance from the origin to plane ABC	$\dfrac{(f_0 \wedge \acute{A} \wedge \acute{B} \wedge \acute{C})^\sim}{\|f \wedge \acute{A} \wedge \acute{B} \wedge \acute{C}\|}$
Sphere $ABCD$	$\acute{A} \wedge \acute{B} \wedge \acute{C} \wedge \acute{D}$

The sphere with center O and radius ρ	$(\acute{O} - \dfrac{\rho^2}{2}f)^{\sim}$
The center of sphere $ABCD$	$(\acute{A} \wedge \acute{B} \wedge \acute{C} \wedge \acute{D})^2 f - 2(\acute{A} \wedge \acute{B} \wedge \acute{C} \wedge \acute{D}) \cdot (f \wedge \acute{A} \wedge \acute{B} \wedge \acute{C} \wedge \acute{D})$
The square of the radius of sphere $ABCD$	$-\dfrac{(\acute{A} \wedge \acute{B} \wedge \acute{C} \wedge \acute{D})^2}{(f \wedge \acute{A} \wedge \acute{B} \wedge \acute{C} \wedge \acute{D})^2}$
The circle on plane ABC, with center O and radius ρ	$(\acute{O} - \dfrac{\rho^2}{2}f) \cdot (f \wedge \acute{A} \wedge \acute{B} \wedge \acute{C})$
The common perpendicular of non-coplanar lines AB and CD	$f \wedge (\acute{A}(f \wedge \acute{B} \wedge \acute{C} \wedge \acute{D}) \cdot N_4$ $-\acute{B}(f \wedge \acute{A} \wedge \acute{C} \wedge \acute{D}) \cdot N_4)$ $\wedge (\acute{C}(f \wedge \acute{A} \wedge \acute{B} \wedge \acute{D}) \cdot N_4$ $-\acute{D}(f \wedge \acute{A} \wedge \acute{B} \wedge \acute{C}) \cdot N_4)$, where $N_4 = f \wedge f_0 \wedge (\acute{A} - \acute{B}) \wedge (\acute{C} - \acute{D})$
The distance between two non-coplanar lines AB, CD	$\dfrac{|f \wedge \acute{A} \wedge \acute{B} \wedge \acute{C} \wedge \acute{D}|}{|(A - B) \wedge (C - D)|}$
The ratio of the signed area of coplanar triangles ABC and DEF	$\dfrac{f \wedge \acute{A} \wedge \acute{B} \wedge \acute{C}}{f \wedge \acute{D} \wedge \acute{E} \wedge \acute{F}}$
The signed volume of tetrahedron $ABCD$	$(f \wedge \acute{A} \wedge \acute{B} \wedge \acute{C} \wedge \acute{D})^{\sim}$

4.2 The affine representations of points, lines and planes

For a geometric problem involving only points, lines and planes, it is better to use $f \wedge \acute{A}$ instead of $\acute{A}$ to represent point A. The reason is, the set $\{f \wedge \acute{A} | A \in \mathcal{R}^3\}$ is affine:

$$\lambda f \wedge \acute{B} + (1 - \lambda)f \wedge \acute{C} = f \wedge \acute{D},$$

where $D = \lambda B + (1 - \lambda)C$. This representation is called the *affine representation* of a point. The *affine representation* of a line (plane) is the representation used in the Wachter model.

We mention only one application here. Let A_r be a Minkowski r-blade, then $f \wedge A_r$ represents a point, line, or plane when $r = 1, 2$, or 3. For two such blades A_r, A_s, where $r < s$, the projection of $f \wedge A_r$ onto $f \wedge A_s$ is

$$P_{f \wedge A_s}(f \wedge A_r) = f \wedge P_{f \wedge A_s}(A_r). \tag{10}$$

It is still a Minkowski r-blade, and represents a geometric entity. When explained geometrically, (10) says that the feet drawn from point A to line BC and plane

BCD are $f \wedge P_{f \wedge B \wedge \dot{C}}(\dot{A})$ and $f \wedge P_{f \wedge \dot{B} \wedge \dot{C} \wedge \dot{D}}(\dot{A})$ respectively, and the projection of line AB onto plane CDE is $f \wedge P_{f \wedge \dot{C} \wedge \dot{D} \wedge \dot{E}}(\dot{A}) \wedge P_{f \wedge \dot{C} \wedge \dot{D} \wedge \dot{E}}(\dot{B})$.

5 Applications

Example 1. Simson triangle.

Let ABC be a triangle, D be a point on the plane. Draw perpendicular lines from D to the three sides AB, BC, CA of triangle ABC. Let C_1, A_1, B_1 be the three feet respectively. Triangle $A_1 B_1 C_1$ is the so-called *Simson triangle*.

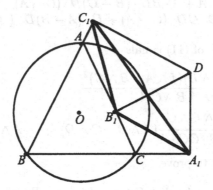

Fig. 2. Simson triangle.

When $A, B, C, D, A_1, B_1, C_1$ are understood to be null vectors, we have the following conclusion:

$$f \wedge A_1 \wedge B_1 \wedge C_1 = \frac{A \wedge B \wedge C \wedge D}{2\rho^2}, \tag{11}$$

where ρ is the radius of circle ABC. (11) can be used to compute the signed area of the Simson triangle.

An immediate corollary of (11) is: when D is constrained on a circle concentric with circle ABC, the area of the Simson triangle is constant. In particular, if D is on circle ABC, then A_1, B_1, C_1 are collinear, which is the classical *Simson's theorem*.

Now we prove (11). Point A_1 is the foot drawn from point D to line BC, therefore,

$$f \wedge A_1 = P_{f \wedge B \wedge C}(f \wedge D). \tag{12}$$

A direct computation gives

$$f \wedge A_1 = f \wedge P_{f \wedge B \wedge C}(D)$$
$$= f \wedge (\frac{B+C}{2} - \frac{D \cdot (B-C)(B-C)}{2B \cdot C}). \tag{13}$$

Similarly we have

$$f \wedge B_1 = f \wedge (\frac{C+A}{2} - \frac{D \cdot (C-A)(C-A)}{2C \cdot A}),$$

$$f \wedge C_1 = f \wedge (\frac{A+B}{2} - \frac{D \cdot (A-B)(A-B)}{2A \cdot B}). \tag{14}$$

Now

$$f \wedge A_1 \wedge B_1 \wedge C_1 = \frac{f \wedge A \wedge B \wedge C}{4A \cdot BB \cdot CC \cdot A} \Gamma, \tag{15}$$

where

$$\Gamma = A \cdot BB \cdot CC \cdot A + A \cdot BD \cdot (B-C)D \cdot (C-A) \\ + D \cdot (A-B)B \cdot CD \cdot (C-A) + D \cdot (A-B)D \cdot (B-C)C \cdot A. \tag{16}$$

The right-hand side of (11) equals

$$-\frac{(A \wedge B \wedge C \wedge D)(f \wedge A \wedge B \wedge C)^2}{2(A \wedge B \wedge C)^2} \\ = \frac{f \wedge A \wedge B \wedge C}{4A \cdot BB \cdot CC \cdot A}(A \wedge B \wedge C \wedge D) \cdot (f \wedge A \wedge B \wedge C), \tag{17}$$

therefore we only need to prove

$$\Gamma = (A \wedge B \wedge C \wedge D) \cdot (f \wedge A \wedge B \wedge C). \tag{18}$$

The only constraint among vectors f, A, B, C, D is that they are in the same 4D space, i.e., $f \wedge A \wedge B \wedge C \wedge D = 0$. So

$$(f \wedge A \wedge B \wedge C \wedge D) \cdot (f \wedge A \wedge B \wedge C \wedge D) = 0. \tag{19}$$

We are going to see that the equality (18) is identical to (19). This can be proved in various ways, for example, we can use affine coordinates of D with A, B, C as the affine basis, and express everything with coordinates, or we can use the method of straightening laws in [31], by which we can keep geometric meaning in the proving procedure. The following are some direct computations without expanding everything to inner products of vectors.

First, we expand the left-hand side of (19) in the following way:

$$((f \wedge A \wedge B \wedge C \wedge D) \cdot f) \cdot (A \wedge B \wedge C \wedge D) \\ = -(f \wedge A \wedge B \wedge C) \cdot (A \wedge B \wedge C \wedge D) \\ + (f \wedge (A \wedge B + B \wedge C + C \wedge A) \wedge D) \cdot (A \wedge B \wedge C \wedge D). \tag{20}$$

From this, using

$$(f \wedge A \wedge B \wedge C) \cdot (A \wedge B \wedge C \wedge D) \\ = -(A \wedge B \wedge C) \cdot (f \cdot (A \wedge B \wedge C \wedge D)) \\ = (A \wedge B \wedge C) \cdot ((A \wedge B + B \wedge C + C \wedge A) \wedge D) - (A \wedge B \wedge C)^2 \tag{21}$$

and

$$(f \wedge (A \wedge B + B \wedge C + C \wedge A) \wedge D) \cdot (A \wedge B \wedge C \wedge D)$$
$$= -(A \wedge B + B \wedge C + C \wedge A) \wedge D) \cdot (f \cdot (A \wedge B \wedge C \wedge D))$$
$$= (D \wedge (A - B) \wedge (B - C))^2 - (A \wedge B \wedge C) \cdot ((A \wedge B + B \wedge C + C \wedge A) \wedge D), \tag{22}$$

we obtain

$$(f \wedge A \wedge B \wedge C \wedge D)^2 = (D \wedge (A - B) \wedge (B - C))^2 + (A \wedge B \wedge C)^2$$
$$-2(A \wedge B \wedge C) \cdot ((A \wedge B + B \wedge C + C \wedge A) \wedge D). \tag{23}$$

Second, using (21), together with

$$(A \wedge B \wedge C)^2 = -2A \cdot BB \cdot CC \cdot A \tag{24}$$

and

$$(D \wedge (A - B) \wedge (B - C))^2 = 2(A \cdot BD \cdot (B - C)D \cdot (C - A)$$
$$+D \cdot (A - B)B \cdot CD \cdot (C - A) + D \cdot (A - B)D \cdot (B - C)C \cdot A), \tag{25}$$

we obtain

$$\Gamma - (A \wedge B \wedge C \wedge D) \cdot (f \wedge A \wedge B \wedge C)$$
$$= \frac{1}{2}(D \wedge (A - B) \wedge (B - C))^2 + (A \wedge B \wedge C)^2)$$
$$-(A \wedge B \wedge C) \cdot ((A \wedge B + B \wedge C + C \wedge A) \wedge D) \tag{26}$$
$$= \frac{1}{2}(f \wedge A \wedge B \wedge C \wedge D)^2.$$

This ends the proof.

Example 2. The direct kinematic problem of a Steward platform.

This example is cited from the work of B. Mourrain [31–33]. Consider a platform attached to the ground by six extensible bars. Changing the lengths of the bars slightly entails a small displacement of the platform so that this mechanism is suited for doing precise movements of adjustment. This parallel robot, also called Steawrd platform or left hand, can be used as flying simulator or hand of a serial robot.

The direct kinematic problem of this robot can be described as follows: let X_i, $i = 1, \ldots, 6$ be the endpoints of the bars on the ground, Z_i, $i = 1, \ldots, 6$ be the points attached to the platform, which is moved by a displacement $D = (R, T)$, where R is a 3D rotation, T is a 3D parallel displacement. The problem is, given the lengths of the six bars $|X_i - Z_i| = l_i$, $i = 1, \ldots, 6$, find the displacement D.

In the Wachter model, the constraint that every X_i is transformed to Z_i by a rigid body motion, where $i = 1, \ldots, 6$, can be expressed by

$$U^{-1}fU = f, \quad \acute{Z}_i = U^{-1}\acute{X}_iU,$$

where U is a spinor in $\mathcal{G}_{4,1}$. The constraint $|X_i - Z_i| = l_i$ is equivalent to

$$\acute{Z}_i \cdot \acute{X}_i = (U^{-1}\acute{X}_iU) \cdot \acute{X}_i = -\frac{l_i^2}{2}.$$

Using the above formulation and results from algebraic geometry, it can be proved that the number of solutions for D is at most 40 [33].

Example 3. The Grassmann model of $\mathcal{R}^n$ within the Wachter model.

In the Wachter model of $\mathcal{R}^n$, let f, f_0 be the two null vectors representing the point at infinity and the origin respectively. Let $F = \{A + f_0 | A \in \mathcal{R}^n\}$. Then F is an affine hyperplane of $\mathcal{R}^n \oplus \langle f_0 \rangle$. The mapping

$$A \mapsto A + f_0, \text{ for } A \in \mathcal{R}^n \tag{27}$$

is one-to-one and onto F. It maps the origin to f_0.

Define an operator ∂ in $\mathcal{G}(\mathcal{R}^n \oplus \langle f_0 \rangle)$ as follows:

$$\partial(x) = -f \cdot x,$$

for $x \in \mathcal{G}(\mathcal{R}^n \oplus \langle f_0 \rangle)$. Then $\partial(A) = 0$ and $\partial(A + f_0) = 1$ for $A \in \mathcal{R}^n$. The Clifford algebra $\mathcal{G}(\mathcal{R}^n \oplus \langle f_0 \rangle)$, together with the mapping (27) and the operator ∂, constitutes a model that is identifiable with the Grassmann model of $\mathcal{R}^n$.

There is another realization of the Grassmann model within the Wachter model, by means of the affine representations of points, lines, planes, etc. In the Wachter model, an $(r-1)$-dimensional plane passing through r points $A_1, \ldots, A_r$ can be represented by

$$\begin{aligned}
f \wedge \acute{A}_1 \wedge \cdots \wedge \acute{A}_r \\
= f \wedge A_1 \wedge \cdots \wedge A_r + f \wedge f_0 \wedge (A_2 - A_1) \wedge \cdots \wedge (A_r - A_1).
\end{aligned} \tag{28}$$

In the set $\{f \wedge x | x \in \mathcal{G}_{n+1,1}\}$ we define the following outer product g:

$$g(f \wedge x, f \wedge y) = f \wedge x \wedge y,$$

for $x, y \in \mathcal{G}_{n+1,1}$. We also define an operator ∂ in it:

$$\partial(f \wedge x) = (f \wedge f_0) \cdot (f \wedge x),$$

for $x \in \mathcal{G}_{n+1,1} - \mathcal{R}$, and $\partial(\lambda f) = 0$ for $\lambda \in \mathcal{R}$. Then $\partial(f \wedge A) = 0$ and $\partial(f \wedge \acute{A}) = 1$ for $A \in \mathcal{R}^n$. The Clifford algebra $\mathcal{G}_{n+1,1}$, together with the affine representations, the outer product g and the operator ∂, provides another realization of the Grassmann model.

Example 4. The degenerate model of the space.

The degenerate model can be derived from the affine representations of points, lines, planes, etc. Let $A_1, \ldots, A_r$ be r points in the space. Then (28) gives a representation of the $(r-1)$-dimensional plane passing through them. The dual of this representation with respect to $f_0 \wedge f$ gives

$$\begin{aligned}
(f \wedge \acute{A}_1 \wedge \cdots \wedge \acute{A}_r)(f \wedge f_0) = \\
(A_2 - A_1) \wedge \cdots \wedge (A_r - A_1) + f \wedge A_1 \wedge \cdots \wedge A_r.
\end{aligned} \tag{29}$$

This is exactly the degenerate model of the space. For example, point A is represented by $1 + f \wedge A$, line AB is represented by $B - A + f \wedge A \wedge B$, plane ABC is represented by $A \wedge B + B \wedge C + C \wedge A + f \wedge A \wedge B \wedge C$.

One advantage of the degenerate model is that only one null vector occurs in the representation, which will simplify computation when no dilation is involved. To compensate for this advantage, we need to redefine the action of a versor on a geometric entity M as follows: let U be a versor inducing a Euclidean transformation. Let $M' = M(f \wedge f_0)$. Then the image of M' under the transformation is

$$Ad_U^*(M') = U^{*-1} M \overline{U}, \tag{30}$$

where the overbar represents an outermorphism that satisfies $\overline{A} = A$ for $A \in \mathcal{R}^3$, and $\overline{f} = -f$. An *outermorphism* g in a Clifford algebra is a linear transformation within the Clifford algebra that satisfies

$$g(a_1 \wedge \ldots \wedge a_r) = g(a_1) \wedge \cdots \wedge g(a_r),$$

for any multivectors $a_1, \ldots, a_r$.

Now we can define a new spinor action "Ad^{**}" as follows:

$$Ad_V^{**}(M) = V^{**-1} M V, \tag{31}$$

where $V = \overline{U}$, the double-star symbol represents an outermorphism that satisfies $A^{**} = -A$ for $A \in \mathcal{R}^3$, and $f^{**} = f$. In this way, the spinor inducing the rotation from direction m to direction n is $m(m+n)$, the spinor inducing the translation along vector l is $1 + f \wedge l/2$.

The geometric relationship between two geometric entities M_1, M_2 is $M_1 M_2^\dagger$.

Example 5. Dual quaternions.

Dual quaternions was first used by W. K. Clifford [6] in geometries. Later on it was used to solve problems in mechanics [5]. The definition of dual quaternions is: let $\mathcal{Q}$ be the set of quaternions, ϵ be a nilpotent algebraic element which commutes with every quaternion. The set $\mathcal{Q} + \epsilon \mathcal{Q}$ is called *dual quaternions*. The set $\mathcal{R} + \epsilon \mathcal{R}$ is called *dual numbers*.

As shown in [17], quaternions are naturally included in the Clifford model of the space: let $\{e_1, e_2, e_3\}$ be an orthonormal basis of $\mathcal{R}^3$. Then the correspondence: $i \mapsto e_1 \wedge e_2, j \mapsto e_2 \wedge e_3, k \mapsto e_1 \wedge e_3$, gives a Clifford algebra isomorphism of $\mathcal{Q}$ and $\mathcal{G}_3^+$.

Dual quaternions is another form of the degenerate model of the space. In the degenerate model, let $\epsilon = f \wedge I_3$, where I_3 is a unit pseudo-scalar of $\mathcal{G}_3$. Then $\mathcal{G}_3^+$ is a vector space spanned by $1, e_1 \wedge e_2, e_2 \wedge e_3, e_1 \wedge e_3$, and $\epsilon \mathcal{G}_3^+$ is a vector space spanned by $f \wedge e_1, f \wedge e_2, f \wedge e_3, f \wedge I_3$. Obviously ϵ commutes with everything in $\mathcal{G}_3^+$. As a consequence, $\mathcal{G}^+(f \wedge I_3) = \mathcal{G}_3^+ + \epsilon \mathcal{G}_3^+$ is isomorphic to dual quaternions as Clifford algebras.

In the degenerate model, the representations of point A, line (A, l) and plane (A, l, m) can be written as

$$
\begin{aligned}
&1 + f \wedge A = 1 + \epsilon A^\perp; \\
&(l + f \wedge A \wedge l)^\perp = l^\perp + \epsilon A^\perp \times l^\perp; \\
&l \wedge m + f \wedge A \wedge l \wedge m = l \wedge m + \epsilon (A \wedge l \wedge m)^\perp.
\end{aligned}
\tag{32}
$$

Here "$\perp$" represents the dual in $\mathcal{G}_3$. From this we get the dual quaternions representation of the space: point A is represented by

$$1 + \epsilon A^{\perp} \in 1 + \epsilon Q_0, \tag{33}$$

where Q_0 is the set of pure quaternions together with zero; line (A, l), where l is a unit vector, is represented by

$$l^{\perp} + \epsilon A^{\perp} \times l^{\perp} \in \hat{Q}_0 + \epsilon Q_0, \tag{34}$$

where $\hat{Q}_0$ is the set of unit pure quaternions; a plane passing through point A and normal to unit vector n is represented by

$$n^{\perp} + \epsilon n^{\perp} \cdot A^{\perp} \in \hat{Q}_0 + \epsilon \mathcal{R}. \tag{35}$$

In the dual quaternions model, the action of versors needs to be redefined. For a point or plane, the action of a versor U is Ad_U^{**}. For a line, the action is $Ad_{\overline{U}}$. More explicitly, the versor generating the reflection with respect to a plane with unit normal n and n-distance δ is

$$U = (n - \delta f)^{\perp} = n^{\perp} + \epsilon \delta \in \hat{Q}_0 + \epsilon \mathcal{R}, \tag{36}$$

the image of geometric entity M, which is either a point, a line, or a plane, under this reflection is

$$\overline{UMU}, \text{ or } \overline{UMU^{-1}}, \text{ or } \overline{UMU}. \tag{37}$$

The spinor generating the rotation from point A to point B centering at the origin is

$$V = -A(A + B) = A^{\perp}(A^{\perp} + B^{\perp}) \in Q, \tag{38}$$

the image of geometric entity M under this rotation is

$$V^{-1}MV. \tag{39}$$

The spinor generating the translation from point A to point B is

$$W = 1 + f\frac{B - A}{2} = 1 + \epsilon\frac{B^{\perp} - A^{\perp}}{2} \in 1 + \epsilon Q_0, \tag{40}$$

the image of geometric entity M, which is either a point, a straight line, or a plane, under this translation is

$$WMW, \text{ or } W^{-1}MW, \text{ or } WMW. \tag{41}$$

As to the geometric relationship between two entities M_1, M_2, if M_1 is not a line but M_2 is, the relationship is represented by $M_1 M_2$, otherwise it is represented by $M_1 M_2^{\dagger}$.

Example 6. Dual vector algebra.

First let us explain what is dual vector algebra. We know that a vector algebra is a three dimensional vector space $\mathcal{V}^3$ equipped with a inner product and a cross product of vectors. Now let ϵ be an algebraic nilpotent element that commutes with everything. The set $\mathcal{V}^3 + \epsilon\mathcal{V}^3$ is called *dual vectors*. Dual vectors, equipped with the inner product and cross product from the vector algebra, form the *dual vector algebra* of $\mathcal{V}^3$.

Let us start from the dual quaternions model of $\mathcal{R}^3$. We know that $\mathcal{Q}_0$, the set of pure quaternions and zero, is isomorphic to $\mathcal{R}^{-3}$ as vector algebra, under the correspondence between $\{i, j, k\}$ and an orthonormal basis of $\mathcal{R}^{-3}$, and the correspondences between the inner products and the cross products in $\mathcal{Q}_0$ and $\mathcal{R}^{-3}$ respectively. As a corollary, $\mathcal{Q}_0 + \epsilon\mathcal{Q}_0$ and the dual vector algebra of $\mathcal{R}^{-3}$ are isomorphic as vector algebras. This isometry identifies the dual quaternions model of $\mathcal{R}^3$ with the dual vector algebra model of $\mathcal{R}^{-3}$: point A is represented by

$$1 + \epsilon A; \tag{42}$$

line (A, l) is represented by

$$l + \epsilon A \times l, \tag{43}$$

where "$\times$" is the cross product in vector algebra; the plane passing through point A and normal to vector n is represented by

$$n + \epsilon n \cdot A. \tag{44}$$

The geometric product of two dual vectors x, y satisfies

$$xy = x \cdot y + x \times y. \tag{45}$$

If we use the inner product in $\mathcal{R}^3$ for vectors, we get the dual vector algebra of $\mathcal{R}^3$.

One application of dual vector algebra in geometry is the computation of two non-coplanar lines [37, 38]. For lines (A, l), (B, m), let n be the unit vector in the direction of $l \times m$, δ be the distance between the two lines, θ be the angle from l to m. Let P be the intersection of the common perpendicular of the two lines with line (A, l). Then

$$(l + \epsilon A \times l)(m + \epsilon B \times m) = e^{\tilde{\theta}\tilde{n}}, \tag{46}$$

where

$$e^{\tilde{\theta}\tilde{n}} = \cos\tilde{\theta} + \tilde{n}\sin\tilde{\theta}, \tag{47}$$

$\tilde{\theta} = \theta + \epsilon\delta$, $\tilde{n} = n + \epsilon P \times n$, and

$$\begin{aligned} \cos\tilde{\theta} &= \cos\theta - \epsilon\delta\sin\theta, \\ \sin\tilde{\theta} &= \sin\theta + \epsilon\delta\cos\theta. \end{aligned} \tag{48}$$

Formula (46) is equivalent to the following result by the Wachter model:

$$\begin{aligned} (f \wedge A \wedge l)&(f \wedge B \wedge m) \\ &= l \cdot m + f \wedge (B - A) \wedge l \wedge m + l \wedge m \\ &\quad + l \cdot mf \wedge (B - A) + f \wedge (A \cdot ml - B \cdot lm). \end{aligned} \tag{49}$$

The general form of (49) in nD geometry is [27]: let I_r, I_s be r-blade, s-blade in $\mathcal{G}_n$ respectively, $0 \le r, s \le n$, then

$$
\begin{aligned}
(f \wedge \acute{A} \wedge I_r)&(f \wedge \acute{B} \wedge I_s) \\
&= ((f \wedge \acute{A})(f \wedge \acute{B})) \wedge (P_{(f \wedge \acute{A})\sim}(I_r) P_{(f \wedge \acute{B})\sim}(I_s)).
\end{aligned}
\tag{50}
$$

References

1. E. Bayro-Corrochano, J. Lasenby and G. Sommer, Geometric algebra: A framework for computing point and line correspondences and projective structure using n-uncalibrated cameras, *Proc. of International conference on Pattern Recognition ICPR'96*, Vienna, Vol. 1, pp. 334–338, 1996.

2. E. Bayro-Corrochano, K. Daniilidis and G. Sommer, Hand-eye calibration in terms of motions of lines using geometric algebra, *Proc. of 10th Scandinavian Conference on Image Analysis*, Lappeenranta, Vol. 1, pp. 397–404, 1997.

3. E. Bayro-Corrochano and J. Lasenby, A unified language for computer vision and robotics, *Algebraic Frames for the Perception-Action Cycle*, G. Sommer and J. J. Koenderink (*eds.*), LNCS 1315, pp. 219–234, 1997.

4. E. Bayro-Corrochano and J. Lasenby, Geometric techniques for the computation of projective invariants using n uncalibrated cameras, *Proc. of Indian Conference on Computer Vision, Graphics and Image Processing*, New Delhi, pp. 95–100, 1998.

5. W. Blaschke, Anwendung dualer quaternionen auf die Kinematik, Annales Acad. Sci. Fennicae, 1–13, 1958.

6. W. K. Clifford, Preliminary sketch of bi-quaternions, Proc. London Math. Soc. 4: 381–395, 1873.

7. A. Crumeyrolle, *Orthogonal and Symplectic Clifford Algebras*, D. Reidel, Dordrecht, Boston, 1990.

8. R. Delanghe, F. Sommen and V. Soucek, *Clifford Algebra and Spinor-Valued Functions*, D. Reidel, Dordrecht, Boston, 1992.

9. O. Faugeras, *Three-dimensional Computer Vision*, MIT Press, 1993.

10. O. Faugeras and B. Mourrain, On the geometry and algebra of the point and line correspondences between N images, *Proc. of Europe-China Workshop on Geometrical Modeling and Invariants for Computer Vision*, R. Mohr and C. Wu (*eds.*), pp. 102–109, 1995.

11. C. Doran, D. Hestenes, F. Sommen and N. V. Acker, Lie groups as spin groups, J. Math. Phys. 34(8): 3642–3669, 1993.

12. T. Havel, Some examples of the use of distances as coordinates for Euclidean geometry, J. Symbolic Comput. 11: 579–593, 1991.

13. T. Havel and A. Dress, Distance geometry and geometric algebra, Found. Phys. 23: 1357–1374, 1993.

14. T. Havel, Geometric algebra and Möbius sphere geometry as a basis for Euclidean invariant theory, *Invariant Methods in Discrete and Computational Geometry*, N. L. White (*ed.*), pp. 245–256, D. Reidel, Dordrecht, Boston, 1995.

15. D. Hestenes, *Space-Time Algebra*, Gordon and Breach, New York, 1966.

16. D. Hestenes and G. Sobczyk, *Clifford Algebra to Geometric Calculus*, D. Reidel, Dordrecht, Boston, 1984.

17. D. Hestenes, *New Foundations for Classical Mechanics*, D. Reidel, Dordrecht, Boston, 1987.

18. D. Hestenes and R. Ziegler, Projective geometry with Clifford algebra, Acta Appl. Math. 23: 25–63, 1991.
19. D. Hestenes, The design of linear algebra and geometry, Acta Appl. Math. 23: 65–93, 1991.
20. D. Hestenes, Invariant body kinematics I: Saccadic and compensatory eye movements, Neural Networks 7 (1): 65–77, 1994.
21. D. Hestenes, Invariant body kinematics II: Reaching and neurogeometry, Neural Networks 7(1): 79–88, 1994.
22. D. Hestenes, H. Li and A. Rockwood, New algebraic tools for classical geometry, *Geometric Computing with Clifford Algebra*, G. Sommer (*ed.*), Springer, 1999.
23. The webpage of D. Hestenes: http://ModelingNTS.la.asu.edu/GC_R&D.html.
24. B. Iversen, *Hyperbolic Geometry*, Cambridge, 1992.
25. J. Lasenby and E. Bayro-Corrochano, Computing 3D projective invariants from points and lines, *Computer Analysis of Images and Patterns*, G. Sommer, K. Daniilidis and J. Pauli (*eds.*), pp. 82–89, 1997.
26. H. Li, Hyperbolic geometry with Clifford algebra, Acta Appl. Math. 48: 317–358, 1997.
27. H. Li, D. Hestenes and A. Rockwood, Generalized homogeneous coordinates for computational geometry, *Geometric Computing with Clifford Algebra*, G. Sommer (*ed.*), Springer, 1999.
28. H. Li, D. Hestenes and A. Rockwood, A universal model for conformal geometries of Euclidean, spherical and double-hyperbolic spaces, *Geometric Computing with Clifford Algebra*, G. Sommer (*ed.*), Springer, 1999.
29. H. Li, D. Hestenes and A. Rockwood, Spherical conformal geometry with geometric algebra, *Geometric Computing with Clifford Algebra*, G. Sommer (*ed.*), Springer, 1999.
30. H. Li, Hyperbolic conformal geometry with Clifford algebra, submitted to Acta Appl. Math. in 1999.
31. B. Mourrain and N. Stolfi, Computational symbolic geometry, *Invariant Methods in Discrete and Computational Geometry*, N. L. White (*ed.*), pp. 107–139, D. Reidel, Dordrecht, Boston, 1995.
32. B. Mourrain and N. Stolfi, Applications of Clifford algebras in robotics, *Computational Kinematics*, J.-P. Merlet and B. Ravani (*eds.*), pp. 41–50, D. Reidel, Dordrecht, Boston, 1995.
33. B. Mourrain, Enumeration problems in geometry, robotics and vision, *Algorithms in Algebraic Geometry and Applications*, L. González and T. Recio (*eds.*), pp. 285–306, Birkhäuser, Basel, 1996.
34. G. Peano, *Geometric Calculus*, 1888 (Translated by L. Kannenberg, 1997)
35. J. Seidel, Distance-geometric development of two-dimensional Euclidean, hyperbolic and spherical geometry I, II, Simon Stevin 29: 32–50, 65–76, 1952.
36. G. Sommer, E. Bayro-Corrochano and T. Bülow, Geometric algebra as a framework for the perception-action cycle, *Proc. of Workshop on Theoretical Foundation of Computer Vision*, Dagstuhl, 1996.
37. E. Study, Geometrie der Dynamen, Leipzig, 1903.
38. A. T. Yang and F. Freudenstein, Application of dual number quaternion algebra to the analysis of spatial mechanisms, J. Appl. Mech. 31: 300–308, 1964.

Decomposing Algebraic Varieties

Dongming Wang

LEIBNIZ–IMAG–CNRS, 46, avenue Félix Viallet, 38031 Grenoble Cedex, France

Abstract. An algebraic variety is a geometric figure defined by the zeros of a set of multivariate polynomials. This paper explains how to adapt two general zero decomposition methods for efficient decomposition of affine algebraic varieties into unmixed and irreducible components. Two devices based on Gröbner bases are presented for computing the generators of the saturated ideals of triangular sets. We also discuss a few techniques and variants which, when properly used, may speed up the decomposition. Experiments for a set of examples are reported with comparison to show the performance and effectiveness of such techniques, variants and the whole decomposition methods. Several theoretical results are stated along with the description of algorithms. The paper ends with a brief mention of some applications of variety decomposition.

1 Introduction

Let $\mathcal{K}$ be a fixed ground field of characteristic 0, $\mathcal{K}[x_1, \ldots, x_n]$ the ring of polynomials in the variables $x_1, \ldots, x_n$ (abbreviated x) with coefficients in $\mathcal{K}$, and $\tilde{\mathcal{K}}$ some extension field of $\mathcal{K}$. For any non-empty polynomial set $\mathbb{P} \subset \mathcal{K}[x]$, Zero($\mathbb{P}$) denotes the set of all common zeros in $\tilde{\mathcal{K}}$ of the polynomials in $\mathbb{P}$. Any element of Zero($\mathbb{P}$) can then be considered as a point in an n-dimensional affine space $\mathbf{A}_{\tilde{\mathcal{K}}}^n$ with coordinates x over $\tilde{\mathcal{K}}$. Let $\mathfrak{V}$ be a collection of points in $\mathbf{A}_{\tilde{\mathcal{K}}}^n$; $\mathfrak{V}$ is called an affine *algebraic variety*, or simply a *variety*, if there exists a polynomial set $\mathbb{P} \subset \mathcal{K}[x]$ such that $\mathfrak{V} = \text{Zero}(\mathbb{P})$. We call $\mathbb{P}$ the *defining set* and $\mathbb{P} = 0$ the *defining equations* of $\mathfrak{V}$. Examples of algebraic varieties are algebraic curves and surfaces in three-dimensional affine space.

A variety $\mathfrak{W}$ is called a *subvariety* of another variety $\mathfrak{V}$, which is denoted as $\mathfrak{W} \subset \mathfrak{V}$, if any point on $\mathfrak{W}$ is also on $\mathfrak{V}$. A variety $\mathfrak{W}$ is called a *true* subvariety of $\mathfrak{V}$ if $\mathfrak{W} \subset \mathfrak{V}$ and $\mathfrak{W} \neq \mathfrak{V}$.

A variety $\mathfrak{V} \subset \mathbf{A}_{\tilde{\mathcal{K}}}^n$ is said to be *irreducible* if it cannot be expressed as the union of two or more true subvarieties of $\mathfrak{V}$. An irreducible subvariety of $\mathfrak{V}$ is said to be *irredundant* if it is not contained in another irreducible subvariety of $\mathfrak{V}$. An irredundant irreducible subvariety of $\mathfrak{V}$ is also called an *irreducible component* of $\mathfrak{V}$. The variety $\mathfrak{V}$ is said to be *unmixed* if all its irreducible components have the same dimension.

We want to decompose any affine algebraic variety defined by a given set of polynomials as the union of a family of irreducible subvarieties. One approach initiated by Wu [22] for this purpose consists of two major steps: first computing an irreducible zero decomposition using the characteristic set method [12, 22] and then determining the defining sets of the subvarieties from the irreducible ascending sets (by means of which the zero decomposition is represented). Wu [23] proposed a method based on Chow bases for determining the prime bases from the obtained irreducible ascending sets. The use of Gröbner bases [2] was proposed in [4] and [14] independently as another method for computing such prime bases. The author has experimented the approach of irreducible variety decomposition based on the computation of characteristic sets and Gröbner bases [15]; 16 examples were given to illustrate the method and its practical performance.

We remark that the technique of determining a prime basis from an irreducible ascending set using Gröbner bases suggested in [4, 14] works as well for determining a defining set $\mathbb{G}$ of the corresponding variety from any (non-trivial) ascending set $\mathbb{A}$; this was shown in [7]. The set $\mathbb{G}$ is actually a finite basis for the saturated ideal $\text{Ideal}(\mathbb{A}) : J^{\infty}$, where $\text{Ideal}(\mathbb{A})$ denotes the ideal generated by the polynomials in $\mathbb{A}$ and J is the product of the initials of the polynomials in $\mathbb{A}$. The same technique of computing a finite basis for an arbitrary saturated ideal was already given in [8]; see also [5, 20]. Using this general technique and a theorem about the dimensionality of any ascending set in [7], an unmixed variety decomposition can be readily obtained from any proper zero decomposition.

In [16, 19, 21] are presented two general methods for decomposing any polynomial system into triangular systems of different kinds, which make use of some early elimination ideas of A. Seidenberg and J. M. Thomas with top-down elimination, splitting and subresultant regular subchain as computational strategies. Other decomposition methods have been proposed in [11] and [9, 10]: the former emphasizes the canonicality of the computed triangular sets with various requirements while the latter aims at computing the variety decomposition represented by regular and irreducible triangular sets. These methods may be used to replace Ritt-Wu's method of characteristic sets for irreducible or unmixed decomposition of algebraic varieties. Note that replacing one method of zero decomposition by another, faster or slower, does not necessarily yield a superior or inferior method for the decomposition of varieties. For the triangular sets computed by different algorithms usually have different forms, that affect the subsequent computation of saturation bases. Therefore, investigating other zero decomposition methods for variety decomposition has practical significance.

One purpose of this paper is to use our zero decomposition methods instead of Ritt-Wu's for decomposing algebraic varieties into unmixed and irreducible components. In Section 3, a different algorithm for constructing the finite bases for saturation ideals is recalled; this method proceeds by computing successively ideal quotients using Gröbner bases. Meanwhile, we explain several variants and techniques which can enhance the practical efficiency in different cases. A number of properties for variety decomposition are presented and show, in particular, the significance of the concept and computation of regular sets. The formal proofs for most of the theoretical results are omitted and can be found from the indicated references. Many of them are contained in the author's forthcoming book [21]. Experiments with our decomposition methods based on Seidenberg's elimination idea and subresultant regular subchains and the previous one using characteristic sets are reported in table form for 22 examples. Our methods are faster in terms of computing time for most of the complete cases. Some of the new techniques, which do not lead to any additional cost, can speed up the decomposition many times when they succeed. The paper ends with a short summary of applications of variety decomposition.

2 Zero Decomposition

Let the variables be ordered as $x_1 \prec \cdots \prec x_n$. For any x_i and $P \in \mathcal{K}[x]$, the *degree* and *leading coefficient* of P in x_i are denoted by $\deg(P, x_i)$ and $\mathrm{lcoef}(P, x_i)$ respectively. If P is not a constant (i.e., $P \notin \mathcal{K}$), we call the smallest index p such that $P \in \mathcal{K}[x_1, \ldots, x_p]$ the *class* of P; x_p the *leading variable* of P, denoted by $\mathrm{lvar}(P)$; $\mathrm{lcoef}(P, x_p)$ the *initial* of P, denoted by $\mathrm{ini}(P)$; and $P - \mathrm{ini}(P)x_p^{\deg(P, x_p)}$ the *reductum* of P, denoted by $\mathrm{red}(P)$. For any two polynomials P and Q with $Q \notin \mathcal{K}$, $\mathrm{prem}(P, Q)$ denotes the *pseudo-remainder* of P with respect to Q in $\mathrm{lvar}(Q)$.

An ordered set $\mathbb{T} = [T_1, \ldots, T_r]$ of r non-constant polynomials is called a *triangular set* if $\mathrm{lvar}(T_1) \prec \cdots \prec \mathrm{lvar}(T_r)$. For any polynomial P,

$$\mathrm{prem}(P, \mathbb{T}) \triangleq \mathrm{prem}(\cdots \mathrm{prem}(P, T_r), \ldots, T_1)$$

is called the *pseudo-remainder* of P with respect to $\mathbb{T}$; the symbol $\triangleq$ reads as "defined to be." For any polynomial set $\mathbb{P}$, $\mathrm{prem}(\mathbb{P}, T_i)$ stands for $\{\mathrm{prem}(P, T_i) \mid P \in \mathbb{P}\}$, $\mathrm{prem}(\mathbb{P}, \mathbb{T})$ for $\{\mathrm{prem}(P, \mathbb{T}) \mid P \in \mathbb{P}\}$, $\mathrm{lvar}(\mathbb{P})$ for $\{\mathrm{lvar}(P) \mid P \in \mathbb{P}\}$, and $\mathrm{ini}(\mathbb{P})$ for $\{\mathrm{ini}(P) \mid P \in \mathbb{P}\}$.

Let $[\mathbb{P}, \mathbb{Q}]$ be a pair of polynomial sets, called a *polynomial system*, with which the zero set $\mathrm{Zero}(\mathbb{P}/\mathbb{Q})$ is of concern, where

$$\mathrm{Zero}(\mathbb{P}/\mathbb{Q}) \triangleq \mathrm{Zero}(\mathbb{P}) \setminus \bigcup_{Q \in \mathbb{Q}} \mathrm{Zero}(Q).$$

We write $\mathrm{Zero}(\mathbb{P})$ for $\mathrm{Zero}(\mathbb{P}/\emptyset)$ and $\mathrm{Zero}(\mathbb{P}/Q)$ for $\mathrm{Zero}(\mathbb{P}/\{Q\})$, etc.

A *triangular system* is a pair $[\mathbb{T}, \mathbb{U}]$ in which $\mathbb{T}$ is a triangular set and $\mathbb{U}$ a polynomial set such that $\mathrm{ini}(T)$ does not vanish on $\mathrm{Zero}(\mathbb{T}/\mathbb{U})$ for any $T \in \mathbb{T}$. A triangular system $[\mathbb{T}, \mathbb{U}]$ is said to be *fine* if $0 \notin \mathrm{prem}(\mathbb{U}, \mathbb{T})$. A triangular set $\mathbb{T}$ is *fine* if $[\mathbb{T}, \mathrm{ini}(\mathbb{T})]$ is fine.

By *zero decomposition* we mean to compute for any given non-empty polynomial set $\mathbb{P} \subset \mathcal{K}[x]$ a finite set Ψ which is either empty (in this case $\mathrm{Zero}(\mathbb{P}) = \emptyset$ in any extension field of $\mathcal{K}$) or of the form $\Psi = \{[\mathbb{T}_1, \mathbb{U}_1], \ldots, [\mathbb{T}_e, \mathbb{U}_e]\}$ such that

$$\mathrm{Zero}(\mathbb{P}) = \bigcup_{i=1}^{e} \mathrm{Zero}(\mathbb{T}_i/\mathbb{U}_i), \tag{1}$$

where each $[\mathbb{T}_i, \mathbb{U}_i]$ is a fine triangular system. In other words, either the emptiness of $\mathrm{Zero}(\mathbb{P})$ is detected or the polynomial set $\mathbb{P}$ is decomposed into finitely many fine triangular systems $[\mathbb{T}_i, \mathbb{U}_i]$. We have been studying three different yet related algorithms reviewed below for such zero decomposition.

Ritt-Wu's algorithm

This is a well-known algorithm attributed to Ritt [12] and Wu [22] which starts by computing a *characteristic set* $\mathbb{C}$ of the given polynomial set $\mathbb{P}$. In the trivial case, $\mathbb{C}$ is contradictory (i.e., $\mathbb{C}$ consists of a single constant) and thus $\text{Zero}(\mathbb{P}) = \emptyset$. Otherwise, the computed $\mathbb{C}$ is a (*weak*) *ascending set* (that is a triangular set with some restriction on the degrees of polynomials) such that

$$\text{Zero}(\mathbb{P}) \subset \text{Zero}(\mathbb{C}), \quad \text{Zero}(\mathbb{C}/\mathbb{I}) \subset \text{Zero}(\mathbb{P}),$$

where $\mathbb{I} = \text{ini}(\mathbb{C})$. In fact, $\text{prem}(\mathbb{P}, \mathbb{C}) = \{0\}$ holds as well. It follows from the above zero relations that

$$\text{Zero}(\mathbb{P}) = \text{Zero}(\mathbb{C}/\mathbb{I}) \cup \bigcup_{I \in \mathbb{I}} \text{Zero}(\mathbb{P} \cup \{I\}).$$

One can proceed further by computing a characteristic set of each $\mathbb{P} \cup \{I\}$ in the same manner. This process will terminate so that a zero decomposition of the form (1) is reached, with $\mathbb{T}_i = \mathbb{C}_i$ and $\mathbb{U}_i = \text{ini}(\mathbb{T}_i)$. In addition, the property $\text{prem}(\mathbb{P}, \mathbb{C}_i) = \{0\}$ holds for all i. See [3, 12, 17, 22] for details.

Seidenberg's algorithm refined

Now let us explain the algorithm described in [16] that computes the zero decomposition (1) by eliminating the variables successively from x_n to x_1 with recursive pseudo-division and by splitting the system whenever pseudo-division is performed. Concretely, for each x_k, $k = n, \ldots, 1$, one proceeds as follows.

The iteration starts with $k = n$. If $\mathbb{P}$ does not contain any polynomial of class k then go for $k - 1$. Otherwise, let $T \in \mathbb{P}$ be of class k and have minimal degree in x_k. Then

$$\mathbb{P} = 0 \quad \Longleftrightarrow \quad \begin{cases} \mathbb{P}^* = 0, I = 0, \text{red}(T) = 0, \text{ or} \\ \text{prem}(\mathbb{P}, T) = 0, T = 0, \quad I \neq 0, \end{cases}$$

where

$$\mathbb{P}^* = \mathbb{P} \setminus \{T\}, \quad I = \text{ini}(T).$$

Therefore, we have

$$\text{Zero}(\mathbb{P}) = \text{Zero}(\mathbb{P}^* \cup \{I, \text{red}(T)\}) \cup \text{Zero}(\text{prem}(\mathbb{P}, T) \cup \{T\}/I).$$

Note that for both $\mathbb{P}^* \cup \{I, \text{red}(T)\}$ and $\text{prem}(\mathbb{P}, T) \cup \{T\}$ either the numbers of polynomials of class k decrease, or so do the minimal degrees of such polynomials in x_k (in comparison with $\mathbb{P}$). The process continues for each split system $[\mathbb{P}', \mathbb{Q}']$ by taking from $\mathbb{P}'$ another polynomial of class k and having minimal degree in x_k (if there exists any) and splitting further. With every $[\mathbb{P}', \mathbb{Q}']$ one proceeds for $k - 1$ when $\mathbb{P}'$ contains none or only one polynomial T' of class k; in the latter case, $\mathbb{Q}'$ is meanwhile replaced by $\text{prem}(\mathbb{Q}', T')$. The system $[\mathbb{P}', \mathbb{Q}']$ is discarded whenever $\mathbb{P}' \cap \mathcal{K} \setminus \{0\} \neq \emptyset$ or $0 \in \mathbb{Q}'$.

Clearly, the above procedure will terminate with a zero decomposition of the form (1).

Subresultant-based algorithm

The third algorithm makes use of the key concept of *subresultant regular subchains* (s.r.s.) and their computation, for which the details can be found in [19]. The algorithm also employs a top-down elimination for x_k, $k = n, \ldots, 1$.

Let us begin with $k = n$. If, trivially, $\mathbb{P}$ does not contain any polynomial of class k, then proceed next for $k-1$. Consider the simple case in which $\mathbb{P}$ contains only one polynomial P of class k; let $I = \text{ini}(P)$. Then

$$\mathbb{P} = 0 \iff \begin{cases} \mathbb{P} = 0, I \neq 0; \text{ or} \\ \mathbb{P} \setminus \{P\} = 0, I = 0, \text{red}(P) = 0. \end{cases}$$

Here two subsystems are produced. For the first, we have obtained a single polynomial P in x_k whose initial is assumed to be non-zero, so the process can continue for next k. For the second, the minimal degree in x_k of the polynomials of class k has decreased. So we can assume that the subsystem may be dealt with by induction.

Now come to the more general case in which $\mathbb{P}$ contains two or more polynomials of class k. Let $P_1, P_2 \in \mathbb{P}$ be two of them with P_2 having minimal degree in x_k and compute the s.r.s. $H_2, \ldots, H_r$ of P_1 and P_2 with respect to x_k. Let $I = \text{lcoef}(P_2, x_k)$ and $I_i = \text{lcoef}(H_i, x_k)$ for $2 \leq i \leq r$. Then, by the properties of s.r.s. we have

$$\mathbb{P} = 0 \iff \begin{cases} \mathbb{P}_2 = 0, I = 0, \text{red}(P_2) = 0, \text{ or} \\ \begin{bmatrix} \mathbb{P}_{12} = 0, H_i = 0, & I \neq 0, \\ I_{i+1} = 0, \ldots, I_r = 0, & I_i \neq 0 \end{bmatrix} \text{ for some } 2 \leq i \leq r, \end{cases}$$

where

$$\mathbb{P}_2 = \mathbb{P} \setminus \{P_2\}, \quad \mathbb{P}_{12} = \mathbb{P} \setminus \{P_1, P_2\}.$$

It follows that

$$\text{Zero}(\mathbb{P}) = \text{Zero}(\mathbb{P}_2 \cup \{I, \text{red}(P_2)\}) \cup \bigcup_{i=2}^{r} \text{Zero}(\mathbb{P}_{12} \cup \{H_i, I_{i+1}, \ldots, I_r\}/II_i).$$

To see the significance of the splitting, one may note that the number of the polynomials of class k in $\mathbb{P}_{12} \cup \{H_i, I_{i+1}, \ldots, I_r\}$ is at least one less than that of the polynomials of class k in $\mathbb{P}$. The process may be continued by iterating the above step for each split system $[\mathbb{P}', \mathbb{Q}']$. The remarks at the end of the preceding subsection should also be added here. We may thus devise another algorithm to compute zero decompositions of the form (1). Such an algorithm is described formally in [21, Sect. 2.3].

A triangular system $[\mathbb{T}, \mathbb{U}]$ or a triangular set $\mathbb{T}$ is said to be *quasi-irreducible* if all the polynomials in $\mathbb{T}$ are irreducible over $\mathcal{K}$. Making use of polynomial factorization over the ground field $\mathcal{K}$, one can compute zero decompositions of the form (1) with each triangular system $[\mathbb{T}_i, \mathbb{U}_i]$ quasi-irreducible.

Computing regular and irreducible triangular systems

Let $\mathbb{T} = [T_1, \ldots, T_r]$ be a fine triangular set with $x_{p_j} = \mathrm{lvar}(T_j)$ and $I_j = \mathrm{ini}(T_j)$ for $1 \leq j \leq r$. $\mathbb{T}$ is said to be *regular* or called a *regular set* if

$$\mathrm{res}(\cdots \mathrm{res}(\mathrm{res}(I_{j+1}, T_j, x_{p_j}), T_{j-1}, x_{p_{j-1}}), \ldots, T_1, x_{p_1}) \neq 0$$

for $2 \leq j \leq r$, where $\mathrm{res}(F, G, x)$ denotes the resultant of F and G with respect to x.

A fine triangular system $[\mathbb{T}, \mathbb{U}]$ is called a *regular system* if $\mathbb{T}$ is regular, $\mathrm{lvar}(\mathbb{T}) \cap \mathrm{lvar}(\mathbb{U}) = \emptyset$, and $\mathrm{ini}(U)$ does not vanish on $\mathrm{Zero}(\mathbb{T}/\mathbb{U})$ for any $U \in \mathbb{U}$.

An algorithm is presented in [21, Sect. 5.1] that computes zero decompositions of the form (1) with each $[\mathbb{T}_i, \mathbb{U}_i]$ a regular system. The main new ingredient of this algorithm is to use the only polynomial P_2 of class k in $\mathbb{P}'$, when obtained, to eliminate all the polynomials of class k from $\mathbb{Q}'$, where $[\mathbb{P}', \mathbb{Q}']$ is the polynomial system under consideration and $\mathrm{ini}(P_2)$ does not vanish on $\mathrm{Zero}(\mathbb{P}'/\mathbb{Q}')$. The elimination is performed with further splitting roughly as follows.

Assume that $\mathbb{Q}'$ contains a polynomial, say P_1, of class k. Compute the s.r.s. $H_2, \ldots, H_r$ of P_1 and P_2 with respect to x_k, and let $I_i = \mathrm{lcoef}(H_i, x_k)$ for $2 \leq i \leq r$. Then, it follows from the properties of s.r.s. that

$$\mathrm{Zero}(\mathbb{P}'/\mathbb{Q}') = \bigcup_{i=2}^{r} \mathrm{Zero}(\mathbb{P}' \setminus \{P_2\} \cup \{L_i, I_{i+1}, \ldots, I_r\}/\mathbb{Q}' \cup \{I_i\}),$$

where L_i is the *pseudo-quotient* of P_2 by H_i with respect to x_k. Observe that, when H_i is of class k, L_i has smaller degree than P_2 in x_k. It is always the case for $2 \leq i \leq r - 1$. If H_r has class smaller than k, then

$$\mathrm{Zero}(\mathbb{P}' \setminus \{P_2\} \cup \{L_r\}/\mathbb{Q}' \cup \{I_r\}) = \mathrm{Zero}(\mathbb{P}'/\mathbb{Q}' \setminus \{P_1\} \cup \{I_r\}),$$

i.e., P_1 may be eliminated from $\mathbb{Q}'$. The above process can be repeated for each split system so that all the polynomials of class k in the second set of the system are eliminated. In this way, we shall obtain from $[\mathbb{P}', \mathbb{Q}']$ finitely many new systems $[\mathbb{P}'_i, \mathbb{Q}'_i]$ such that each $\mathbb{P}'_i$ contains either none or exactly one polynomial of class k, while in the latter case $\mathbb{Q}'_i$ does not contain any polynomial of class k.

A fine triangular set $\mathbb{T}$ as above is said to be *irreducible* if each T_j, as a polynomial of x_{p_j}, is irreducible over $\mathcal{K}(x_1, \ldots, x_{p_j-1})$, where $x_{p_1}, \ldots, x_{p_{j-1}}$ are adjoined as algebraic elements by the equations $T_1 = 0, \ldots, T_{j-1} = 0$ in the case $j > 1$.

A fine triangular system $[\mathbb{T}, \mathbb{U}]$ is said to be *irreducible* if $\mathbb{T}$ is irreducible.

If $\mathbb{T}$ is reducible, one can find a k such that

$$\mathbb{T}^{\{k-1\}} = [T_1, \ldots, T_{k-1}]$$

is irreducible and the polynomial T_k has an irreducible factorization of the form

$$DT_k \doteq F_1 \cdots F_t$$

over the extension field $\mathcal{K}(x_1, \ldots, x_{p_k-1})$ with the algebraic elements $x_{p_1}, \ldots,$ $x_{p_{k-1}}$ adjoined by $T_1 = 0, \ldots, T_{k-1} = 0$, where

$$D \in \mathcal{K}[x_1, \ldots, x_{p_k-1}], \quad F_j \in \mathcal{K}[x_1, \ldots, x_{p_k}]$$

all have degree smaller than $\deg(T_i, x_{p_i})$ for $1 \leq i \leq k-1$ and the dot equality means that $\mathrm{prem}(DT_k - F_1 \cdots F_t, \mathbb{T}^{\{k-1\}}) = 0$ (see [21, Sect. 7.5]).

Therefore, for each triangular system $[\mathbb{T}_i, \mathbb{U}_i]$ in (1) with $\mathbb{T}_i$ reducible, say having the same form as $\mathbb{T}$ above, one may get a further zero decomposition of the form

$$\mathrm{Zero}(\mathbb{T}_i/\mathbb{U}_i) = \mathrm{Zero}(\{D\} \cup \mathbb{T}_i/\mathbb{U}_i) \cup \bigcup_{j=1}^{t} \mathrm{Zero}(\mathbb{T}_{ij}/\mathbb{U}_i \cup \{D\}), \qquad (2)$$

where each $\mathbb{T}_{ij}$ is an ordered polynomial set obtained from $\mathbb{T}_i$ by replacing T_k with F_j.

Let $\mathbb{U}_{ij} = \mathrm{prem}(\mathbb{U}_i \cup \{D\}, \mathbb{T}_{ij})$. If $\mathbb{U}_{ij}$ contains 0 for some j, then the corresponding component in (2) can be simply removed. It is easy to see that for those components in which $\mathbb{U}_{ij}$ does not contain 0, $[\mathbb{T}_{ij}, \mathbb{U}_{ij}]$ is still a fine triangular system and its irreducibility can be further verified.

The polynomial set $\{D\} \cup \mathbb{T}_i$ may no longer be in triangular form, but it can be further triangularized by applying the same algorithm as for $\mathbb{P}$. In this way we shall finally arrive at a zero decomposition of the form (1) with each triangular set $\mathbb{T}_i$ irreducible.

In case Ritt-Wu's algorithm is not used, the computed $\mathbb{U}_i$ may contain many more polynomials than actually needed. The redundant polynomials can be deleted by using the following theorem [16, 21], which is of both theoretical interest and practical importance.

Theorem 1. *Let $\mathbb{P}$ be any polynomial set in $\mathcal{K}[x]$ having a zero decomposition of the form (1) with each $[\mathbb{T}_i, \mathbb{U}_i]$ a regular or irreducible triangular system. Then*

$$\mathrm{Zero}(\mathbb{P}) = \bigcup_{i=1}^{e} \mathrm{Zero}(\mathbb{T}_i/\mathrm{ini}(\mathbb{T}_i)). \qquad (3)$$

Let $\mathbb{P}$ have a zero decomposition of the form (3) with each $\mathbb{T}_i$ a fine triangular set and $\Psi = \{\mathbb{T}_1, \ldots, \mathbb{T}_e\}$. We set $\Psi = \emptyset$ when $\mathrm{Zero}(\mathbb{P}) = \emptyset$ is verified. Ψ is called a *W-characteristic series* of $\mathbb{P}$ if $\mathrm{prem}(\mathbb{P}, \mathbb{T}_i) = \{0\}$ for each i. Ψ is called a *regular series* of $\mathbb{P}$ if all $\mathbb{T}_i$ are regular. In general, a regular series is not necessarily a W-characteristic series of $\mathbb{P}$ (i.e., there is no guarantee for $\mathrm{prem}(\mathbb{P}, \mathbb{T}_i) = \{0\}$). However, it is proved in [21] that there exists an integer q such that $\mathrm{prem}(P^q, \mathbb{T}_i) = 0$ for all $P \in \mathbb{P}$ when $\mathbb{T}_i$ is regular.

3 Decomposition of Varieties

In the decomposition (3), $\mathrm{Zero}(\mathbb{T}_i/\mathrm{ini}(\mathbb{T}_i))$ is not necessarily an algebraic variety. We need to construct from each $\mathbb{T}_i$ a finite set $\mathbb{P}_i$ of polynomials, called a *saturation basis* of $\mathbb{T}_i$, that defines a subvariety of $\mathrm{Zero}(\mathbb{P})$.

Ideal saturation for triangular sets

Let $\mathfrak{I}$ be an ideal and F a polynomial in $\mathcal{K}[\boldsymbol{x}]$. The *saturation* of $\mathfrak{I}$ with respect to F is the ideal

$$\mathfrak{I} : F^{\infty} \triangleq \{P \in \mathcal{K}[\boldsymbol{x}] \mid F^q P \in \mathfrak{I} \text{ for some integer } q > 0\}.$$

For any non-empty polynomial set $\mathbb{P} \subset \mathcal{K}[\boldsymbol{x}]$, $\mathrm{Ideal}(\mathbb{P})$ denotes the ideal generated by (the polynomials in) $\mathbb{P}$. A finite basis for $\mathfrak{I} : F^{\infty}$ can be computed by using Gröbner bases [2] according to the following lemma.

Lemma 1. *Let $\mathfrak{I}$ be an ideal generated by $\mathbb{P}$ and F a polynomial in $\mathcal{K}[\boldsymbol{x}]$; $F_1, \ldots, F_t$ be t factors of F such that $F_1 \cdots F_t \neq 0 \iff F \neq 0$;*

$$\mathbb{P}^* = \mathbb{P} \cup \{zF - 1\}, \quad \mathbb{P}^\star = \mathbb{P} \cup \{z_i F_i - 1 \mid 1 \leq i \leq t\},$$

where $z, z_1, \ldots, z_t$ are new indeterminates; and $\mathbb{G}^*, \mathbb{G}^\star$ be the Gröbner bases of $\mathbb{P}^*$ in $\mathcal{K}[\boldsymbol{x}, z]$ and of $\mathbb{P}^\star$ in $\mathcal{K}[\boldsymbol{x}, z_1, \ldots, z_t]$ with respect to the purely lexicographical term ordering (plex) determined with $x_l \prec z$ and $x_l \prec z_j$, respectively. Then

$$\begin{aligned}
\mathfrak{I} : F^{\infty} &= \mathrm{Ideal}(\mathbb{P}^*) \cap \mathcal{K}[\boldsymbol{x}] = \mathrm{Ideal}(\mathbb{G}^* \cap \mathcal{K}[\boldsymbol{x}]) \\
&= \mathrm{Ideal}(\mathbb{P}^\star) \cap \mathcal{K}[\boldsymbol{x}] = \mathrm{Ideal}(\mathbb{G}^\star \cap \mathcal{K}[\boldsymbol{x}]).
\end{aligned}$$

Proof. See [8, 7, 20].

In fact, for the Gröbner bases computation any compatible elimination ordering in which $x_1^{i_1} \cdots x_n^{i_n} \prec z$ (or z_j) does. As we have mentioned in Section 1, the above technique of computing saturation bases was introduced independently by several researchers. The result stated in Lemma 1 appeared first in [8]. The special case of the lemma, in which $\mathbb{P}$ is an irreducible ascending set $\mathbb{A}$ and F is the product of the initials of the polynomials in $\mathbb{A}$, was proved in [4, 14]. A proof for a slightly different version of the lemma may be found in [7]. The lemma in the above form is stated and proved in [20, 21]. What is explained below is another method for determining a finite basis for any $\mathfrak{I} : F^{\infty}$; this method has been presented in [5].

Let $\mathfrak{I}$ and $\mathfrak{J}$ be two ideals in $\mathcal{K}[\boldsymbol{x}]$. The infinite set of polynomials

$$\mathfrak{I} : \mathfrak{J} \triangleq \{F \in \mathcal{K}[\boldsymbol{x}] \mid FG \in \mathfrak{I} \text{ for all } G \in \mathfrak{J}\}$$

is called the *ideal quotient* of $\mathfrak{I}$ by $\mathfrak{J}$. It is easy to show that in $\mathcal{K}[\boldsymbol{x}]$ the quotient of two ideals is an ideal (see, e.g., [5, p. 193]). For any polynomial F, we write $\mathfrak{I} : F$ instead of $\mathfrak{I} : \mathrm{Ideal}(\{F\})$.

Lemma 2. *Let $\mathfrak{I}$ be an ideal and F a polynomial in $\mathcal{K}[\boldsymbol{x}]$, and let k be an integer ≥ 1. Then*

$$\mathfrak{I} : F^{\infty} = \mathfrak{I} : F^k \iff \mathfrak{I} : F^k = \mathfrak{I} : F^{k+1}.$$

As a consequence, the minimal k can be determined by computing $\mathfrak{I} : F^i$ with i increasing from 1.

Proof. Exercise in [5, p. 196].

This lemma provides an algorithm for determining a basis for $\mathfrak{I} : F^\infty$ by computing the bases for the ideal quotients $\mathfrak{I} : F^k$ with k increasing from 1. The basis is obtained when $\mathfrak{I} : F^k = \mathfrak{I} : F^{k+1}$ is verified for some k; in this case $\mathfrak{I} : F^k = \mathfrak{I} : F^\infty$.

A basis for the ideal quotient $\mathfrak{I} : G$ can be computed by using Gröbner bases in the following way. Let $\mathfrak{I}$ be generated by $F_1, \ldots, F_s$. Compute the Gröbner basis $\mathbb{G}$ of

$$\{zF_1, \ldots, zF_s, (1-z)G\}$$

in $\mathcal{K}[\boldsymbol{x}, z]$ with respect to an elimination term ordering with $x_i \prec z$ and let

$$\mathbb{H} = \{H/G \mid H \in \mathbb{G} \cap \mathcal{K}[\boldsymbol{x}]\},$$

where z is a new indeterminate. Then $\mathfrak{I} : G = \text{Ideal}(\mathbb{H})$ and thus $\mathbb{H}$ is a finite basis for $\mathfrak{I} : G$. See [5,8] for the correctness of this method.

Let $\mathbb{T}$ be any triangular set in $\mathcal{K}[\boldsymbol{x}]$. The *saturation* of $\mathbb{T}$ is the ideal

$$\text{sat}(\mathbb{T}) \triangleq \text{Ideal}(\mathbb{T}) : J^\infty,$$

where $J = \prod_{T \in \mathbb{T}} \text{ini}(T)$.

Let $\mathbb{P}$ be a finite basis for $\text{sat}(\mathbb{T})$; the following relation is obvious

$$\text{Ideal}(\mathbb{T}) \subset \text{sat}(\mathbb{T}) = \text{Ideal}(\mathbb{P}).$$

Theorem 2. *Let* $\mathbb{T} \subset \mathcal{K}[\boldsymbol{x}]$ *be any triangular set. Then* $\mathbb{T}$ *is regular if and only if*

$$\text{sat}(\mathbb{T}) = \{P \in \mathcal{K}[\boldsymbol{x}] \mid \text{prem}(P, \mathbb{T}) = 0\}.$$

Proof. See [1] and [21, Sect. 6.2].

For any irreducible triangular set $\mathbb{T}$, Theorem 5 below asserts that $\text{sat}(\mathbb{T})$ is a prime ideal. For any $F \in \mathcal{K}[\boldsymbol{x}]$, if $\text{prem}(F, \mathbb{T}) \neq 0$, then $F \notin \text{sat}(\mathbb{T})$ according to Theorem 2 and thus $\text{sat}(\mathbb{T}) : F^\infty = \text{sat}(\mathbb{T})$ by definition. This result is generalized in the following lemma for regular sets.

Lemma 3. *Let* $\mathbb{T}$ *be a regular set and* F *any polynomial in* $\mathcal{K}[\boldsymbol{x}]$. *If* $\text{res}(F, \mathbb{T}) \neq 0$, *then* $\text{sat}(\mathbb{T}) : F^\infty = \text{sat}(\mathbb{T})$.

Proof. See [21, Sect. 6.2].

Proposition 1. *Let* $[\mathbb{T}, \mathbb{U}]$ *be a regular system in* $\mathcal{K}[\boldsymbol{x}]$ *and* $V = \prod_{U \in \mathbb{U}} U$. *Then*

$$\text{Ideal}(\mathbb{T}) : V^\infty = \text{sat}(\mathbb{T}). \tag{4}$$

Proof. See [21, Sect. 6.2].

As a consequence of (4), we have $\text{Zero}(\text{Ideal}(\mathbb{T}) : V^\infty) = \text{Zero}(\text{sat}(\mathbb{T}))$.

Unmixed decomposition

Refer to the zero decomposition (3) which provides a representation of the variety $\mathfrak{V}$ defined by $\mathbb{P}$ in terms of its subvarieties determined by $\mathbb{T}_i$. As we have mentioned, each $\mathrm{Zero}(\mathbb{T}_i/\mathrm{ini}(\mathbb{T}_i))$ is not necessarily an algebraic variety; it is a *quasi-algebraic variety*. From the results above, we have seen that a corresponding set of polynomials may be obtained by determining, from each $\mathbb{T}_i$, a saturation basis with Gröbner bases computation.

Theorem 3. *Let* $\mathbb{P}$ *be a non-empty polynomial set in* $\mathcal{K}[\boldsymbol{x}]$ *and* $\{\mathbb{T}_1, \ldots, \mathbb{T}_e\}$ *a W-characteristic series or a regular series of* $\mathbb{P}$. *Then*

$$\mathrm{Zero}(\mathbb{P}) = \bigcup_{i=1}^{e} \mathrm{Zero}(\mathrm{sat}(\mathbb{T}_i)). \tag{5}$$

Proof. See [7] and [21, Sect. 6.2]. $\quad\blacksquare$

Let $|S|$ denote the number of elements in the set S. The following result used by Chou and Gao [3] provides a useful criterion for removing some redundant subvarieties in the decomposition (5) without computing their defining sets.

Lemma 4. *Let* $\mathbb{P}$ *and* $\mathbb{T}_i$ *be as in Theorem 3. If* $|\mathbb{T}_j| > |\mathbb{P}|$, *then*

$$\mathrm{Zero}(\mathrm{sat}(\mathbb{T}_j)) \subset \bigcup_{\substack{1 \le i \le e \\ i \ne j}} \mathrm{Zero}(\mathrm{sat}(\mathbb{T}_i));$$

thus $\mathrm{Zero}(\mathrm{sat}(\mathbb{T}_j))$ *can be deleted from* (5).

Proof. As $|\mathbb{T}_j| > |\mathbb{P}|$, the dimension of $\mathbb{T}_j$ is smaller than $n - |\mathbb{P}|$. By the affine dimension theorem (see [3] or [21, Sect. 6.1]) and Theorem 4, $\mathrm{Zero}(\mathrm{sat}(\mathbb{T}_j))$ is a redundant component of $\mathrm{Zero}(\mathbb{P})$. $\quad\blacksquare$

A triangular set $\mathbb{T}$ is said to be *perfect* if $\mathrm{Zero}(\mathbb{T}/\mathrm{ini}(\mathbb{T})) \ne \emptyset$ in some suitable extension field of $\mathcal{K}$. The following result is due to Gao and Chou [7].

Theorem 4. *Let* $\mathbb{T}$ *be any triangular set in* $\mathcal{K}[\boldsymbol{x}]$. *If* $\mathbb{T}$ *is not perfect then* $\mathrm{sat}(\mathbb{T}) = \mathcal{K}[\boldsymbol{x}]$; *if* $\mathbb{T}$ *is perfect then* $\mathrm{Zero}(\mathrm{sat}(\mathbb{T}))$ *is an unmixed variety of dimension* $n - |\mathbb{T}|$.

Recall that any regular or irreducible triangular set $\mathbb{T}$ is perfect, so the variety $\mathrm{Zero}(\mathrm{sat}(\mathbb{T}))$ is unmixed of dimension $n - |\mathbb{T}|$.

In (5), for each i let $\mathbb{P}_i$ be a finite basis for $\mathrm{sat}(\mathbb{T}_i)$ which can be determined by computing Gröbner bases according to Lemma 1 or by computing $\mathrm{Ideal}(\mathbb{T}_i) : J_i^k$ with k increasing from 1. If $\mathrm{sat}(\mathbb{T}_i) = \mathcal{K}[\boldsymbol{x}]$, then the constant 1 is contained in (the Gröbner basis of) $\mathbb{P}_i$. Let us assume that such $\mathbb{P}_i$ is simply removed. Thus, a variety decomposition of the following form is obtained:

$$\mathrm{Zero}(\mathbb{P}) = \bigcup_{i=1}^{e} \mathrm{Zero}(\mathbb{P}_i). \tag{6}$$

Apparently, $\text{Zero}(\mathbb{P}) = \emptyset$ when $e = 0$ (i.e., all such $\mathbb{P}_i$ have been removed). By Theorem 4, each $\mathbb{P}_i$ defines an unmixed algebraic variety.

Let $\mathfrak{V}_i = \text{Zero}(\mathbb{P}_i)$; then the decomposition (6) can be rewritten as

$$\mathfrak{V} = \mathfrak{V}_1 \cup \cdots \cup \mathfrak{V}_e.$$

This decomposition may be contractible; that is, some variety may be a subvariety of another. Some of the redundant subvarieties may be easily removed by using Lemma 4. The following lemma points out how to remove some other redundant components.

Lemma 5. *Let $\mathbb{G}$ be a Gröbner basis and $\mathbb{P}$ an arbitrary polynomial set in $\mathcal{K}[\boldsymbol{x}]$. If every polynomial in $\mathbb{P}$ has normal form 0 modulo $\mathbb{G}$, then $\text{Zero}(\mathbb{G}) \subset \text{Zero}(\mathbb{P})$.*

Proof. Since every polynomial in $\mathbb{P}$ has normal form 0 modulo $\mathbb{G}$, $\text{Ideal}(\mathbb{P}) \subset \text{Ideal}(\mathbb{G})$. It follows that $\text{Zero}(\mathbb{G}) \subset \text{Zero}(\mathbb{P})$.

The method for decomposing an algebraic variety into unmixed components explained above can be described in the following algorithmic form.

Algorithm UnmVarDec (Input: $\mathbb{P}$; Output: Ψ). Given a non-empty polynomial set $\mathbb{P} \subset \mathcal{K}[\boldsymbol{x}]$, this algorithm computes a finite set Ψ of polynomial sets $\mathbb{P}_1, \ldots, \mathbb{P}_e$ such that the decomposition (6) holds and each $\mathbb{P}_i$ defines an unmixed algebraic variety.

U1. Compute a W-characteristic series or a regular series Φ of $\mathbb{P}$ and set $\Psi \leftarrow \emptyset$.

U2. While $\Phi \neq \emptyset$ do:

    U2.1. Let $\mathbb{T}$ be an element of Φ and set $\Phi \leftarrow \Phi \setminus \{\mathbb{T}\}$. If $|\mathbb{T}| > |\mathbb{P}|$ then go to U2.

    U2.2. Compute a finite basis for $\text{sat}(\mathbb{T})$ according to Lemma 1 or 2, let it be given as a Gröbner basis $\mathbb{G}$ and set $\Psi \leftarrow \Psi \cup \{\mathbb{G}\}$.

U3. While $\exists \mathbb{G}, \mathbb{G}^* \in \Psi$ such that the normal form of G modulo $\mathbb{G}^*$ is 0 for all $G \in \mathbb{G}$ do:

    Set $\Psi \leftarrow \Psi \setminus \{\mathbb{G}^*\}$.

The termination of the algorithm is obvious. The variety decomposition (6) and the unmixture of each $\text{Zero}(\mathbb{P}_i)$ is guaranteed by Lemma 1 and Theorem 4.

Note that the unmixed decomposition (6) computed by Algorithm UnmVarDec is not necessarily irredundant. To remove *all* redundant components, extra computation is required. The removal of redundant subvarieties by examining the containment relations among the corresponding Gröbner bases has the drawback that one component can be removed only if the corresponding Gröbner basis has already been computed. The following lemma provides another criterion for removing redundant components.

Lemma 6. *Let $\mathbb{T}$ be a regular set in $\mathcal{K}[\boldsymbol{x}]$ and $\mathbb{P}$ a finite basis for $\text{sat}(\mathbb{T})$. If $\mathbb{P}^*$ is a polynomial set such that $\text{prem}(\mathbb{P}^*, \mathbb{T}) = \{0\}$, then $\text{Zero}(\mathbb{P}) \subset \text{Zero}(\mathbb{P}^*)$.*

Proof. Since $\mathbb{T}$ is regular and $\text{prem}(\mathbb{P}^*, \mathbb{T}) = \{0\}$, $\mathbb{P}^* \subset \text{sat}(\mathbb{T})$ by Theorem 2. It follows that

$$\text{Zero}(\mathbb{P}) = \text{Zero}(\text{sat}(\mathbb{T})) \subset \text{Zero}(\mathbb{P}^*).$$

Irreducible decomposition

Now we come to decompose an arbitrary algebraic variety defined by a polynomial set into a family of irreducible subvarieties. This is done with an analogy to the unmixed decomposition of $\text{Zero}(\mathbb{P})$, requiring in addition that each triangular set $\mathbb{T}_i$ in the zero decomposition is irreducible. Then any finite basis for $\text{sat}(\mathbb{T}_i)$ will define an irreducible variety with any generic zero of $\mathbb{T}_i$ as its generic point.

Note that any point $\boldsymbol{\xi}$ of an algebraic variety $\mathfrak{V}$ over some extension of $\mathcal{K}$, which is such that every polynomial annulled by $\boldsymbol{\xi}$ vanishes on $\mathfrak{V}$, is called a *generic point* of $\mathfrak{V}$. A *generic zero* of an irreducible triangular set $\mathbb{T}$ is a zero $(\bar{x}_1, \ldots, \bar{x}_n)$ of $\mathbb{T}$ in which $\bar{x}_i$ takes an indeterminate when $x_i \notin \text{lvar}(\mathbb{T})$. An ideal $\mathfrak{I} \subset \mathcal{K}[\boldsymbol{x}]$ is said to be *prime* if whenever $F, G \in \mathcal{K}[\boldsymbol{x}]$ and $FG \in \mathfrak{I}$, either $F \in \mathfrak{I}$ or $G \in \mathfrak{I}$.

Theorem 5. *For any irreducible triangular set* $\mathbb{T} \subset \mathcal{K}[\boldsymbol{x}]$, *the ideal* $\text{sat}(\mathbb{T})$ *is prime.*

Proof. Let $\boldsymbol{\xi}$ be a generic zero of $\mathbb{T}$; then

$$\text{prem}(P, \mathbb{T}) = 0 \iff P(\boldsymbol{\xi}) = 0$$

for any $P \in \mathcal{K}[\boldsymbol{x}]$ by Lemma 3 in [22]. Let $FG \in \text{sat}(\mathbb{T})$. Then $\text{prem}(FG, \mathbb{T}) = 0$ according to Theorem 2, so

$$F(\boldsymbol{\xi})G(\boldsymbol{\xi}) = 0.$$

It follows that either $F(\boldsymbol{\xi}) = 0$ or $G(\boldsymbol{\xi}) = 0$; that is, either $\text{prem}(F, \mathbb{T}) = 0$ or $\text{prem}(G, \mathbb{T}) = 0$. By Theorem 2, we have either $F \in \text{sat}(\mathbb{T})$ or $G \in \text{sat}(\mathbb{T})$. Therefore, $\text{sat}(\mathbb{T})$ is prime. ∎

When $\text{sat}(\mathbb{T})$ is prime, its finite basis is called a *prime basis* of $\mathbb{T}$ and denoted by $\text{PB}(\mathbb{T})$. Then the variety defined by $\text{PB}(\mathbb{T})$ should have any generic zero of $\mathbb{T}$ as its generic point.

Proposition 2. *Let* $\mathbb{T}_1$ *and* $\mathbb{T}_2$ *be two irreducible triangular sets in* $\mathcal{K}[\boldsymbol{x}]$ *which have the same set of generic zeros. Then* $\text{sat}(\mathbb{T}_1) = \text{sat}(\mathbb{T}_2)$.

Proof. See [21, Sect. 6.2]. ∎

Proposition 3. *Let* $\mathbb{T}_1$ *and* $\mathbb{T}_2$ *be two triangular sets in* $\mathcal{K}[\boldsymbol{x}]$ *with* $\text{lvar}(\mathbb{T}_1) = \text{lvar}(\mathbb{T}_2)$, *and* $\mathbb{T}_2$ *be irreducible. If* $\text{prem}(\mathbb{T}_2, \mathbb{T}_1) = \{0\}$, *then* $\mathbb{T}_1$ *is also irreducible and has the same set of generic zeros as* $\mathbb{T}_2$; *thus* $\text{sat}(\mathbb{T}_1) = \text{sat}(\mathbb{T}_2)$.

Proof. See [21, Sect. 6.2]. ∎

Proposition 3 generalizes a result in [3]; in the same paper the following is also proved.

Proposition 4. *Let* $\mathbb{T}_1$ *and* $\mathbb{T}_2$ *be two triangular sets in* $\mathcal{K}[\boldsymbol{x}]$, *of which* $\mathbb{T}_1$ *is irreducible. If* $\text{prem}(\mathbb{T}_2, \mathbb{T}_1) = \{0\}$ *and* $0 \notin \text{prem}(\text{ini}(\mathbb{T}_2), \mathbb{T}_1)$, *then* $\text{sat}(\mathbb{T}_2) \subset \text{sat}(\mathbb{T}_1)$.

Let each $\mathbb{T}_i$ in (5) be irreducible. Then we have the following zero decomposition

$$\text{Zero}(\mathbb{P}) = \bigcup_{i=1}^{e} \text{Zero}(\text{PB}(\mathbb{T}_i)).$$

Now each $\text{PB}(\mathbb{T}_i)$, which can be exactly determined by using Gröbner bases, defines an irreducible algebraic variety and we have thus accomplished an irreducible decomposition of the variety $\mathfrak{V}$ defined by $\mathbb{P}$.

This decomposition is not necessarily minimal. All the redundant subvarieties can be removed by using Proposition 4 and Lemma 6 or 5, so one can get a *minimal* irreducible decomposition.

The following algorithm is modified from UnmVarDec and terminates obviously. Its correctness follows from the above discussions and the remarks on early removal of redundant components in Section 4 (that ensures the decomposition minimal).

Algorithm IrrVarDec (Input: $\mathbb{P}$; Output: Ψ). Given a non-empty polynomial set $\mathbb{P} \subset \mathcal{K}[\boldsymbol{x}]$, this algorithm computes a finite set Ψ of polynomial sets $\mathbb{P}_1, \ldots, \mathbb{P}_e$ such that the decomposition (6) holds, it is minimal, and each $\mathbb{P}_i$ defines an irreducible algebraic variety.

I1. Compute an irreducible W-characteristic series Φ of $\mathbb{P}$ and set $\Phi \leftarrow \{\mathbb{T} \in \Phi \mid |\mathbb{T}| \le |\mathbb{P}|\}, \Psi \leftarrow \emptyset$.

I2. While $\Phi \ne \emptyset$ do:

    I2.1. Let $\mathbb{T}$ be an element of Φ of highest dimension and set $\Phi \leftarrow \Phi \setminus \{\mathbb{T}\}$.

    I2.2. Compute a finite basis $\mathbb{G}$ for $\text{sat}(\mathbb{T})$ according to Lemma 1 or 2 and set $\Psi \leftarrow \Psi \cup \{\mathbb{G}\}$.

    I2.3. While $\exists \mathbb{T}^* \in \Phi$ such that $\text{prem}(\mathbb{G}, \mathbb{T}^*) = \{0\}$ do:
        Set $\Phi \leftarrow \Phi \setminus \{\mathbb{T}^*\}$.

4   Examples, Experiments and Techniques

Here examples and experiments are presented mainly to illustrate irreducible decomposition which usually is more difficult.

Three illustrative examples

Example 1. Consider the polynomial set $\mathbb{P} = \{P_1, P_2, P_3\}$, where

$$P_1 = -x_2 x_3 x_4 - x_4 - x_1 x_3 + x_2^2 - x_2 - x_1,$$
$$P_2 = x_2 x_3^3 - 2x_1 x_3^2 + x_3^2 + x_1 x_2 x_3 - 2x_2 x_3 + x_1,$$
$$P_3 = 2x_2 x_3 x_4 + 2x_4 + x_2 x_3^3 - 2x_1 x_3^2 + x_3^2 + x_1 x_2 x_3 - 2x_2 x_3 + 2x_1 x_3 + 2x_2 + x_1$$

are all irreducible over $\boldsymbol{Q}$ (the field of rational numbers). $\mathbb{P}$ can be decomposed into three fine quasi-irreducible triangular systems $[\mathbb{T}, \{x_2, I\}], [\mathbb{T}', \{x_1\}], [\mathbb{T}'', \emptyset]$ such that

$$\text{Zero}(\mathbb{P}) = \text{Zero}(\mathbb{T}/x_2 I) \cup \text{Zero}(\mathbb{T}'/x_1) \cup \text{Zero}(\mathbb{T}''),$$

where
$$\mathbb{T} = \begin{bmatrix} x_2^2 - x_1, \\ x_2 x_3^3 - 2x_1 x_3^2 + x_3^2 + x_1 x_2 x_3 - 2x_2 x_3 + x_1, \\ I x_4 + x_1 x_2 x_3 + x_1 \end{bmatrix},$$

$$\mathbb{T}' = [x_2^2 - x_1, I], \quad \mathbb{T}'' = [x_1, \ldots, x_4]$$

and $I = x_1 x_3 + x_2$.

Let the three polynomials in the above $\mathbb{T}$ be denoted T_1, T_2, T_3 respectively. $\mathbb{T}$ is not regular because $\mathrm{res}(\mathrm{res}(I, T_2, x_3), T_1, x_2) = 0$. In fact, $\mathbb{T}$ is reducible: one may verify that T_2 has an irreducible factorization

$$T_2 \doteq \frac{x_2}{x_1} I (x_3 - x_2)^2$$

over $\mathbf{Q}(x_1, x_2)$ with x_2 having minimal polynomial T_1. Therefore,

$$\mathrm{Zero}(\mathbb{T}/x_2 I) = \mathrm{Zero}([T_1, x_3 - x_2, T_3]/x_2 I) = \mathrm{Zero}(\mathbb{T}^*/x_2 I),$$

where

$$\mathbb{T}^* = [T_1, x_3 - x_2, x_4 + x_2].$$

It is easy to see that both $\mathbb{T}'$ and $\mathbb{T}''$ are irreducible. So we have, after applying Theorem 1 and removing a redundant component, an irreducible zero decomposition of the form

$$\mathrm{Zero}(\mathbb{P}) = \mathrm{Zero}(\mathbb{T}^*) \cup \mathrm{Zero}(\mathbb{T}'/x_1).$$

The three polynomials in $\mathbb{T}^*$ all have constant initials, so $\mathbb{T}^*$ defines by itself an irreducible algebraic variety. To determine a prime basis of $\mathbb{T}'$, we compute the Gröbner basis $\mathbb{G}$ of $\mathbb{T}' \cup \{zx_1 - 1\}$ over $\mathbf{Q}$ under plex with $x_1 \prec \cdots \prec x_4 \prec z$. It is found that

$$\mathbb{G} = \mathbb{T}' \cup \{x_2 x_3 + 1, z - x_3^2\},$$

so the prime basis wanted is

$$\mathbb{P}' = \mathbb{T}' \cup \{x_2 x_3 + 1\}.$$

Hence we obtain finally the following irreducible decomposition of the variety defined by $\mathbb{P}$

$$\mathrm{Zero}(\mathbb{P}) = \mathrm{Zero}(\mathbb{T}^*) \cup \mathrm{Zero}(\mathbb{P}').$$

Finally we remark that the triangular set $\mathbb{T}'$ is irreducible, even though the second polynomial I in $\mathbb{T}'$ may be written as $I = x_2(x_2 x_3 + 1)$ over $\mathbf{Q}(x_1, x_2)$ with x_2 having minimal polynomial $x_2^2 - x_1$. Let $\bar{\mathbb{T}} = [x_2^2 - x_1, x_2 x_3 + 1]$. It follows from the factorized from of I and Theorem 1 that

$$\mathrm{Zero}(\mathbb{T}'/x_1) = \mathrm{Zero}(\bar{\mathbb{T}}/x_2) = \mathrm{Zero}(\bar{\mathbb{T}});$$

the last equality holds because $x_2 x_3 + 1 = 0$ implies that $x_2 \neq 0$. Therefore, $\mathbb{P}'$ can be replaced by $\bar{\mathbb{T}}$ (in fact, $\mathbb{P}'$ is a plex Gröbner basis of $\bar{\mathbb{T}}$). Simplifications of this kind to avoid the computation of saturation bases can be achieved by some other algorithms.

Example 2. Let the algebraic variety $\mathfrak{V}$ be defined by $\mathbb{P} = \{P_1, P_2, P_3\}$, where

$$P_1 = 3x_3x_4 - x_2^2 + 2x_1 - 2,$$
$$P_2 = 3x_1^2x_4 + 4x_2x_3 + 6x_1x_3 - 2x_2^2 - 3x_1x_2,$$
$$P_3 = 3x_3^2x_4 + x_1x_4 - x_2^2x_3 - x_2.$$

With $x_1 \prec \cdots \prec x_4$, $\mathbb{P}$ may be decomposed into 2 irreducible triangular sets $\mathbb{T}_1$ and $\mathbb{T}_2$ such that

$$\mathrm{Zero}(\mathbb{P}) = \mathrm{Zero}(\mathbb{T}_1/I) \cup \mathrm{Zero}(\mathbb{T}_2/x_2),$$

where

$$\mathbb{T}_1 = [T_1, T_2, Ix_4 - 2x_2^2 - 3x_1x_2],$$
$$\mathbb{T}_2 = [x_1, 2x_3 - x_2, 3x_2x_4 - 2x_2^2 - 4];$$

$$T_1 = 2x_2^4 - 12x_1^2x_2^3 + 9x_1x_2^3 - 9x_1^4x_2^2 + 8x_1x_2^2 - 8x_2^2 + 24x_1^3x_2 - 24x_1^2x_2 + 18x_1^5 - 18x_1^4,$$
$$T_2 = Ix_3 - x_2^2;$$

$$I = 2x_2 + 3x_1^2.$$

To obtain an irreducible decomposition of $\mathfrak{V}$, we determine the prime bases of $\mathbb{T}_1$ and $\mathbb{T}_2$ by computing the respective Gröbner bases $\mathbb{G}_1$ and $\mathbb{G}_2$ of

$$\mathbb{T}_1 \cup \{zI - 1\} \quad \text{and} \quad \mathbb{T}_2 \cup \{x_2z - 1\}$$

according to Lemma 1. The Gröbner bases may be found to consist of 8 and 4 polynomials respectively. Let $\mathbb{V}_i = \mathbb{G}_i \cap \mathcal{K}[x_1, \ldots, x_4]$ and $\mathfrak{V}_i = \mathrm{Zero}(\mathbb{V}_i)$ for $i = 1, 2$. We have

$$\mathbb{V}_1 = \begin{bmatrix} T_1, \\ 27x_1^4x_3 - 27x_1^3x_3 + 2x_2^3 - 15x_1^2x_2^2 + 9x_1x_2^2 + 8x_1x_2 - 8x_2 + 12x_1^3 - 12x_1^2, \\ T_2, \\ 12x_1x_3^2 - 12x_3^2 - 9x_1^2x_3 - 2x_1x_2^2 + 3x_2^2 + 4x_1^2 - 4x_1, \\ x_1x_4 - 2x_1x_3 + 2x_3 - x_2, \\ x_2x_4 + 3x_1^2x_3 - 3x_1x_3 - x_2^2, \\ P_1 \end{bmatrix}$$

and $\mathbb{V}_2 = \mathbb{T}_2$ such that $\mathfrak{V} = \mathfrak{V}_1 \cup \mathfrak{V}_2$. One can check with ease that this decomposition is minimal.

Example 3. As a more complicated example, consider the algebraic variety defined by the following five polynomials

$$P_1 = x_4x_2 + x_1 + x_2x_6 + 3x_5,$$
$$P_2 = 54x_4x_5 + 9x_4x_2x_6 - 9x_1x_6 - 9x_2x_7 - 18x_3x_2 - 2x_2^3,$$
$$P_3 = 18x_3x_5 - 9x_4^2x_5 + 3x_3x_2x_6 + 3x_4x_6x_1 + 3x_4x_7x_2 - 3x_1x_7 - 3x_3x_1 - 2x_2^2x_1,$$
$$P_4 = 3x_3x_1x_6 + 3x_3x_2x_7 + 3x_4x_1x_7 - 18x_4x_3x_5 - 2x_2x_1^2,$$
$$P_5 = 9x_3x_1x_7 - 27x_3^2x_5 - 2x_1^3.$$

Let $\mathbb{P} = \{P_1, \ldots, P_5\}$ and the variable ordering be $\omega_1: x_1 \prec \cdots \prec x_7$. Under ω_1, $\mathbb{P}$ can be decomposed into 9 irreducible triangular sets $\mathbb{T}_1, \ldots, \mathbb{T}_9$ such that

$$\text{Zero}(\mathbb{P}) = \bigcup_{i=1}^{9} \text{Zero}(\mathbb{T}_i/\text{ini}(\mathbb{T}_i)).$$

For $i = 6, \ldots, 9$, the triangular set $\mathbb{T}_i$ contains more than 5 polynomials and thus need not be considered for the variety decomposition by Lemma 4. Let $\mathbb{V}_i$ be the prime basis of $\mathbb{T}_i$ under the ordering ω_1 for $i = 3, 4, 5$. Obviously $\mathbb{T}_3$ already defines an irreducible variety, so $\mathbb{V}_3 = \mathbb{T}_3$. It remains to determine the prime bases from $\mathbb{T}_1, \mathbb{T}_2, \mathbb{T}_4$ and $\mathbb{T}_5$ according to Lemma 1 or 2. One may find that $\mathbb{V}_4 = \mathbb{T}_4$ and $\mathbb{V}_5$ is the same as the set obtained by replacing the last polynomial in $\mathbb{T}_5$ with

$$9x_7 + 9x_4x_6 - 2x_2^2.$$

A prime basis of $\mathbb{T}_1$ under ω_1 contains 20 polynomials. To reduce the number of elements, we compute a Gröbner basis of this prime basis with respect to another variable ordering $\omega_2: x_4 \prec x_2 \prec x_6 \prec x_3 \prec x_1 \prec x_7 \prec x_5$. The new basis $\mathbb{V}_1$ consists of 10 polynomials as follows

$$\mathbb{V}_1 = \begin{bmatrix} 81x_3^3 + 72x_2^2x_3^2 + 16x_4^4x_3 + 90x_4^2x_2^2x_3 + 4x_4^2x_2^4 + 18x_4^4x_2^2, \\ 6x_4x_2^2x_1 + 9x_4^3x_1 - 9x_2x_3^2 - 4x_2^3x_3 + 9x_4^2x_2x_3 + 2x_4^2x_2^3 + 9x_4^4x_2, \\ 9x_3x_1 + 4x_2^2x_1 + 9x_4^2x_1 + 18x_4x_2x_3 + 2x_4x_2^3 + 9x_4^3x_2, \\ x_1^2 + x_4x_2x_1 - x_2^2x_3, \\ 9x_4^3x_7 - 6x_4x_2x_6x_1 - 12x_4^2x_2x_1 + 9x_6x_3^2 + 18x_4x_3^2 + 4x_2^2x_6x_3 \\ \quad - 9x_4^2x_6x_3 + 8x_4x_2^2x_3 - 2x_4^2x_2^2x_6 - 2x_4^3x_2^2, \\ 9x_2x_7 + 9x_6x_1 + 18x_4x_1 + 18x_2x_3 + 9x_4x_2x_6 + 2x_2^3 + 18x_4^2x_2, \\ 9x_3x_7 + 9x_4^2x_7 - 4x_2x_6x_1 - 8x_4x_2x_1 + 18x_3^2 - 9x_4x_6x_3 + 2x_2^2x_3 \\ \quad - 2x_4x_2^2x_6 - 2x_4^3x_2^2, \\ 9x_1x_7 - 6x_2^2x_1 - 18x_4^2x_1 + 9x_2x_6x_3 - 18x_4x_2x_3 - 4x_4x_2^3 - 18x_4^3x_2, \\ 81x_7^2 + 81x_4x_6x_7 - 162x_4^2x_7 + 108x_2x_6x_1 + 216x_4x_2x_1 - 324x_3^2 \\ \quad - 81x_6^2x_3 + 162x_4x_6x_3 - 72x_2^2x_3 + 54x_4x_2^2x_6 - 4x_2^4 + 36x_4^2x_2^2, \\ P_1 \end{bmatrix}.$$

As for $\mathbb{T}_2$, the difficult case, let T_i denote the ith polynomial of $\mathbb{T}_2$ and I_i the initial of T_i for $1 \leq i \leq 5$. The non-constant initials are

$$I_2, \quad I_3, \quad \text{and} \quad I_4 = I_5 = x_2.$$

Thus, it is necessary to determine a prime basis from $\mathbb{T}_2$ by computing a Gröbner basis of the enlarged polynomial set, for instance, $\mathbb{T}_2 \cup \{z_1I_4 - 1, z_2I_3 - 1, z_3I_2 - 1\}$ or $\mathbb{T}_2 \cup \{zI_2I_3I_4 - 1\}$. Nevertheless, the Gröbner basis cannot be easily computed in either case. We have tried some of the most powerful Gröbner bases packages without success. For this reason, we normalize $\mathbb{T}_2$ to get another triangular set

$\mathbb{T}_2^*$ (see [21, Sect. 5.2] for the details of normalization): it is obtained from $\mathbb{T}_2$ by replacing T_2 and T_3 respectively with

$$T_2^* = -4x_1^3x_2x_4 + 81x_3^4 + 9x_2^2x_3^2 - 9x_1^2x_3^2 + 6x_1^2x_2^2x_3 - 2x_1^4,$$

$$T_3^* = 972x_1^7x_5 + 729(2x_2^4 + 27x_1^2)x_2^2x_3^5 + 81(2x_2^8 + 9x_1^2x_2^4 - 81x_1^4)x_3^4$$
$$- 648x_1^2(x_2^4 + 9x_1^2)x_2^2x_3^3 + 9x_1^2(8x_2^8 + 180x_1^2x_2^4 + 81x_1^4)x_3^2$$
$$- 36x_1^4(2x_2^4 + 27x_1^2)x_2^2x_3 + 2x_1^4(4x_2^8 + 90x_1^2x_2^4 + 243x_1^4).$$

$\mathbb{T}_2^*$ and $\mathbb{T}_2$ have the same set of generic zeros, so the prime bases constructed from them define the same irreducible algebraic variety. $\mathbb{T}_2^*$ possesses the property that the initials of its polynomials only involve the parameters x_1 and x_2.

A prime basis of $\mathbb{T}_2^*$ can be easily determined by computing the corresponding Gröbner basis with respect to the variable ordering ω_1 or ω_2 according to Lemma 1. The basis under ω_2 contains 9 elements and is as follows

$$\mathbb{V}_2 = \begin{bmatrix} 81x_4^3x_6^2 + 16x_2^4x_6 + 108x_4^2x_2^2x_6 + 324x_4^4x_6 + 20x_4x_2^4 + 144x_4^3x_2^2 + 324x_4^5, \\ 144x_2^2x_3 + 729x_4^2x_3 + 81x_4^3x_6 + 16x_2^4 + 144x_4^2x_2^2 + 405x_4^4, \\ 4x_6x_3 + 5x_4x_3 + x_4^2x_6 + x_4^3, \\ 4x_2x_1 + 27x_4x_3 + 2x_4x_2^2 + 9x_4^3, \\ 18x_6x_1 + 36x_4x_1 - 18x_2x_3 + 9x_4x_2x_6 - 2x_2^3, \\ 972x_4x_3x_1 + 324x_4^3x_1 - 1296x_2x_3^2 - 405x_4^2x_2x_3 + 81x_4^3x_2x_6 + 16x_2^5 \\ \quad + 108x_4^2x_2^3 + 243x_4^4x_2, \\ 144x_1^2 + 1296x_3^2 - 81x_4^2x_3 - 81x_4^3x_6 - 16x_2^4 - 144x_4^2x_2^2 - 405x_4^4, \\ 6x_7 + 18x_3 + 3x_4x_6 + 2x_2^2 + 12x_4^2, \\ P_1 \end{bmatrix}.$$

It is easy to verify that both $\text{Zero}(\mathbb{V}_4)$ and $\text{Zero}(\mathbb{V}_5)$ are subvarieties of $\text{Zero}(\mathbb{V}_1)$. Therefore, the variety defined by $\mathbb{P}$ is decomposed into three irreducible subvarieties defined by $\mathbb{V}_1, \mathbb{V}_2$ and $\mathbb{V}_3$. Symbolically,

$$\text{Zero}(\mathbb{P}) = \text{Zero}(\mathbb{V}_1) \cup \text{Zero}(\mathbb{V}_2) \cup \text{Zero}(\mathbb{V}_3),$$

where $\text{Zero}(\mathbb{V}_i)$ is irreducible for $i = 1, 2, 3$.

Experimental results

We first provide timing statistics for irreducible variety decomposition on 22 examples using the three zero decomposition algorithms reviewed in Section 2. The first 16 examples are given in [15]. Here are added to our test set the last six examples, for which the defining polynomials of the varieties as well as the variable orderings used for the computation are listed in the appendix.

The heading entries Ritt-Wu, Seidenberg refined and Subresultant-based in Table 1 indicate the respective zero decomposition algorithms used for the irreducible variety decomposition. The experiments were made with Maple V.5

running on an i686 under Linux 2.0.36 at LIP6, Université Paris VI. The timings are given in CPU seconds and include the time for garbage collection. The symbol >2000 indicates that the computation was interrupted manually after about 2000 CPU seconds, and the cases marked with *** were rejected by Maple for "object too large."

Ex	Ritt-Wu	Seidenberg refined	Subresultant-based
1	.05	.02	.02
2	.52	.45	.36
3	1.01	.69	.66
4	.43	.11	.07
5	.93	.66	2.43
6	2.03	11.73	113.06
7	.63	.49	.41
8	9.53	1.15	2.16
9	17.19	4.92	4.8
10	>2000	>2000	>2000
11	3.81	5.06	***
12	7.58	2.38	7.05
13	17.11	5.62	8.35
14	33.54	16.64	19.68
15	29.47	5.56	4.18
16	***	***	***
17	18.35	1.54	.72
18	>2000	>2000	>2000
19	***	44.39	31.74
20	13.91	6.96	4.45
21	>2000	>2000	>2000
22	>2000	>2000	>2000

Table 1. Irreducible variety decomposition for 22 examples

From Table 1 one sees that irreducible variety decomposition using algorithms based on Seidenberg's elimination idea and s.r.s. is faster than that using characteristic sets for most of the complete test cases. Example 6 is one of the few notable examples for which our algorithms are slower than Ritt-Wu's for the irreducible zero decomposition as already pointed out in [18]. The difference of computing times with the three zero decomposition algorithms is often not significant because the involved saturation bases computation may take a large part of the total computing time. It is also possible that computing the saturation bases from the triangular sets requires more time than that from the ascending sets. This actually happens with Example 9 (for which the irreducible zero decomposition by either algorithm does not take much time) in Maple V.3.

As we have explained previously, there are two different ways for computing the prime/saturation bases of triangular sets. There are also a few variants for the algorithms which may affect the computation. Let $\mathbb{T} = [T_1, \ldots, T_r]$ be an irreducible triangular set in $\mathcal{K}[x]$, $F_1, \ldots, F_t$ be all the distinct irreducible factors (over $\mathcal{K}$) of $\prod_{i=1}^{r} \text{ini}(T_i)$, and $F = F_1 \cdots F_t$. We consider the following four variants for the construction of a prime basis $\mathbb{V}$ of $\mathbb{T}$:

V1. Compute the Gröbner basis $\mathbb{G}$ of $\mathbb{T} \cup \{zF - 1\}$ with respect to plex[1] with $x_1 \prec \cdots \prec x_n \prec z$ and take $\mathbb{G} \cap \mathcal{K}[x]$ for $\mathbb{V}$, where z is a new indeterminate;

V2. Compute the Gröbner basis $\mathbb{G}$ of $\mathbb{T} \cup \{z_1F_1 - 1, \ldots, z_tF_t - 1\}$ with respect to plex with $x_1 \prec \cdots \prec x_n \prec z_1 \prec \cdots \prec z_t$ and take $\mathbb{G} \cap \mathcal{K}[x]$ for $\mathbb{V}$, where $z_1, \ldots, z_t$ are new indeterminates;

V3. Compute the generating sets $\mathbb{V}_k$ for $\text{Ideal}(\mathbb{T}) : F^k$, $k = 1, 2, \ldots$, until $\text{Ideal}(\mathbb{T}) : F^k = \text{Ideal}(\mathbb{T}) : F^{k+1}$ and take the last $\mathbb{V}_k$ for $\mathbb{V}$, where the quotient ideals are computed using plex Gröbner bases with $x_1 \prec \cdots \prec x_n \prec z$ according to what is explained in the paragraph after Lemma 2;

V4. Normalize $\mathbb{T}$ to obtain an equivalent triangular set $\bar{\mathbb{T}}$ (see [21, Sect. 5.2] and Example 3) and then compute a prime basis of $\bar{\mathbb{T}}$ according to V2 above.

The following table shows the computational effect of the four variants.

Ex	V1	V2	V3	V4
3	.41	.46	1.51	.24
9	11.18	11.98	63.11	5.18
13	3.09	—	64.5	2.43
14	16.19	11.22	246.62	12.24
18	>2000	>2000	>2000	>2000
19	109.64	23.45	>2000	>2000
20	2.6	—	23.49	.79
21	>2000	>2000	>2000	>2000
22	>2000	>2000	>2000	>2000

Table 2. Variants for prime bases computation

The dash — in Table 2 denotes the cases in which the variant V2 coincides with V1 (i.e., $t = 1$). It occurs that the method V3 performs worst for these examples. For the two better variants V2 and V4, it is not evident which one is superior to the other. An additional message is that, in Maple V.3 running on an Alpha station 600 at MMRC, Academia Sinica, the Gröbner bases were computed in 602.62 and 1371.7 CPU seconds respectively with V1 and V2 for Example 18, and in 201.68 CPU seconds with V4 for Example 21. This makes it even difficult to predetermine which of the three variants V1, V2 and V4 to take for a concrete problem.

As the computation of saturation bases represents a major time-consuming step, it is of interest to try other Gröbner bases implementations for our test ex-

---

[1] We have tried the new Groebner package with the elimination term order lexdeg in Maple V.5. For most of our examples, notably Ex 9 and Ex 19, the computation using lexdeg is much slower than that using plex.

amples. We did so using the Gröbner basis packages in Reduce 3.4 and Risa/Asir[2] in comparison with **grobner** in Maple V.2, and the Gb package of J.-C. Faugère in comparison with **Groebner** in Maple V.5. For Example 14, 14 Gröbner bases have to be computed and the variety is decomposed into 6 irreducible components. The detailed timings obtained on one and the same machine at ISIS in August 1993 for Maple V.2, Reduce 3.4 and Risa/Asir, and on the above-mentioned i686 at LIP6 for Maple V.5 and Gb 3.1156 are exhibited below.

No	1	2	3	4	5	6	7
Maple V.2	7.32	11.55	5.12	5.83	23.08	3.43	27.17
Reduce 3.4	.58	1.92	.51	.73	1.72	.32	9.08
Risa/Asir	1.5	2.91	.78	1.4	4.89	.5	7.74
Maple V.5	.74	.59	.37	.53	3.08	.22	.77
Gb 3.1156	.03	0	.01	0	.09	.01	.02

No	8	9	10	11	12	13	14
Maple V.2	4.75	3.03	3.07	121.93	4.08	2.77	3698.38
Reduce 3.4	.38	.23	.52	14.65	.32	.38	944.18
Risa/Asir	.75	.56	.69	14.97	.74	.74	243.76
Maple V.5	.32	.19	.17	6.31	.22	.23	4.31
Gb 3.1156	.01	.02	.01	.05	0	.01	.02

Table 3. Gröbner bases computation for Example 14

The remarkable difference of timings shown in Table 3 was somewhat to our surprise, and indicates the inefficiency of the Gröbner basis packages in Maple and the interest in implementing our decomposition algorithms using another computer algebra system or a low-level programming language for effective applications. We are still working with Maple mainly because of our experience and the existence of an elimination library we have developed. Although Gb took little time to compute the 14 Gröbner bases for Example 14, it did not succeed in computing the two Gröbner bases needed for Example 18. We are going to try Faugère's FGb, a new version of Gb, and other implementations of Gröbner bases such as those in Magma and CoCoA for large examples.

Techniques for practical efficiency

The algorithms for variety decomposition discussed in the previous sections employ several time-consuming subalgorithms. Improvements on each subalgorithm have positive consequence for the overall performance. Roughly speaking, the

[2] This is an experimental system developed by the computer algebra group at ISIS, Fujitsu Laboratories Limited, Japan. M. Noro from this group informed the author that the computing time with Risa/Asir in Table 3 may be much reduced by setting some switches on. Moreover, by using another C implementation of a Gröbner basis package in Risa/Asir, the 14 Gröbner bases can be computed in 3 CPU seconds.

main algorithms consist of two essentially independent steps: one computes the zero decomposition and the other constructs the saturation bases. If the zero decomposition cannot be completed, the variety decomposition fails. When the first step is successful, we come to the second step of constructing the saturation bases. Two different algorithms with variants for this step have been explained in Section 3. Our experiments have shown the dependence of their efficiency on problems and the potential of using other Gröbner packages. The construction of saturation bases is unnecessary in some cases, for instance, when in (1) $e = 1$ and $\mathbb{T}_1$ is irreducible (and thus the given variety is irreducible), or all $\mathbb{T}_i$ are of dimension 0, or all the initials of the polynomials in $\mathbb{T}_i$ are constants. This happened for about half of the cases in our test. Construction of the prime bases requires expensive Gröbner bases computation; for this we simply use the grobner/Groebner packages in Maple without modification. For four of the test examples we failed in computing the Gröbner bases within 2000 CPU seconds for each or because of "object too large."

For zero decomposition, we continue studying algorithmic and implementational strategies to speed up each substep. Some of the strategies and techniques have been discussed in the papers where the zero decomposition algorithms are proposed and experimented. In what follows, we mention a few removing techniques, some of which have not been given before.

EARLY REMOVAL OF REDUNDANT COMPONENTS. The zero decomposition methods reviewed in Section 2 may produce many redundant branches/components. It is always desirable to remove such redundant components as early as possible. In fact, there are a number of techniques for doing so, see [3, 7, 15, 17, 18] for example. One of the most effective techniques suggested by Chou and Gao [3, 7] for variety decomposition is the application of the affine dimension theorem (i.e., $\mathbb{T}_i$ is redundant if $|\mathbb{T}_i| > |\mathbb{P}| = s$, see Lemma 4) when a W-characteristic series $\mathbb{T}_i$ of $\mathbb{P}$ is computed. For being more efficient, we have used this technique dynamically (i.e., whether the final $\mathbb{T}_i$ will contain more than s polynomials is examined during the computation of $\mathbb{T}_i$) in the implementation of our zero decomposition algorithms based Seidenberg's idea and s.r.s. This has taken the advantage of top-down elimination: for a triangular system $[\mathbb{T}, \mathbb{U}]$, it is known that either $|\mathbb{T}| > s$ or $\mathrm{Zero}(\mathbb{T}/\mathbb{U}) = \emptyset$ as soon as the last s polynomials of $\mathbb{T}$ have been obtained and at least one more polynomial of $\mathbb{T}$ can be obtained (it is the case when there still exist polynomials for elimination to produce $\mathbb{T}$). In other words, a redundant triangular system may be removed before it is completely computed. This technique is particularly useful for removing such redundant triangular systems $[\mathbb{T}, \mathbb{U}]$ of lower dimension for which the first few polynomials in $\mathbb{T}$ are large and difficult to compute.

For another technique, consider the last step U3 of UnmVarDec which aims at removing some redundant components. This step does not appear in IrrVarDec due to the following observation. Since $\mathbb{T}$ taken from Ψ in step I2.1 has the highest dimension, the remaining triangular sets $\mathbb{T}^*$ in Ψ are all those whose dimensions are not higher than that of $\mathbb{T}$; thus $\mathrm{Zero}(\mathrm{sat}(\mathbb{T}^*))$ are all the candidates of subvarieties of $\mathrm{Zero}(\mathrm{sat}(\mathbb{T}))$. After the saturation basis $\mathbb{G}$ of $\mathbb{T}$ is computed in step I2.2, we immediately compute prem($\mathbb{G}, \mathbb{T}^*$) in step I2.3, and remove such $\mathbb{T}^*$ when prem($\mathbb{G}, \mathbb{T}^*$) = $\{0\}$ is verified. Therefore, in the case where some $\mathbb{T}^*$ is successfully removed, its saturation basis must not have been computed.

This simple technique deducts the time for computing the prime bases that are redundant, and does not lead to any additional cost. The redundancy occurs often for examples from geometry theorem proving where the corresponding varieties are of higher dimension. We applied the technique to Example 14 and got a further surprise. Among the 14 Gröbner bases, 8 (including the one whose computation takes much time) correspond to redundant subvarieties and need not be computed according to this technique. The previous total decomposition time 2663.2 CPU seconds (with Maple V.2 on a SUN SparcServer 690/51) is reduced to 113.5 seconds. In other words, the technique speeds up the decomposition by a factor of 23.

However, the above-explained technique cannot be employed when the pseudo-division verification is replaced by the normal form reduction.

REMOVAL OF COMPUTED SUBVARIETIES. Now we describe a technique to remove a subvariety from the given variety as soon as the subvariety is computed. This is similar to the case of polynomial factorization in which an irreducible factor can readily be removed from the polynomial being factorized when the factor is found. However, the removal of subvarieties appears much more difficult computationally. The removing technique can be incorporated into the decomposition algorithms according to the following theorem.

Theorem 6. *Let $\mathbb{P}$ and $\mathbb{Q} = \{F_1, \ldots, F_t\}$ be two polynomial sets in $\mathcal{K}[\boldsymbol{x}]$ with* Zero($\mathbb{Q}$) $\subset$ Zero($\mathbb{P}$) *and $\mathfrak{I}$ be the ideal generated by*

$$\mathbb{P} \cup \{zF_1 + \cdots + z^t F_t - 1\} \quad in \quad \mathcal{K}[\boldsymbol{x}, z] \tag{7}$$

or by

$$\mathbb{P} \cup \{z_1 F_1 + \cdots + z_t F_t - 1\} \quad in \quad \mathcal{K}[\boldsymbol{x}, z_1, \ldots, z_t], \tag{8}$$

where $z, z_1, \ldots, z_t$ are new indeterminates. Then

$$\text{Zero}(\mathbb{P}) = \text{Zero}(\mathbb{Q}) \cup \text{Zero}(\mathfrak{I} \cap \mathcal{K}[\boldsymbol{x}]).$$

Proof. Consider the case in which

$$\mathfrak{I} = \text{Ideal}(\mathbb{P} \cup \{zF_1 + \cdots + z^t F_t - 1\}).$$

Let $\bar{\boldsymbol{x}} \in$ Zero($\mathbb{P}$). For any $P \in \mathfrak{I} \cap \mathcal{K}[\boldsymbol{x}]$, there exists a polynomial $Q \in \mathcal{K}[\boldsymbol{x}, z]$ such that

$$P - Q(zF_1 + \cdots + z^t F_t - 1) \in \text{Ideal}(\mathbb{P}) \subset \mathcal{K}[\boldsymbol{x}, z].$$

Hence

$$P(\bar{\boldsymbol{x}}) = Q(\bar{\boldsymbol{x}}, z)[zF_1(\bar{\boldsymbol{x}}) + \cdots + z^t F_t(\bar{\boldsymbol{x}}) - 1] \tag{9}$$

for arbitrary z. Suppose that $\bar{\boldsymbol{x}} \notin$ Zero($\mathbb{Q}$). Then there exists some j such that $F_j(\bar{\boldsymbol{x}}) \neq 0$. So there is a $\bar{z} \in \tilde{\mathcal{K}}$ such that $\bar{z}F_1(\bar{\boldsymbol{x}}) + \cdots + \bar{z}^t F_t(\bar{\boldsymbol{x}}) - 1 = 0$. Plunging $\bar{z}$ into (9), we get $P(\bar{\boldsymbol{x}}) = 0$. Therefore, Zero($\mathbb{P}$) $\subset$ Zero($\mathbb{Q}$) $\cup$ Zero($\mathfrak{I} \cap \mathcal{K}[\boldsymbol{x}]$).

To show the opposite, let $\bar{\boldsymbol{x}} \in \text{Zero}(\mathfrak{I} \cap \mathcal{K}[\boldsymbol{x}])$. Obviously, for $z \in \mathcal{K}[\boldsymbol{x}, z]$ and any $P \in \mathbb{P}$

$$\text{Zero}(\mathbb{P}) \subset \text{Zero}(P(zF_1 + \cdots + z^t F_t)).$$

By Hilbert's Nullstellensatz, there exists an exponent $q > 0$ such that

$$P^q (zF_1 + \cdots + z^t F_t)^q \in \text{Ideal}(\mathbb{P}) \subset \mathcal{K}[\boldsymbol{x}, z].$$

It follows that

$$P^q + P^q [(zF_1 + \cdots + z^t F_t)^{q-1} + (zF_1 + \cdots + z^t F_t)^{q-2} + \cdots + 1]$$
$$\cdot (zF_1 + \cdots + z^t F_t - 1) \in \text{Ideal}(\mathbb{P}) \subset \mathcal{K}[\boldsymbol{x}, z],$$

so that $P^q \in \mathfrak{I}$. Since P does not involve z, $P^q \in \mathfrak{I} \cap \mathcal{K}[\boldsymbol{x}]$. Hence, $P^q(\bar{\boldsymbol{x}}) = 0$ and thus $P(\bar{\boldsymbol{x}}) = 0$. This proves that $\text{Zero}(\mathfrak{I} \cap \mathcal{K}[\boldsymbol{x}]) \subset \text{Zero}(\mathbb{P})$.

The case in which $\mathfrak{I} = \text{Ideal}(\mathbb{P} \cup \{z_1 F_1 + \cdots + z_t F_t - 1\})$ is proved analogously, observing that if $F_1(\bar{\boldsymbol{x}}), \ldots, F_t(\bar{\boldsymbol{x}})$ are not all 0, then there exist $\bar{z}_1, \ldots, \bar{z}_t$ such that $\bar{z}_1 F_1(\bar{\boldsymbol{x}}) + \cdots + \bar{z}_t F_t(\bar{\boldsymbol{x}}) - 1 = 0$, and $P^q(z_1 F_1 + \cdots + z_t F_t)^q \in \text{Ideal}(\mathbb{P}) \subset \mathcal{K}[\boldsymbol{x}, z_1, \ldots, z_t]$ for some integer $q > 0$.

This theorem suggests a way to remove any subvariety $\text{Zero}(\mathbb{Q})$ from the given variety $\text{Zero}(\mathbb{P})$ by determining a finite basis $\mathbb{H}$ for the ideal $\mathfrak{I} \cap \mathcal{K}[\boldsymbol{x}]$. The latter can be done, for instance, by computing a Gröbner basis of (7) or of (8) with respect to plex determined by $x_j \prec z$ or $x_j \prec z_l$ together with its elimination property. Thus, decomposing $\text{Zero}(\mathbb{P})$ is reduced to decomposing $\text{Zero}(\mathbb{Q})$ and $\text{Zero}(\mathbb{H})$. The author has tested this technique. Nevertheless, the Gröbner bases computation in this case is too inefficient and we had no gain from the experiments. One can make use of the technique only when a more effective procedure for determining the finite bases is available.

In fact, the removal of $\text{Zero}(\mathbb{Q})$ from $\text{Zero}(\mathbb{P})$ corresponds to computing the quotient $\text{Ideal}(\mathbb{P}) : \text{Ideal}(\mathbb{Q})$. The latter can be done by a possibly more efficient algorithm described in [5, pp. 193–195].

Applications

Decomposing given algebraic varieties into irreducible or unmixed components is a fundamental task in classical algebraic geometry and has various applications in modern geometry engineering. Among such applications we can mention two: one in computer-aided geometric design where the considered geometric objects are desired to be decomposed into *simpler* subobjects and the other in automated geometry theorem proving where the configuration of the geometric hypotheses needs to be decomposed in order to determine for which components the geometric theorem holds true.

An irreducible decomposition of the algebraic variety defined by $\mathbb{P}$ corresponds to a prime decomposition of the radical of the ideal $\mathfrak{I}$ generated by $\mathbb{P}$. From the latter a primary decomposition of $\mathfrak{I}$ can be constructed, see [6, 13] for details. This points out an application of variety decomposition to computational polynomial ideal theory.

In various mathematical problems such as the qualitative study of differential equations (where Example 3 comes), necessary and sufficient conditions are often established and given as polynomial equalities in different ways and by different researchers. Variety decomposition provides an effective way to examine the relationships such as equivalence and containment among different sets of conditions.

Acknowledgments. This work was supported by FWF and CEC under ESPRIT BRP 6471 (MEDLAR II), by ISIS of Fujitsu Laboratories Limited under VRSP, and by CNRS/CAS under a cooperation project between LEIBNIZ and CICA. I wish to thank T. Nakagawa, T. Takeshima, K. Yokoyama, and L. Yang who organized my visit at ISIS and at CICA where part of this paper was written. I am also grateful to an anonymous referee whose comments have helped improve the paper, in particular, clarify the redundancy issue about the decomposition computed by Algorithm UnmVarDec.

References

1. Aubry, P., Lazard, D., Moreno Maza, M. (1999): On the theories of triangular sets. *J. Symb. Comput.* **28**: 105–124.
2. Buchberger, B. (1985): Gröbner bases: An algorithmic method in polynomial ideal theory. In: *Multidimensional systems theory* (N. K. Bose, ed.), D. Reidel Publ. Co., Dordrecht Boston, pp. 184–232.
3. Chou, S.-C., Gao, X.-S. (1990): Ritt-Wu's decomposition algorithm and geometry theorem proving. In: *Proceedings CADE-10*, Kaiserslautern, July 24–27, 1990, Springer-Verlag, Berlin Heidelberg, pp. 207–220 (Lecture notes in computer science, vol. 449) [also as Tech. Rep. TR-89-09, Department of Computer Science, The Univ. of Texas at Austin, USA].
4. Chou, S.-C., Schelter, W. F., Yang, J.-G. (1990): An algorithm for constructing Gröbner bases from characteristic sets and its application to geometry. *Algorithmica* **5**: 147–154.
5. Cox, D., Little, J., O'Shea, D. (1992): *Ideals, varieties, and algorithms.* Springer-Verlag, New York Berlin.
6. Eisenbud, D., Huneke, C., Vasconcelos, W. (1992): Direct methods for primary decomposition. *Invent. Math.* **110**: 207–235.
7. Gao, X.-S., Chou, S.-C. (1993): On the dimension of an arbitrary ascending chain. *Chinese Sci. Bull.* **38**: 799–804.
8. Gianni, P., Trager, B., Zacharias, G. (1988): Gröbner bases and primary decomposition of polynomial ideals. *J. Symb. Comput.* **6**: 149–167.
9. Kalkbrener, M. (1993): A generalized Euclidean algorithm for computing triangular representations of algebraic varieties. *J. Symb. Comput.* **15**: 143–167.
10. Kalkbrener, M. (1994): Prime decompositions of radicals in polynomial rings. *J. Symb. Comput.* **18**: 365–372.
11. Lazard, D. (1991): A new method for solving algebraic systems of positive dimension. *Discrete Appl. Math.* **33**: 147–160.
12. Ritt, J. F. (1950): *Differential algebra.* Am. Math. Soc., New York.
13. Shimoyama, T., Yokoyama, K. (1996): Localization and primary decomposition of polynomial ideals. *J. Symb. Comput.* **22**: 247–277.

14. Wang D. (1989): A method for determining the finite basis of an ideal from its characteristic set with application to irreducible decomposition of algebraic varieties. *Math. Mech. Res. Preprints* **4**: 124–140.
15. Wang, D. (1992): Irreducible decomposition of algebraic varieties via characteristic sets and Gröbner bases. *Comput. Aided Geom. Des.* **9**: 471–484.
16. Wang, D. (1993): An elimination method for polynomial systems. *J. Symb. Comput.* **16**: 83–114.
17. Wang, D. (1995): An implementation of the characteristic set method in Maple. In: *Automated practical reasoning: Algebraic approaches* (J. Pfalzgraf, D. Wang, eds.). Springer-Verlag, Wien New York, pp. 187–201.
18. Wang, D. (1996): Solving polynomial equations: Characteristic sets and triangular systems. *Math. Comput. Simulation* **42**(4-6): 339–351.
19. Wang, D. (1998): Decomposing polynomial systems into simple systems. *J. Symb. Comput.* **25**: 295–314.
20. Wang, D. (1998): Unmixed and prime decomposition of radicals of polynomial ideals. *ACM SIGSAM Bull.* **32**(4): 2–9.
21. Wang, D. (1999): *Elimination methods.* Springer-Verlag, Wien New York (in press).
22. Wu, W.-t. (1984): Basic principles of mechanical theorem proving in elementary geometries. *J. Syst. Sci. Math. Sci.* **4**: 207–235.
23. Wu, W.-t. (1989): On the generic zero and Chow basis of an irreducible ascending set. *Math. Mech. Res. Preprints* **4**: 1–21.
24. Yang, L., Zhang, J.-Z. (1994): Searching dependency between algebraic equations: An algorithm applied to automated reasoning. In: *Artificial intelligence in mathematics* (Johnson, J., McKee, S., Vella, A., eds.). Oxford Univ. Press, Oxford, pp. 147–156.

Appendix. List of Examples 17–22. In the following list, $\mathbb{P}$ is the polynomial set that defines the algebraic variety to be decomposed and ω is the variable ordering used for the computation. These examples were chosen quite randomly yet with moderate difficulty so that the algorithms under discussion can be tested.

Example 17. Let $\mathbb{P} = \{x^{31} - x^6 - x - y, x^8 - z, x^{10} - t\}$ with $\omega\colon t \prec z \prec y \prec x$. The variety defined by $\mathbb{P}$ is irreducible and is of dimension 1 in the 4-dimensional space. The set of polynomials is taken from a paper by C. Traverso and L. Donati (Proc. ISSAC '89, pp. 192–198).

Example 18. Let $\mathbb{P} = \{P_1, \ldots, P_4\}$, where

$$P_1 = (x - u)^2 + (y - v)^2 - 1,$$
$$P_2 = v^2 - u^3,$$
$$P_3 = 2v(x - u) + 3u^2(y - v),$$
$$P_4 = (3wu^2 - 1)(2wv - 1)$$

with $\omega\colon x \prec y \prec u \prec v \prec w$. The irreducible decomposition of the variety defined by $\mathbb{P}$ could not be completed within 2000 CPU seconds. The set of polynomials was communicated by P. Vermeer (Department of Computer Science, Purdue Univ., USA).

Example 19. Let $\mathbb{P} = \{P_1, P_2, P_3\}$, where

$$P_1 = (x_1 - y_1)^2 + (x_2 - y_2)^2 - (x_1' - y_1')^2 - (x_2' - y_2')^2,$$
$$P_2 = (x_1' - \zeta_1)^2 + (x_2' - \zeta_2)^2 - (x_1 - z_1)^2 - (x_2 - z_2)^2,$$
$$P_3 = (y_1' - \zeta_1)^2 + (y_2' - \zeta_2)^2 - (y_1 - z_1)^2 - (y_2 - z_2)^2$$

with ω: $x_1 \prec x_2 \prec y_1 \prec y_2 \prec z_1 \prec z_2 \prec x_1' \prec x_2' \prec y_1' \prec y_2' \prec \zeta_1 \prec \zeta_2$. The variety defined by $\mathbb{P}$ may be decomposed into two irreducible components. The set of polynomials is taken from a paper by J. Pfalzgraf (Ann. Math. Artif. Intell. **13**: 173–193).

Example 20. Let $\mathbb{P} = \{P_1, \ldots, P_6\}$, where

$$P_1 = 2x_1 - 1,$$
$$P_2 = x_2^2 + x_1^2 - 1,$$
$$P_3 = x_4^2 + x_3^2 - u_2^2 - u_1^2,$$
$$P_4 = 2u_2 x_4 + 2u_1 x_3 - u_2^2 - u_1^2,$$
$$P_5 = x_6^2 + x_5^2 - 2x_5 - u_2^2 - u_1^2 + 2u_1,$$
$$P_6 = 2u_2 x_6 + 2u_1 x_5 - 2x_5 - u_2^2 - u_1^2 + 1$$

with ω: $u_1 \prec u_2 \prec x_1 \prec \cdots \prec x_6$. The variety defined by $\mathbb{P}$ may be decomposed into 9 irreducible components. The set of polynomials comes from an algebraic formulation of Steiner's theorem given in a paper by the author (Proc. CADE-12, LNAI **814**, pp. 386–400).

Example 21. Let $\mathbb{P} = \{P_1, \ldots, P_6\}$, where

$$P_1 = (3u_1^2 - 1)x_2 + u_1(u_1^2 - 3)x_1 - u_1(3u_1^2 - 1),$$
$$P_2 = (3u_2^2 - 1)x_2 + u_2(u_2^2 - 3)x_1 - u_2(3u_2^2 - 1),$$
$$\begin{aligned}
P_3 = &-x_1 F_1 x_4^3 + 3(F_2 x_3 + x_1 F_3)x_4^2 \\
&+ 3[x_1 F_1 x_3^2 - 2(x_2^4 - u_2 x_2^3 - 2u_1 x_2^3 - u_2 x_1^2 x_2 - 2u_1 x_1^2 x_2 + F_4)x_3 \\
&\quad - x_1(x_2^4 + 2x_1^2 x_2^2 + u_1^2 u_2 x_2 - F_4)]x_4 \\
&- F_2 x_3^3 - 3x_1 F_3 x_3^2 + 3(x_2^5 + 2x_1^2 x_2^3 + x_1^4 x_2 - u_1^2 x_1^2 x_2 - F_5)x_3 \\
&- x_1(2u_1 u_2 x_2^3 + u_1^2 x_2^3 - 2u_1^2 u_2 x_2^2 + 2u_1 u_2 x_1^2 x_2 - F_5),
\end{aligned}$$
$$P_4 = -(x_2 + u_1 x_1 - u_1)x_4 + (u_1 x_2 - x_1 - u_1^2)x_3 + u_1 x_2 + u_1^2 x_1 - u_1^2,$$
$$P_5 = -(x_2 + u_2 x_1 - u_2)x_6 + (u_2 x_2 - x_1 - u_2^2)x_5 + u_2 x_2 + u_2^2 x_1 - u_2^2,$$
$$\begin{aligned}
P_6 = &-(x_1 G_1 x_4 - G_2 x_3 - x_1 G_3)x_6 + (G_2 x_4 + x_1 G_1 x_3 - G_4)x_5 \\
&+ x_1 G_3 x_4 - G_4 x_3 - x_1(u_2 x_2^2 + u_1 x_2^2 - 2u_1 u_2 x_2 + u_2 x_1^2 + u_1 x_1^2)
\end{aligned}$$

with

$$F_1 = 3x_2^2 - 2u_2 x_2 - 4u_1 x_2 - x_1^2 + 2u_1 u_2 + u_1^2,$$
$$F_2 = x_2^3 - u_2 x_2^2 - 2u_1 x_2^2 - 3x_1^2 x_2 + 2u_1 u_2 x_2 + u_1^2 x_2 + u_2 x_1^2 + 2u_1 x_1^2 - u_1^2 u_2,$$
$$F_3 = 2x_2^3 - u_2 x_2^2 - 2u_1 x_2^2 + 2x_1^2 x_2 - u_2 x_1^2 - 2u_1 x_1^2 + u_1^2 u_2,$$
$$F_4 = 2u_1 u_2 x_2^2 + u_1^2 x_2^2 - u_1^2 u_2 x_2 - x_1^4 + 2u_1 u_2 x_1^2 + u_1^2 x_1^2,$$
$$\begin{aligned}
F_5 = &u_2 x_2^4 + 2u_1 x_2^4 - 2u_1 u_2 x_2^3 - u_1^2 x_2^3 + 2u_2 x_1^2 x_2^2 + 4u_1 x_1^2 x_2^2 + u_1^2 u_2 x_2^2 \\
&- 2u_1 u_2 x_1^2 x_2 - 2u_1^2 x_1^2 x_2 + u_2 x_1^4 + 2u_1 x_1^4 - u_1^2 u_2 x_1^2,
\end{aligned}$$

$$G_1 = 2x_2 - u_2 - u_1,$$
$$G_2 = x_2^2 - u_2 x_2 - u_1 x_2 - x_1^2 + u_1 u_2,$$
$$G_3 = x_2^2 + x_1^2 - u_1 u_2,$$
$$G_4 = x_2^3 - u_2 x_2^2 - u_1 x_2^2 + x_1^2 x_2 + u_1 u_2 x_2 - u_2 x_1^2 - u_1 x_1^2$$

and ω: $u_1 \prec u_2 \prec x_1 \prec \cdots \prec x_6$. The variety defined by $\mathbb{P}$ may be decomposed into 12 irreducible components (in Maple V.3 on an Alpha station 600) when the technique of normalization is used. The set of polynomials comes from an algebraic formulation of Morley's theorem given in a paper by the author (Proc. CADE-12, LNAI **814**, pp. 386–400).

Example 22. Let $\mathbb{P} = \{P_1, \ldots, P_7\}$, where

$$P_1 = 4u_1^2 u_2^4 x_1^2 - (2u_2^2 u_3 + 2u_1^2 u_2^2 - E)(2u_1^2 u_2^2 u_3 - 2u_2^2 + u_1^2 E),$$
$$P_2 = 4u_2^4(2u_1^2 x_1 - 2u_1^2 u_3 - A)x_2^2 + C^2[4I x_2 - 2u_1^2 D^2 x_1 - F],$$
$$P_3 = -4u_2^4(2u_1^2 x_1 + 2u_1^2 u_3 + A)x_3^2 + C^2[4I x_3 + 2u_1^2 D^2 x_1 - F],$$
$$P_4 = u_1 u_2 x_4 - u_1^2 u_2^2 + 1,$$
$$P_5 = u_1 u_2 x_5 + u_2^2 - u_1^2,$$
$$P_6 = 2u_1^2[u_2^2(x_5 + x_4) - CD]x_6 - u_1^2 E(x_5 + x_4) + BCD,$$
$$P_7 = I x_7 - [2u_1^2 u_2^2 x_5 - (u_2^2 - u_1^2)(u_1^2 u_2^2 + 1)]x_6 + u_1^2 E x_5 - u_2^4 + u_1^4$$

with

$$I = (u_2^2 - u_1^2)(u_1^2 u_2^2 - 1), \quad F = 2[u_1^2 u_2^4 - 2(B + u_1^2)u_2^2 + u_1^2]u_3 - AD^2;$$
$$A = u_1^4 - 1, \quad B = u_1^4 + 1, \quad C = u_2^2 - 1, \quad D = u_2^2 + 1, \quad E = u_2^4 + 1$$

and ω: $u_1 \prec u_2 \prec u_3 \prec x_1 \prec \cdots \prec x_7$. The irreducible decomposition of the variety defined by $\mathbb{P}$ could not be completed within 2000 CPU seconds. The set of polynomials comes from an algebraic formulation of Thébault-Taylor's theorem given in Yang and Zhang (1994).

An Application of Automatic Theorem Proving in Computer Vision

Didier Bondyfalat, Bernard Mourrain, and Théodore Papadopoulo

INRIA, Saga and Robotvis
2004 route des lucioles, BP. 93
06902 Sophia Antipolis, France
First.Last@sophia.inria.fr

Abstract. Getting accurate construction of tridimensional CAD models is a field of great importance: with the increasing complexity of the models that modeling tools can manage nowadays, it becomes more and more necessary to construct geometrically accurate descriptions. Maybe the most promising technique, because of its full generality, is the use of automatic geometric tools: these can be used for checking the geometrical coherency and discovering geometrical properties of the model. In this paper, we describe an automatic method for constructing the model of a given geometrical configuration and for discovering the theorems of this configuration. This approach motivated by 3D modeling problems is based on characteristic set techniques and generic polynomials in the bracket algebra.

1 Introduction

The construction of tridimensional CAD models is of growing importance: with the increase of power of the computers and networks, it is now a common experience to navigate within a virtual 3D world. One of the biggest challenges to obtain the best navigation is to get models that are both photo-realistic and accurate. One particular technique, derived from computer vision, makes possible the construction of a 3D model of a scene, starting from simple pictures of it [10,9] (Fig. 1).

In order to achieve this result, several stages are necessary: first, special image features (i.e. corner points) must be extracted from the pictures and matched across the various views. These matched features are then used to calibrate the cameras, which consists in finding the parameters defining their states (such as their focal distances) and positions in the Euclidean space. Once the cameras are calibrated, reconstructing a 3D feature is just a matter of triangulation provided that the object is visible in at least two pictures. Using such a scheme, it is quite easy to obtain a realistic 3D model since the photometric information of the pictures can be used to augment the pure geometric model.

However, in most cases, these techniques are based only on the viewing geometry (i.e. the geometry of light rays captured by the cameras). Since most

Fig. 1. The 3D model

human made objects (e.g. buildings, tools, ...) are usually geometrically structured (coplanar or aligned points, orthogonal or parallel lines, ...), a very rich source of information is thus ignored, whereas it could be used to "stabilize" results obtained from the noisy data. So far a priori knowledge of the scene to be reconstructed is used only through CAD models describing it in terms of geometrical primitive blocks such as cubes, pyramids, cones, spheres, ...Each primitive block being parameterized, it is then possible to "optimize" the 3D model so as to obtain the best 3D structure predicting the observed images [9, 16]. In many cases, creating such CAD models is a tedious task and a weaker description of the geometric structure (in terms of raw geometrical properties) would be more convenient.

Let us illustrate our claim by a picture (Fig. 2): the scene is a district of a city. As it can be seen, buildings are never completely visible so that it would be difficult to model those as polyhedrons. On the contrary, it is fairly easy to associate to each frontage a polygon and to each edge a segment, as well as polyhedrons to some details. These primitives can then be constrained using relations such as orthogonality (e.g. of frontages), parallelism and angle (for windows). Of course, not all properties need to be introduced one by one and higher level of abstraction can be used: for example, rectangles can be associated to each window.

The problem which then arises is organizing these descriptions in such a way that it can be used by reconstruction algorithms. If we want a fairly generic system, this definitely requires some form of geometrical reasoning. The potential importance of using such reasoning methods for computer vision has been recognized for many years [12], as it is natural to think that geometry plays a

Fig. 2. Geometrical features in a picture of buildings

significant role in the analysis and comprehension of at least human made scenes. Taking account of it would greatly improve the quality and the generality of the current systems. However, the complexity of the usual scenes makes the use of current tools difficult. Consequently, those have only been used so far in simple limited situations, e.g. to constrain the position of the camera directly from the observations of certain known geometric patterns [13, 22].

In this paper, we intend to show how some techniques inspired by automated geometric reasoning can be used as a way to structure the a priori geometrical knowledge of a given 3D situation in such a way that it can be used to derive better 3D models when using computer vision techniques. The next section describes the computer vision problem we are interested in and presents the geometrical tools that are needed in order to solve it. Section 3 then analyzes the inherent difficulties of using the tools that have been developed within the framework of automatic theorem proving. Section 4 proposes the two new tools that have been developed to cope with our problem. Finally, Section 5 shows results obtained with those tools.

2 Building Geometrically Constrained 3D Models from Images

3D models such as the one shown in Fig. 1 are built uniquely from a set of images (Fig. 3). As sketched briefly in the introduction, three main steps are used in order to achieve this results[1]:

Fig. 3. Different views of "place des arcades", Valbonne

- First, characteristic points are extracted from each image.
- Those points are then matched across the images: points corresponding to different views of a same 3D point are associated.
- Point matches are then used to establish a set of 3×3 matrices known as fundamental matrices that encompass the projective structure of the rigid scene for each two images. The fundamental matrices are then combined in order to establish the Euclidian positions and properties of each of the cameras. Note that this whole process is based only on the weak assumption that

---

[1] The whole process is depicted with point primitives for the sake of simplicity, it is however more general and can be used with other kinds of primitives such as lines.

the scene is Euclidian and rigid. Once this geometry has been established, it is easy to obtain a 3D point for each of the point matches by intersecting the corresponding rays. An initial 3D model has thus been obtained.

However, this scheme still has to face with many problems:

- Extracted points are always corrupted with noise. This noise has many sources: electronic or optic effects, sampling, discretization and even biases when automatic feature extraction is used.
- Incorrect matches can occur: even though a certain number of such matches can be (and are) detected during the fundamental matrix computation, some of those may still influence the final result of the algorithm.
- Computing the camera positions and properties from fundamental matrices is inherently non-linear. In difficult situations, this gives rise to unstabilities that perturb the final result.

Of course these effects are maximized if using fully automatic methods and the trend has been to use manual or semi-automatic methods instead. Nevertheless, such inaccuracies always remain. This is all the more annoying that the process sketched above is not *directly* optimizing the quality of the 3D model with respect to the observed data. Consequently, one additional step is usually performed to complete the above process. This step called *bundle adjustment* refines the 3D structure and cameras parameters in such a way that the quadratic errors between observed and predicted points in the images are minimized. To formalize things better, let us call $\mathbf{M}_j$ a given 3D points and $\mathbf{m}_i^j$ its observed image obtained with the i-th camera P_i. Denoting by $P_i(\mathbf{M}_j)$ the predicted image point corresponding to $\mathbf{m}_i^j$, and assuming that there are N cameras and m 3D points, the bundle adjustment process minimizes the following criterion:

$$\sum_{i=1}^{N} \sum_{j=1}^{m} d(\mathbf{m}_i^j, P_i(\mathbf{M}_j))^2 \,,$$

where $d(\mathbf{m}_1, \mathbf{m}_2)$ represents the distance between $\mathbf{m}_1$ and $\mathbf{m}_2$. Note that since the 3D points $\mathbf{M}_j$ are computed from the data, the parameters over which the minimization is carried on are just the $P_i, i = 1, \ldots, N$. In practise, this minimization is carried on by doing alternating sequences of 3D reconstructions (fixing the camera parameters) and of camera calibrations (fixing the 3D point positions). Figures 1 and 4, show the results obtained using the data of Fig. 3. The 3D-model is shown from different points of view. With the orthographic top view, we can see that the roofs are not rectangular and that the frontages are not planar. These models are not verifying exactly some of the geometrical properties of the scene that would be expected [2, 1].

2.1 Incorporating Constraints in the Reconstruction Process

Of course, one way to improve this is to use the knowledge of geometrical properties of the scene: this would both stabilize the calibration and give models

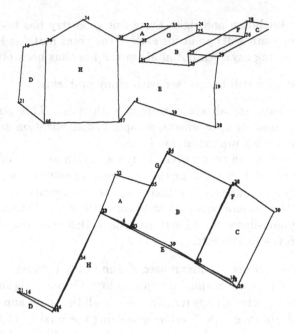

Fig. 4. A wire frame view of the 3D-model

which respect the 3D geometry of the scene. To do so, we consider an enriched input consisting of:

- A set $\tilde{P}$ of 3D primitives which consists of the reconstructed 3D points that have been used so far are used but also of logical grouping of those to introduce higher-level kinds of primitives such as 3D planes or lines. As before, each 3D point is attached to a set of matched image points.
- A set $\tilde{C}$ of geometric constraints attached to these primitives (incidences, angles, ...).

This input has to be provided by the user. Using the image points associated with $\tilde{P}$, we compute a set of camera positions and properties as well as an initial 3D model as above. But we now want to introduce the known constraints $\tilde{C}$ in the bundle adjustment process. This can be achieved in two different ways:

Using a Parameterization. One technique would be to exhibit somehow a parameterization $\mathbf{p}$ of the scene or more precisely of the set of 3D primitives $\tilde{P}$ obeying to the constraints $\tilde{C}$. Such a parameterization can usually be defined by a subset $\mathcal{M}$ of the 3D points of $\tilde{P}$ and by a set of expressions $M_i(\mathbf{p})$ derived from $\tilde{C}$ defining all the other 3D points of $\tilde{P}$ in terms of those of $\mathcal{M}$. By construction, $\tilde{P}$ is then completely defined in such a way that the constraints $\tilde{C}$ are satisfied. Using such a tool, it is easy to modify the bundle adjustment process to take account of the constraints. It is sufficient to just modify the 3D reconstruction part by:

– Reconstruct only the points of $\mathcal{M}$ from the image data.
– Deduce the other 3D points using the rules $M_i(\mathbf{p})$.

The cost function that is minimized then becomes:

$$\sum_{i=1}^{N}\sum_{j=1}^{m} d(\mathbf{m}_i^j, P_i(\mathbf{M}_j(\mathbf{p})))^2.$$

This scheme has been implemented. The result obtained for the above scene is shown in Fig. 7. As the points of $\mathcal{M}$ play a special role in the method, problems can be expected if those points are very erroneous due to image measurements as those errors will be propagated to the complete model. This is one weakness of the method. Its major advantages are its simplicity (once the parameterization is computed) and the correctness of its results as it is guaranteed that the 3D constraints $\tilde{C}$ are satisfied *at each time instant*.

Using an Optimization Method. An alternative to the previous method is to rely on optimization techniques to incrementally improve the geometric 3D structure at each reconstruction step. In its principle, implementing this method is not very difficult. Each constraint of $\tilde{C}$ can be expressed as an algebraic expression that must cancel out. Minimizing the residuals of those expression with respect to the 3D point coordinates is just a matter of gradient descent.

As such, this method suffers from problems similar to those arising in the method presented in the previous section: although all the 3D primitives seem to play a similar role, this is still not true as geometrical constraints explicitly written in $\tilde{C}$ have a particular status with respect to all the various geometric properties of the figure that are consequences of those. The net effect of this is that again, some primitives will play a particular role. To understand better this problem, let us consider a set of lines $l_1, l_2, \ldots, l_k$. We want to state that all these lines are parallel by writing that l_j is parallel to l_1 for $j = 2, \ldots, k$. In this scheme, l_1 is clearly granted with a central role that is completely artificial. If it turns up that l_1 is precisely a primitive that has been reconstructed erroneously whereas $l_2, \ldots l_k$ are recovered correctly, then this error will tend to propagate to all the other lines. This effect would not have appeared if all the parallelism constraints had been explicitly stated in $\tilde{C}$.

This suggests that it might be extremely important to be able to generate all the consequences of the given set of constraints $\tilde{C}$. Using all those in the minimization process would reduce the risk of propagating errors inconsiderately.

In the following sections, two techniques are presented. The first one attempts to compute a parameterization $\mathbf{p}$ given $\tilde{P}$ and $\tilde{C}$ and has been used with the program TOTALCALIB to produce the result shown in Fig. 7. A comparison with the result given in Fig. 4 clearly shows the improvement of the geometric 3D structure. The second technique shows how consequences of $\tilde{C}$ can be systematically computed. The modified bundle adjustment of Section 2.1 has not yet been implemented, but, as explained above, using this technique should greatly improve the stability of such a method.

3 Drawbacks of the Classical Geometric Reasoning Tools in Computer Vision

In order to do automatic geometrical reasoning, it is necessary to translate the geometry into an algebraic form. Nowadays, two kinds of techniques are opposed: one uses a reference frame (a coordinate system), the other, on the contrary, is free from the choice of a reference frame, using more intrinsic quantities, i.e. invariants. We briefly summarize these approaches and evaluate them in the context of computer vision problems.

3.1 Using a Reference Frame

Here, we give short descriptions of some provers which use a coordinate system to formulate the proofs. We refer to [21] for a more complete treatment of the algebraic methods that are used. The most well-known techniques for theorem proving are probably Wu's method [4, 20, 19] and Gröbner basis method [6, 23]. In these methods, the geometrical objects are defined by several parameters (for a point, its coordinates; for a line in the planar case, the 3 coefficients of its equation; ...). The geometrical properties are represented by polynomial equations into the parameters of the objects. Roughly speaking, proving a geometric theorem is equivalent to showing that a polynomial vanishes on a certain manifold, which corresponds to the generic configurations of the theorem. More precisely [25, 26, 6], this "reduces" to show that the polynomial representing the conclusion of a theorem is a member of the radical ideal generated by the polynomials representing its hypotheses divided (or saturated) by the ideal generated by its degeneracy conditions.

Characteristic set methods transform the initial system of equations into one (or several) triangular set(s), by which any valid geometric property of the configuration can be reduced to zero. These triangular sets are obtained as fixed points of an autoreduction process of the initial equations.

Gröbner basis methods compute a basis of the initial ideal saturated by the degenerate conditions. This also allows the reduction to zero of any polynomial which corresponds to a geometric property of the configuration.

3.2 Using Geometric Invariants

Another class of methods uses invariant calculus and are of particular interest in computer vision (for these invariants quantities may be computed directly in the images). The theorem provers of this family use intrinsic computations, that are usually more compact and easier to interpret in natural language. We briefly review some of these provers (see also [21]).

In the prover **Cinderella** [7, 17], the geometrical properties are represented by polynomials in brackets (more precisely by binomials of determinants), many geometric properties can be written in this way. In this case, the reduction of the conclusion is performed by simple linear algebra on integer matrices representing the exponents of the variables involved in these binomials.

In the works of [5, 14], hypotheses are also translated into polynomials of the bracket algebra (extended in **GEX** [5] by other primitives like the area of quadrilateral, inner products of vectors, angles). The methods are essentially based on rewriting rules on points. In the system **GEO** [14], only collinearity relations are considered. Rewriting rules on points are constructed formally in the exterior algebra, taking into account the generic degree of freedom of each point.

Clifford algebras which are exterior algebras extended with a quadratic form and a related product (see [8]) have also been used [18, 22]. Theorems are proved by reduction modulo algebraic expression associated with the hypotheses.

Other primitives like distance between points have also been considered. In [11] for instance, the geometrical properties are translated into polynomials in the distances between points, and straightening laws are applied to get a normal form and test the validity of some theorems.

3.3 A Comparison of these Methods in our Context

We come back now to our problem of 3D modelisation and compare first the existing provers in this context.

Wu's and Wang's methods are ordering the points and using a system of co-ordinates, according the theorem to be proved. The method has the advantage to deal uniformly with different kinds of geometric constraints (collinearity, orthogonality, angles, ...). The interpretation of the polynomial calculus in terms of geometric relations is not obvious. Moreover, as experimented on the scene "place des arcades" (Fig. 1), the computation is very expensive on problems involving many points or objects. In this scene, 102 parameters and 101 equations are necessary to describe the geometry of the scene. With such a number of parameters, it is not possible to use characteristic set or Gröbner basis techniques, for we have an explosion of the computation time and memory resource during the computation of these polynomial sets. Notice that this explosion is essentially due to the inter-reduction of the polynomials.

The Richter-Gebert Crapo method has the advantage to reduce the proof to simple linear algebra on integer matrices. Usually the proofs are very short. However, there is no way to assert that a property is generically false (we have only sufficient conditions). Combining both points, lines, planes in a same computation seems to be difficult. The extension to the 3D case is possible but expensive.

The method of Chou-Gao-Zhang has the advantage to produce fast and simple proofs. It requires also a constructive order on the points and it has some difficulties with circles (quadratic rewriting rules instead of linear ones for lines and points). Its generalization to the 3D case has not yet been really investigated.

In the context of 3D modelisation, none of these approaches yield a satisfactory answer. The main difficulties that we encounter are:

- The definition of a natural constructive order on the objects,
- The construction of reductor sets that can handle many geometric objects

– The automatic generation of theorems.

We describe our treatment of these points in the next sections.

4 Application of Theorem Proving to Computer Vision

4.1 Building a Constructive Order

Our configurations are defined by features (points, lines, planes, ...) and the constraints between them without any information on their construction or their degree of freedom. We describe here an automatic tool, which finds the degree of freedom for all the features and which sorts the features so that we can build the configuration. Such an order will be called *constructive order*. We start with some definitions:

– The degree of freedom of an object is the dimension of the manifold to which this object belongs. In other words, it is the number of parameters of the object which can be instantiated randomly.
– A generic component of a configuration is a component where all the objects are distinct and the dimension of the intersection of two objects is the largest possible, taking into account all the constraints on the two objects.
– An object will be called *constructible* iff its degree of freedom is equal to zero.
– A constraint will be called *active* iff only one of the two objects which are involved in it, is built.

The Algorithm. It is an iterative algorithm which starts with an empty list of built objects and an empty list of active constraints. The initial degree of freedom of each object is computed (3 for points and planes, 4 for lines). The algorithm stops when all the objects are built. Here are its main steps:

1. Search one unbuilt object with minimal degree of freedom.
2. Compute the rewriting rules for it, using its constraints with the previous objects.
3. Build it, that is add it to the list of constructed objects.
4. Compute the new set of active constraints.
5. Compute the new degrees of freedom of all the unbuilt objects.

We discuss now the problems to be solved during these steps.

How to Choose the Object to be Built. In step (1), we have to choose one object of minimal degree of freedom. In many cases, we will have many choices and we choose the object among the objects of smallest dimension. If this refinement is not sufficient, we choose the object, which induces the maximum number of active constraints. If there still remains several candidates, we choose any of them.

Constructing Sparse Reductor Sets. A main obstacle of the provers described in Section 3 is their inherent complexity, due to the quick growth of the size of the polynomials, with the number of geometric objects. In the applications that we consider, the number of objects (e.g. about 50 points, lines or planes in the scene "place des arcades") is much larger than in classical theorems (see [3, 5]), although the geometric constraints may be simpler. It implies that in characteristic set methods, we deal with many variables and since the polynomials are interreduced, the output polynomials of the triangular sets will involve all these variables and will thus be huge. The modification that we propose to such methods consists in replacing dense characteristic sets by sparse triangular sets of polynomials, which also allow to reduce every valid geometric property to zero. These sets will be called hereafter *sparse reductor sets*.

Our modification is specific to geometrical problems, where object configurations are described by a few kind of geometric constraints (collinearity, coplanarity, orthogonality, cocyclicity ...). If a point X is described by such a set of hypotheses, the possible configurations for X belong to a finite number of classes, going from the generic intersection case to a finite number of degenerate cases. For instance, if X is on two projective lines, either these lines are intersecting properly and X is their intersection or they are identical and X is any point on the line. Once this so called *geometric type* is known, we can easily derive a reductor set describing the position of X. As the position of X depends explicitly on few previous points (those which actually appear in the hypotheses on X), this yields a reductor set which is sparse. Let us illustrate now this approach, which has already been used in [15] for computations in the exterior algebra.

We consider here an approach using a reference frame, in which we express the hypotheses by polynomials

$$\begin{cases} H_1(X, A_1, \ldots, A_{k_1}) = 0 \\ \vdots \\ H_l(X, U_1, \ldots, U_{k_l}) = 0 \end{cases} \tag{1}$$

where H_i are polynomials in the coordinates of the objects, X is the point to be built and $A_1, \ldots, U_k$ are the built points. As the geometric constraints on X involve specific and simple polynomials in the coordinates of X (usually linear or quadratic), the possible *geometric type* of intersection can effectively be analyzed by checking the vanishing (or non-vanishing) of certain resultant polynomials in the coordinates of the previous points. For instance, if a point X is in the intersection of the linear spaces $A = \langle A_1, \ldots, A_{k_1} \rangle$ and $B = \langle B_1, \ldots, B_{k_2} \rangle$ The intersection of the linear spaces can be defined formally as the last non-zero term of the expansion:

$$D(A, B) = D_{k_1}(A, B) \oplus \cdots \oplus D_{k_1+k_2-n}(A, B).$$

(see [15] for more details).

Checking the vanishing (or non-vanishing) of these resultant polynomials can be performed by using the reductor set associated with these previous points.

It is also possible to do it by sampling techniques (choosing random values for the free parameters and computing the other). This is the method that we have used to construct a sparse reductor set for the scene "place des arcades", because using directly characteristic sets was intractable in this case. In our application, we may even skip this checking step, and use external informations, that we known from the initial model.

How to Avoid to the Redundant Constraints. During the update of the degrees of freedom in step (5), we have to look to all the active constraints on each unbuilt object and to compute the degree of freedom induced by these constraints. This degree of freedom can be computed by reduction of the constraints on the new point by a triangular set (in order to remove redundant hypotheses). In practise, we do not use this technique because it requires too much memory and time resources. We prefer to use a random instantiation of the degrees of the parameters defining each built object (its degree of freedom). Once the previous points are instantiated, the constraints defining the object to be built defines a polynomial system in 3 or 4 variables (3 for points and planes, 4 for lines). The computation of a triangular set for such a system is easy and fast. This yields the degree of freedom of (or the dimension of the variety described by) the new point. If this leads to an impossible construction (an empty variety), we stop the computation, backtrack to the last choice step (1) and we choose another point.

Backtracking. The refinements of the choice described in step (1) are based on heuristics and but may lead to a degenerated component. Thus during the algorithm, if we detect that one constraint is not satisfied or that we are not in a generic component (e.g. two objects are the same), we backtrack to step (1). For that, we maintain a list of possible choices at each step of the algorithm and process this list until a non-degenerate component is built. If this is not possible, we stop and output "failure".

Example 1 We consider the following hypotheses, defining a subconfiguration of "place des arcades" (Fig. 5):

```
seq(point(p.i),i=1..4):
seq(line(l.i),i=1..2):
map(x->plane(x),[H1,H2,H3,H11]):

map(OnObject,[p2,p3],H1):
map(OnObject,[p1,p2,p3],H2):
map(OnObject,[p1,p2],H11):
OnObject(p4,H3):
map(OnObject,[p1,p2],l1):
map(OnObject,[p2,p3],l2):
PerpendicularConstraint(l1,l2):
PerpendicularConstraint(H1,H2):
PerpendicularConstraint(H2,H3):
```

`PerpendicularConstraint(H11,H3):`

If we execute the algorithm on this scene, the following order is obtained:
$[P_2, l_1, P_1, l_2, P_3, H_2, H_1]$.

At this stage, it remains to be built p_4, H_3 et H_{11}. The degrees of freedom and the dimensions are the same for H_{11} and H_3; H_3 activates two constraints whereas H_{11} activates that only one constraint. Thus the algorithm chooses the plane H_3, adds $[H_3, H_{11}]$ to the list "component". This choice leads us to a degenerated component of the configuration (in this case H_{11} is equal with H_2). Thus, the algorithm backtracks, deletes H_3 of the list $[H_3, H_{11}]$, and constructs H_{11}, and so on. The constructive order found in this example is
$[P_2, l_1, P_1, l_2, P_3, H_2, H_1, H_{11}, H_3, p_4]$.

4.2   Theorem Discovering

Up to recently, theorem provers were used only to check the conclusion of a theorem. However, in our context as explained in Section 2.1, the theorems are not inputs of the problem and have to searched in the geometric configurations, in order to improve the modelisation process. This task can be performed by trial and error, using a geometric prover but it is tedious and requires geometric intuition. We are interested in a more systematic approach.

From an algebraic point of view, any polynomial of the ideal associated with the generic configurations is a theorem. If we are looking for relations between specific points, we intersect this ideal with the subring of coordinates of these points. This is the approach used in [23], the elimination process being carried on by Gröbner basis computations.

Our approach proceeds differently: a (sparse) reductor set of the scene is computed and used to recover the theorems. It is based on the following simple remark:

Remark 2 The theorems of a configuration correspond to homogeneous invariant relations of small degree in the bracket algebra.

The principle. We begin by introducing some notations in the bracket algebra: Let Ξ be a subset of $\mathbb{E}$, a **letter** of the bracket algebra on Ξ is either an element of Ξ or the linear form $\mathcal{H}$ defining the hyperplane at infinity, or the bilinear form $\mathcal{Q}$ defining the metric of the Euclidean space.

Definition 3. *We denote by* **support of a monomial** *of the bracket algebra on* Ξ, *the set of letters (counted with their multiplicity) appearing in the monomial.*

The support of $\mathcal{H}(A)\,\mathcal{H}(C)\,\mathcal{Q}(A,B)\,|A,B,C||A,D,E|$ is

$$Supp(\mathcal{H}(A)\,\mathcal{H}(C)\,\mathcal{Q}(A,B)\,|A,B,C||A,D,E|) = \mathcal{H}^2, \mathcal{Q}, A^3, B, C^2, D, E.$$

We denote by **homogeneous component** the set of all the monomials with the same support.

According to Remark 2, the theorems correspond to relations between monomials of a same homogeneous component. We thus construct all the monomials of a same homogeneous component up to a given degree and consider a linear combination of these monomials with indeterminate coefficients. This *generic polynomial* is then reduced by the reductor set either using a reference frame or invariants and the remaining system is solved in order to determine the coefficients which yield the properties of the configuration.

Algorithm 4 *For a geometrical configuration C:*

1. *Choose a* **reference frame** *and compute a* **reductor set** *RS for the hypotheses of C.*
2. *For each* **homogeneous component** *in the* **bracket algebra**, *of C:*
   - *compute a* **generic polynomial** $\mathcal{P}$ *in the formal vector parameter $u = [u_1, \cdots, u_k]$ (where k is the number of monomials in the homogeneous component).*
   - *compute the* **remainder** R *of the reduction of P by RS.*
   - *compute the vectors u such that $R = 0$ and find their geometric interpretation.*

Notice that the last step reduces to solving a linear system in u, for the polynomial P is linear in u and the reduction does not change this property. Moreover, theorems correspond to integer solutions of this system. The task of interpretation of the geometric relations is based on a list of binomial relations corresponding to basic geometrical properties. We do not have yet a systematic way to do it. This will be discussed further in the conclusion. Note however that having such an interpretation is not necessary for the application described in Section 2.1.

5 Examples

This section details two complete examples of the techniques just described. These experimentations were performed mostly in MAPLE. The graphs were obtained with the tool DOT. The wire-frame models of Fig. 7 were obtained with the tool TOTALCALIB of the Robotvis project.

5.1 A Complete Example: "La place des arcades" Valbonne France

In this section, we detail the case of the parameterization of the 3D scene depicted in figure (Fig. 3).

Geometrical Constraints and Results after Processing. From the image sequence of Fig. 3, we can infer basic relations between the geometric objects: some frontages are parallel or perpendicular, the roofs of the building are rect-

angular ... Below are written the constraints that have been used (MAPLE file).
See Fig. 5 for the notations.

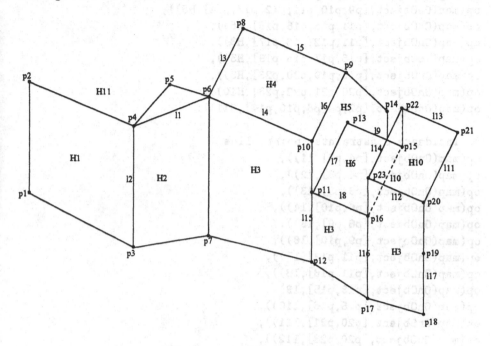

Fig. 5. A model built by hand

```
## Definition of the objects in the model
seq(point(p.i),i=1..23),
map(x->plane(x),[H1,H2,H3,H4,H5,H6,H10,H11]),
seq(line(1.i),i=1..17):

## Incidence constraints: line, plane
OnObject(11,H2),
OnObject(11,H11),
OnObject(12,H1),
OnObject(17,H5),
OnObject(18,H3),
OnObject(18,H6),
OnObject(112,H3),
OnObject(112,H10),
OnObject(111,H10),
OnObject(115,H3),
OnObject(115,H5):

## Incidence constraints: point, plane
op(map(OnObject,[p1,p2,p3,p4],H1)),
op(map(OnObject,[p3,p4,p5,p6,p7],H2)),
```

```
op(map(OnObject,[p6,p7,p10,p11,p12],H3)),
op(map(OnObject,[p6,p8,p9,p10],H4)),
op(map(OnObject,[p9,p10,p11,p12,p13,p14],H5)),
op(map(OnObject,[p11,p13,p15,p16],H6)),
op(map(OnObject,[p11,p12,p16,p17],H3)),
op(map(OnObject,[p16,p17,p18,p19],H3)),
op(map(OnObject,[p16,p19,p20,p23],H3)),
op(map(OnObject,[p20,p21,p22,p23],H10)),
op(map(OnObject,[p2,p4,p6,p10,p14],H11)):

## Incidence constraints: point, line
op(map(OnObject,[p4,p6],11)),
op(map(OnObject,[p4,p3],12)),
op(map(OnObject,[p6,p8],13)),
op(map(OnObject,[p6,p10],14)),
op(map(OnObject,[p8,p9],15)),
op(map(OnObject,[p9,p10],16)),
op(map(OnObject,[p11,p13],17)),
op(map(OnObject,[p11,p16],18)),
op(map(OnObject,[p13,p15],19)),
op(map(OnObject,[p15,p16],110)),
op(map(OnObject,[p20,p21],111)),
op(map(OnObject,[p20,p23],112)),
op(map(OnObject,[p21,p22],113)),
op(map(OnObject,[p22,p23],114)),
op(map(OnObject,[p10,p11,p12],115)),
op(map(OnObject,[p16,p17,p23],116)),
op(map(OnObject,[p18,p19,p20],117)):

## Constraints of parallelism: plane, plane
ParallelConstraint(H5,H2),
ParallelConstraint(H4,H6),
ParallelConstraint(H6,H10):

## Constraints of parallelism: line, line
ParallelConstraint(13,16),
ParallelConstraint(14,15),
ParallelConstraint(17,110),
ParallelConstraint(18,19),
ParallelConstraint(111,114),
ParallelConstraint(112,113):

## Constraints of orthogonality: plane, plane
PerpendicularConstraint(H11,H3),
PerpendicularConstraint(H11,H2),
```

```
PerpendicularConstraint(H1,H2),
PerpendicularConstraint(H3,H2):
```

Constraints of orthogonality: line, line
```
PerpendicularConstraint(l1,l2),
PerpendicularConstraint(l4,l6),
PerpendicularConstraint(l7,l9),
PerpendicularConstraint(l11,l12):
```

Algorithm 1 is used to obtain a constructive order (using the function **Ordre**) which gives the following list of objects sorted in the increase order of construction:

```
[p6, l1, p4, l2, p3, H2, H1, H11, H3, p10, H5, l4, l15, p11, p2,
p7, p12, p14, l6, l3, p8, H4, l5, p9, H6, l8, p13, l7, l9, p16,
l10, p15, H10, l12, p20, l11, p23, l14, l16, p21, l13, p22, p17,
p18, l17, p19, p1, p5].
```

This constructive order can be represented by the dependence graph (Fig. 6) where each node is associated to one object (ellipse nodes represent points, rectangular nodes, planes and diamond nodes, lines). The color of node encode the degree of freedom of the object represented by the node (darkgrey for three, grey for two, lightgrey for one and white for zero). Each edge is associated to one geometrical constraint. The orientation of the edge shows the dependence between the nodes. Now, we can compute a triangular set of the polynomial equations of the geometrical constraints. The variable order is described below. The independent variables appear in the first list, and in the second, are the variables which are not free.

```
[p6[1],p6[2],p6[3],l1[1],l1[2],p4[1],l2[1],p3[1],p10[1],p11[2],
p2[1], p7[1],p12[1],p14[1],l6[1],p8[1],p13[1],p16[1],H10[1],
p20[1],p23[1], p21[1],p17[1],p18[1],p18[2],p19[1],p1[1],
p1[2],p5[1],p5[2]]
```

```
[l1[3], l1[4], p4[2], p4[3], l2[2], l2[3], l2[4], p3[2], p3[3],
H2[1], H2[2], H2[3], H1[1], H1[2], H1[3], H11[1], H11[2], H11[3],
H3[1], H3[2], H3[3], p10[2], p10[3], H5[1], H5[2], H5[3], l4[1],
l4[2], l4[3], l4[4], l15[1], l15[2], l15[3], l15[4], p11[2],
p11[3], p2[2], p2[3], p7[2], p7[3], p12[2], p12[3], p14[2],
p14[3], l6[2], l6[3], l6[4], l3[1], l3[2], l3[3], l3[4], p8[2],
p8[3], H4[1], H4[2], H4[3], l5[1], l5[2], l5[3], l5[4], p9[1],
p9[2], p9[3], H6[1], H6[2], H6[3], l8[1], l8[2], l8[3], l8[4],
p13[2], p13[3], l7[1], l7[2], l7[3], l7[4], l9[1], l9[2],
l9[3], l9[4], p16[2], p16[3], l10[1], l10[2], l10[3], l10[4],
p15[1], p15[2], p15[3], H10[2], H10[3], l12[1], l12[2], l12[3],
l12[4], p20[2], p20[3], l11[1], l11[2], l11[3], l11[4], p23[2],
p23[3], l14[1], l14[2], l14[3], l14[4], l16[1], l16[2], l16[3],
```

224

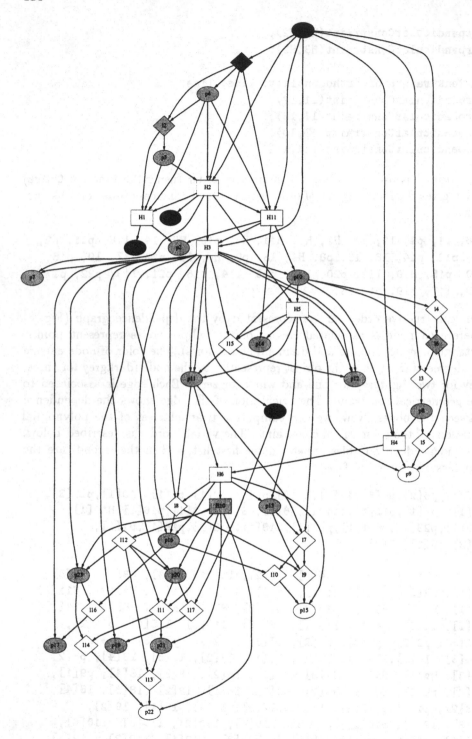

Fig. 6. Dependence graph of "place des arcades"

116[4], p21[2], p21[3], 113[1], 113[2], 113[3], 113[4], p22[1],
p22[2], p22[3], p17[2], p17[3], p18[3], 117[1], 117[2], 117[3],
117[4], p19[2], p19[3], p1[3], p5[3]]

Below, is the triangular set computed with respect the variable order described
above. This triangular set has 131 polynomials. Here are the first and last polynomials of this set:

[p6[3] 11[3] - p6[1] + 11[1],
p6[2] - 11[2] - p6[3] 11[4],
11[4] p4[1] - 11[4] 11[1] - 11[3] p4[2] + 11[3] 11[2],
p4[2] - 11[2] - p4[3] 11[4],

   .   .
   .   .
   .   .

H3[3] p19[2] - H3[3] 117[2] + 117[4] p19[1] H3[1]
                            + 117[4] p19[2] H3[2] + 117[4],
p19[1] H3[1] + p19[2] H3[2] + p19[3] H3[3] + 1,
p1[1] H1[1] + p1[2] H1[2] + p1[3] H1[3] + 1,
p5[1] H2[1] + p5[2] H2[2] + p5[3] H2[3] + 1].

The corrected model after adjustment is shown in Fig. 7.

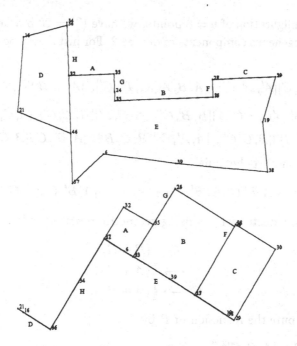

Fig. 7. A constrained wire-frame reconstruction.

5.2 Examples of Theorem Discovering

We illustrate now the method proposed for discovering theorems in three different contexts. The first context is related to characteristic sets, the second to direct rewriting rules on points and the last one to computations in the exterior algebra.

Example 5 (The median configuration) Let ABC be a triangle and A', B', C' be the middle points of $[BC]$, $[AC]$, $[AB]$, respectively.

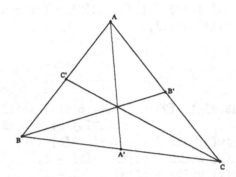

Fig. 8. The median configuration

For this configuration of $n = 6$ points, we have $\binom{n}{3} = 20$ brackets and $5\binom{n}{5} + \binom{n}{6} = 31$ homogeneous components of degree 2. For instance choose one of these 31 components:

$$|A, B, C||A', B', C'|, \ |A, B, A'||C, B', C'|, \ |A, B, B'||C, A', C'|,$$

$$|A, B, C'||C, A', B'|, |A, C, A'||B, B', C'|, |A, C, B'||B, A', C'|, |A, C, C'||B, A', B'|,$$

$$|A, A', B'||B, C, C'|, \ |A, A', C'||B, C, B'|, \ |A, B', C'||B, C, A'| \ .$$

We form the generic polynomial

$$P = u_1|A, B, C||A', B', C'| + \cdots + u_{10}|A, B', C'||B, C, A'| \ ,$$

and compute a reductor set RS using geometric rewriting rules:

$$A' \longrightarrow \tfrac{1}{2}(B + C) \ ,$$
$$B' \longrightarrow \tfrac{1}{2}(A + C) \ ,$$
$$C' \longrightarrow \tfrac{1}{2}(A + B) \ .$$

Then, we compute the reduction of P by RS:

$$R = |A, B, C|^2 \left(-u_3 + u_2 - u_9 + 2u_1 - u_5 + u_8 - u_7\right) \ .$$

We solve the diophantine equation (linear system over $\mathbb{Z}$):

– For the solution $u_5 = 1$, $u_9 = -1$, $u_1 = u_2 = u_3 = u_4 = u_6 = u_7 = u_8 = u_{10} = 0$, we obtain this bracket equation:

$$|A, C, A'||B, B', C'| - |A, A', C'||B, C, B'| = 0 .$$

We can recognize the property the lines (AA'), (BB'), (CC') are concurrent.

– Another solution is $u_2 = u_5 = 1$, $u_1 = u_3 = u_4 = u_6 = u_7 = u_8 = u_9 = u_{10} = 0$. Thus, we obtain this bracket equation:

$$|A, A', B||B', C', C| + |A, A', C||B', C', B| = 0 ,$$

which means that the lines $(B'C')$ and (BC) are parallel.
The proof: A' is the meet point of (AA') and (BC). I is the meet point of $(B'C')$ and (BC), thus we have:

$$(A', I; B, C) = \frac{|A, A', B||B', C', C|}{|A, A', C||B', C', B|} ,$$

which by hypothesis is equal to -1. So the points B, C, A', I are in harmonic division. But, the fact that A' is the middle point of $[BC]$ implies that I is at infinity. Consequently, (BC) and $(B'C')$ are parallel.

Example 6 (Configuration of a rectangle) Let $ABCD$ a parallelogram such as $AC \equiv BD$. Choosing a reference frame, $A(0, 0)$, $B(x_1, 0)$, $C(x_2, x_3)$, $D(x_4, x_5)$, the hypotheses are:

$$\begin{array}{lll}
(AB) \parallel (CD) & -x_1(x_2 - x_4) = 0 , \\
(AD) \parallel (BC) & x_4 x_3 - x_4 x_1 - x_5 x_2 = 0 , \\
AC \equiv BD & x_2^2 + x_3^2 - x_4^2 - (-x_5 + x_1)^2 = 0 .
\end{array}$$

We compute a characteristic set CS:

$$x_1^2 x_2 (x_1 - x_3) ,$$
$$x_1(x_2 - x_4) ,$$
$$x_5 x_1^2 x_2 .$$

We add the points of the absolute conics I and J, so we have 6 points and 20 brackets and 31 homogeneous components. We choose the homogeneous component of support: A^2, B, D, I, J.

$$P = u_1|A, B, D||A, I, J| + u_2|A, B, I||A, D, J| + u_3|A, B, J||A, D, I| .$$

Then, we compute the pseudo-remainder R of P by CS:

$$R = x_1^2 x_2^2 (2u_1 + u_2 - u_3),$$

and solve the diophantine equation:

$$2u_1 + u_2 - u_3 = 0 .$$

One solution is $u_1 = 0$ and $u_2 = u_3 = 1$,

$$|A, B, I||A, D, J| + |A, B, J||A, D, I| = 0 \, .$$

This equation says that the lines (AB) and (AD) are perpendicular.

Example 7 (Configuration of "place des arcades") We want to find the properties involving orthogonality relations in the sub-configuration of Fig. 9. First, we defined a constructive order using the method described in Section 4.1.

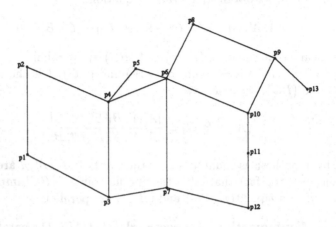

Fig. 9. A sub-configuration of "place des arcades"

Secondly, we construct a reductor set of the configuration, based on rewriting rule in the exterior algebra. We have used here the dual operator *, which computes the normal to a plane. We assume here that the orthogonality defined by duality is a same that the orthogonality defined by the quadric Q. We obtain the following rewriting rules:

$$l_3 \rightarrow p_8 \wedge p_6$$
$$p_8 \rightarrow H_2 \vee l_5$$
$$l_5 \rightarrow (l_4 \vee \mathcal{H}) \wedge p_9$$
$$H_4 \rightarrow p_6 \wedge p_9 \wedge p_{10}$$
$$l_6 \rightarrow p_9 \wedge p_{10}$$
$$p_9 \rightarrow x_5 \, p_{10} + x_6 \, p_{12} + (1 - x_5 - x_6)((U_1 \wedge U_2) \vee H_5)$$
$$p_{11} \rightarrow x_3 \, p_{10} + (1 - x_3)(l_7 \vee \mathcal{H})$$
$$l7 \rightarrow H_5 \vee H_6$$
$$H_1 \rightarrow p_3 \wedge p_4 \wedge (H_2)^*$$
$$p_3 \rightarrow x_2 \, p_4 + (1 x_2)(l_2 \vee \mathcal{H})$$
$$l_2 \rightarrow ((l_1 \vee \mathcal{H})^* \vee \mathcal{H}) \wedge p_4$$
$$p_4 \rightarrow x_1 \, p_6 + (1 - x_1)(l_1 \vee \mathcal{H})$$
$$l_1 \rightarrow H_2 \vee H_3$$
$$l_4 \rightarrow H_6 \vee H_3$$
$$p_{10} \rightarrow H_3 \vee H_5 \vee H_6$$

$$H_5 \rightarrow (H_2 \vee \mathcal{H}) \wedge p_{12}$$
$$H_2 \rightarrow p_6 \wedge p_7 \wedge (H_6)^*$$
$$H_6 \rightarrow p_6 \wedge p_7 \wedge p_{12} .$$

In this case, the classical techniques using coordinates failed due to the computational memory resource.

Then, we compute the different homogeneous components involving $\mathcal{Q}^2, \mathcal{H}$ and concentrate on the homogeneous component in the points p_4, p_6, p_8, p_9. In this case, we have only one generic component

$$u_1((p_4 \wedge p_6) \vee \mathcal{H})^* \wedge ((p_8 \wedge p_9) \vee \mathcal{H}) + u_2((p_4 \wedge p_8) \vee \mathcal{H})^* \wedge ((p_6 \wedge p_9) \vee \mathcal{H})$$
$$+u_3((p_4 \wedge p_9) \vee \mathcal{H})^* \wedge ((p_6 \wedge p_8) \vee \mathcal{H}) + u_4((p_6 \wedge p_8) \vee \mathcal{H})^* \wedge ((p_4 \wedge p_9) \vee \mathcal{H})$$
$$+u_5((p_6 \wedge p_9) \vee \mathcal{H})^* \wedge ((p_4 \wedge p_8) \vee \mathcal{H}) + u_6((p_8 \wedge p_9) \vee \mathcal{H})^* \wedge ((p_4 \wedge p_6) \vee \mathcal{H}) .$$

and check that the only valid relation is for $u_2 = \cdots = u_6 = 0$, that is (p_4, p_6) orthogonal to (p_8, p_9).

6 Conclusion

This paper describes an application of theorem proving to 3D modelisation problems in computer vision. We combine the construction of a reductor set with the analysis of the graph of constraints, in order to compute a correct parameterization of our 3D model. This parameterization is then used in a minimization process to fit the parameters defining the model with its images in the camera(s). We propose a new technique for sorting points and constructing sparse reductor sets. This method uses a coordinate system and mixes symbolic and numeric computations. This enables us to treat systems that were intractable by Gröbner basis or characteristic set methods, but many improvements may still be considered.

We also focus on discovering theorems, which can be used in the numerical steps in the 3D reconstruction, described in Section 2.1. Our approach uses the bracket algebra formalism, which has the advantage to deal with invariant quantities (like determinants of planar points) which can be useful for direct computations from the images. This formalism also allows the decomposition of algebraic expressions into homogeneous components that are used to find the geometric properties of the configuration. This task can be easily parallelized, for the computation in each component is independent of the others once the reductor set is available. Indeed, in practice, we are running several reduction processes (one for each homogeneous component) in parallel on different workstations. This homogeneity also permits to restrict our search to a subset of points of the configuration or to special classes of relations (projective, affine, Euclidean). A drawback of the approach is the fast growth of the number of components with the number of geometric features and the degree of the monomials. Fortunately, interesting properties in a figure correspond to relations of small degree. An interesting feature of our approach is that reductor sets are computed only once.

Among the points that need improvements, we would like to avoid as much as possible the backtracking in the construction of the order on the points. One solution would be e.g. to add new and trival connections in the graph that can be inferred directly from the geometric hypotheses. Interpretation of relations obtained with Algorithm 4 in terms of geometric properties may be automatised using Cayley factorization techniques [24]. As a final remark, the numerical quality of the parameterization of the model also needs a better understanding and deeper analysis, as it is used in a numerical optimization process.

Acknowledgments. This work is partially supported by the European project CUMULI (LTR 21 914). Many thanks to S. Bougnoux that provided us with the basic tools and results for the reconstruction of the scene "place des arcades".

References

1. D. Bondyfalat and S. Bougnoux. Imposing Euclidean Constraints during Self-Calibration Processes. In *Proc. ECCV98 Workshop SMILE*, 1998.
2. S. Bougnoux. From Projective to Euclidean Space under any Practical Situation, a Criticism of Self-Calibration. In *Proc. of the 6th International Conference on Computer Vision*, pages 790–796, Bombay, India, Jan. 1998, IEEE Computer Society Press.
3. S.-C. Chou. *Mechanical Geometry Theorem Proving*. Reidel, Dordrecht Boston, 1988.
4. S.-C. Chou and X.-S. Gao. Ritt-Wu's Decomposition Algorithm and Geometry Theorem Proving. In *Proc. CADE-10*, pages 202–220, Kaiserslautern, Germany, 1990.
5. S.-C. Chou, X.-S. Gao, and J.-Z. Zhang. *Machine Proofs in Geometry: Automated Production of Readable Proofs for Geometry Problems*, volume 6 of *Series on Applied Mathematics*. World Scientific, 1994.
6. S.-C. Chou, W. F. Schelter, and J.-G. Yang. Characteristic Sets and Gröbner Bases in Geometry Theorem Proving. In *Resolution of Equations in Algebraic Structures*, pages 33–92. Academic Press, 1989.
7. H. Crapo and J. Richter-Gebert. Automatic Proving of Geometric Theorems. In *Invariant Methods in Discrete and Computational Geometry*, pages 167–196, 1995.
8. A. Crumeyrolle. *Orthogonal and Simplectic Clifford Algebra*. Kluwer Academic Plublishers, 1990.
9. P. E. Debevec, C. J. Taylor, and J. Malik. Modeling and Rendering Architecture from Photographs: A Hybrid Geometry- and Image-Based Approach. In *Proc. SIGGRAPH*, pages 11–20, New Orleans, Aug. 1996.
10. O. Faugeras. *Three-Dimensional Computer Vision: A Geometric Viewpoint*. MIT Press, 1993.
11. T. Havel. Some Examples of the Use of Distances as Coordinates for Euclidean Geometry. *Journal of Symbolic Computation*, 11: 579–593, 1991.
12. D. Kapur and J. L. Mundy, editors. *Geometric Reasoning*. MIT Press, Cambridge, 1989.
13. D. Kapur and J. L. Mundy. Wu's Method and its Application to Perspective Viewing. In *Geometric Reasoning* [12], pages 15–36.
14. B. Mourrain. *Approche Effective de la Théorie des Invariants des Groupes Classiques*. PhD thesis, Ecole Polytechnique, Sept. 1991.

15. B. Mourrain. Géométrie et Interprétation Générique ; un Algorithme. In *Effective Methods in Algebraic Geometry (MEGA'90)*, volume 94 of *Progress in Math.*, pages 363–377, Castiglioncello (Italy), 1991. Birkhäuser.

16. P. Poulin, M. Ouimet, and M.-C. Frasson. Interactively Modeling with Photogrammetry. In *Proc. of Eurographics Workshop on Rendering 98*, pages 93–104, June 1998.

17. J. Richter-Gebert. Mechanical Theorem Proving in Projective Geometry. *Annals of Mathematics and Artificial Intelligence*, 13: 139–172, 1995.

18. S. Stifter. Geometry Theorem Proving in Vector Spaces by Means of Gröbner Bases. In M. Bronstein, editor, *Proc. ISSAC'93*, ACM Press, pages 301–310, Jul. 1993.

19. D. Wang. *CharSets 2.0 — A Package for Decomposing Polynomial Systems.* Preprint, LEIBNIZ – Institut IMAG, Feb. 1996.

20. D. Wang. Zero Decomposition for Theorem Proving in Geometry. In *Proc. IMACS International Conference on Applications of Computer Algebra*, Albuquerque, USA, May 1995.

21. D. Wang. Geometry Machines: From AI to SMC. In *Proc. AISMC-3*, volume 1138 of *LNCS*, pages 213–239, 1996.

22. D. Wang. Clifford Algebraic Calculus for Geometric Reasoning with Application to Computer Vision, volume 1360 of *Springer's LNAI*, pages 115–140. Springer, 1997.

23. D. Wang. Gröbner Bases Applied to Geometric Theorem Proving and Discovering. In B. Buchberger and F. Winkler, editors, *Gröbner Bases and Applications*, pages 281–301. Cambridge Univ. Press, 1998.

24. N. White. Multilinear Cayley Factorization. *Journal of Symbolic Computation*, 11(5 & 6): 421–438, May/June 1991.

25. W.-t Wu. Basic Principles of Mechanical Theorem Proving in Elementary Geometries. *J. Automated Reasoning*, 2: 221–252, 1986.

26. W.-t Wu. *Mechanical Theorem Proving in Geometries: Basic Principles* (translated from Chinese by X. Jin and D. Wang). Texts and Monographie in Symbolic Computation. Springer, 1994.

Automated Geometry Diagram Construction and Engineering Geometry*

Xiao-Shan Gao

Institute of Systems Science, Academia Sinica, Beijing 100080, P. R. China
xgao@mmrc.iss.ac.cn

Abstract. This paper reviews three main techniques for automated geometry diagram construction: synthetic methods, numerical computation methods, and symbolic computation methods. We also show how to use these techniques in parametric mechanical CAD, linkage design, computer vision, dynamic geometry, and CAI (computer aided instruction). The methods and the applications reviewed in this paper are closely connected and could be appropriately named as *engineering geometry*.

1 Introduction

In the area of automated geometry reasoning, most of the efforts have been concentrated on automated geometry theorem proving and discovering. These efforts lead to the classic work of Tarski [61], Gelernter [29], and Wu [68]. On the other hand, automated geometry diagram construction (AGDC) is more or less overlooked. Actually, AGDC has been studied in the CAD community under a different name: geometric constraint solving (GCS) and with a different perspective: engineering diagram drawing. GCS is the central topic in much of the current work of developing intelligent or parametric CAD systems [2, 5, 21, 33, 43, 53, 59, 63] and interactive constraint-based graphic systems [3, 32, 60]. The main advantage of using the constraint approach is that the resulting systems accept declarative descriptions of diagrams or engineering drawings, while for conventional CAD systems the users need to specify how to draw the diagrams. As a result, parametric systems are more powerful and user friendly.

Three main approaches have been developed for automated diagram construction: the synthetic approach, the numerical approach, and the symbolic approach.

In the *synthetic approach*, a pre-treatment is carried out to transform the constraint problem into a step-by-step constructive form which is easy to draw. A majority of the work is to transform the problem to a constructive form that can be drawn with ruler and compass. Once a constraint diagram is transformed into constructive form, all its solutions can be computed efficiently. Two main techniques used in the synthetic approach are the *rule-based search* and the *graph analysis*. The basic idea of the rule-based search [2, 5, 21, 59, 63] is to

---

* This work was supported in part by an Outstanding Youth Grant from the Chinese NSF and the National "973" Project.

represent the knowledge of geometric construction as deductive rules and use various search techniques from AI to find a step-by-step construction procedure for a constrained diagram. The basic idea of the graph analysis method [33, 46, 43, 53, 62] is to represent the geometric conditions as a graph and use various techniques from graph theory to attack the problem.

In the *numerical approach*, geometric constraints are translated into algebraic equations and various numerical techniques are used to solve these equations. The first generation constraint-based systems, such as Sketchpad [60] and ThingLab [3], used numerical relaxation as the last resort of solving constraint problems. Newton iteration method was used in Juno-2 [32] and in the Variational Geometry [51]. The main advantage of numerical methods is their fast speed and generality. On the other hand, it also needs further improvements in several aspects: like finding stable solutions during animation and finding all solutions of equation systems.

In the *symbolic approach* [42, 22], we also transform the geometric constraints into algebraic equations. Instead of using numerical methods to solve the algebraic equations directly, we first use general symbolic methods such as Wu-Ritt's characteristic set method [68], the Gröbner basis method [7], the Dixon resultant method [39], and Collins' cylindrical algebraic decomposition method [15] to change the equation set to new forms which are easy to solve, and then solve the new equations numerically. The advantage of the symbolic computational approach is that it may provide complete methods for many problems in AGDC. On the other hand, current symbolic methods are still too slow for real time computation.

The above AGDC techniques can be applied to a variety of engineering problems: parametric mechanical design, linkage design, robotics, computer vision, and geometric modeling. They can also be used in dynamic geometry, physics, and CAI (computer aided instruction). We call the methods and the applications reviewed in this paper *engineering geometry*.

This paper is an attempt to give an overview of the research on engineering geometry with emphasis on the work done by the extended Wu group. The reader may find details for other related work based on the given references. In the next section, we introduce two synthetic approaches: the global propagation and the graph decomposition approach. In Section 3, we introduce the application of the numerical optimization methods to AGDC. In Section 4, we introduce the symbolic approach based on Wu-Ritt's characteristic set method. In Section 5, we introduce dynamic geometry as a tool of generating diagrams. In Section 6, we discuss applications of these techniques.

2   Synthetic Approaches to AGDC

Synthetic approaches provide a geometric way of solving algebraic equations. They can generate all solutions of a diagram efficiently and with high precision. The solutions provided by this kind of approaches are stable and have geometric meanings. So in most drawing systems, synthetic approaches are first used to solve the diagram construction problem. If synthetic approaches fail, other approaches will be used.

In this section, we will give a brief introduction to two synthetic approaches: the global propagation approach [21] and Owen's graph decomposition approach [53]. Other important synthetic approaches include Hoffmann's graph analysis approach [33], Kramer's freedom of degree analysis approach [43], Latheam and Middleditch's approach based on weighted graphs [46], Lee's optimized rule-based method [47], and rule-based methods proposed in [1,64].

2.1 The Global Propagation Method

Local Propagation is a basic AI technique to solve constraint problems [57,49]. It is to find unknown quantities one by one from the set of known quantities. For instance, from the relation $5F - 9c = 160$ about two kinds of measurement of temperature c and F, we may find the following propagational rules:

1. If c is known, we may compute F with the formula: $F = \frac{9}{5}c + 32$.
2. If F is known, we may compute c with the formula: $c = \frac{5(F-32)}{9}$.

For AGDC, local propagation is closely related to the *method of locus intersection*. For instance, to draw a triangle ABC with sides $AB = 4, BC = 5$, and $CA = 6$, we first draw an arbitrary point as point A. To draw point B, we first draw a circle (locus) with center A and radius 4, and point B is an arbitrary point on this circle. To draw point C, we first draw two circles (loci) with centers A and B and radii 6 and 5 respectively, and C is the intersection of these two circles.

It is clear that the power of local propagation is limited. It can not solve constraint problems with *loops*, i.e., more than one geometric objects are constrained or referred to simultaneously in the problem. *Global propagation* is proposed to solve constraint problems with loops [21]. Like the *local propagation method*, it tries to determine a geometric object with the locus intersection method. But the global propagation uses not only the constraints involving this object but also *implicit information* derived from other constraints. The global information needed in the propagation comes from a *geometry information base (GIB)* built before the construction begins. The GIB for a configuration is actually a database containing all the properties of the configuration that can be deduced by using a fixed set of geometric axioms [14].

Figure 1 is an example from [33]. Starting from point B, it is easy to construct B, C, X. Next, we want to construct D. Since $|XD|$ is known, we need only to know the direction of XD which can be determined by the following *global propagation of line directions*: $XD \perp DE$, $\angle(DEA) = a_2$, $AE \perp AB$, and $\angle(ABC) = a_1$.

Generally speaking, to determine the direction of a line L, we first consider some *basic situations*:

1. L passes through two known points.
2. L cuts a fixed length between a pair of parallel lines.

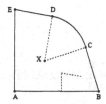

The constraints:
$|BC| = d1, |ED| = d2,$
$|XC| = d3, |XD| = d3$
ANGLE(ABC) = a1, ANGLE(DEA) = a2
PERP(X,C,C,B), PERP(X,D,D,E), PERP(B,A,A,E)

Fig. 1. An example to show the propagation of line direction

3. L cuts a pair of segments on the sides of a known angle such that the ratio of the segments is known.
4. L cuts a pair of segments between three known lines passing through a point such that the ratio of the segments is known.

Secondly, line directions can be propagated as follows:

$$l_1 \; \tau_1 \; l_2 \; \tau_2 \; \ldots \; \tau_{s-1} \; l_s,$$

where l_i are lines and τ_i represent the fact that l_i and l_{i+1} form a known angle. If the direction of l_s is determined by one of the basic situations, the direction of line l_1 is known by the above propagation. This process clearly demonstrates the global feature of the propagation.

An important technique in the global propagation is to use *LC transformation*. An LC transformation is a one-to-one map from the Euclidean plane to itself which transforms a line to a line and a circle to a circle. We mainly use two kinds of LC transformations.

The first LC transformation is as follows. Suppose that segment AB has a fixed direction and a fixed length. Then point A is on a line or a circle if and only if point B is on a line or a circle. An example is given in Fig. 2. Suppose that points A and C have been constructed. Next, we construct point B. Since $\angle ABC = a_1$, B is on a circle C_1. Similarly, point D is on a circle C_2. Since BD has a fixed direction and a fixed length, point B must be on another circle C_3, which is the transformation of circle C_2. B is the intersection of circles C_2 and C_3. To construct C_3, we first construct the center X_1 of circle C_2 and then move it for a distance of $|BD|$ along the direction of BD to obtain the center X_2 of C_3. The radius of C_3 is the same as that of C_2.

The second LC transformation is related to ratio constraints. If points P, Q, R are collinear, PQ/RQ is known, and one of them, say Q, is already known, then point P is on a line or a circle if and only if point R is on a line or a circle. Consider the constraint problem in Fig. 3. Suppose that we have constructed points A and B. Since $\angle(ACB) = a_1$, point C is on a circle C_1. We construct the center X_2 of circle C_1. Since M is the midpoint of BC, by the second LC transformation, point M is on another circle C_2. Hence M is the intersection of circle C_2 and the circle with A as center and with d_2 as radius.

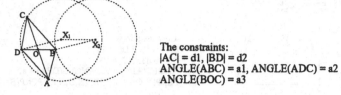

The constraints:
$|AC| = d1$, $|BD| = d2$
ANGLE(ABC) = a1, ANGLE(ADC) = a2
ANGLE(BOC) = a3

Fig. 2. An example to show the first kind of LC transformation

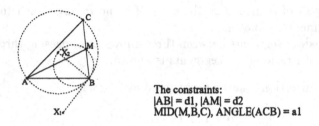

The constraints:
$|AB| = d1$, $|AM| = d2$
MID(M,B,C), ANGLE(ACB) = a1

Fig. 3. An example to show the second kind of LC transformation

On the top level, the method works as follows [21]:

1. For a constraint problem

$$[[Q_1, \ldots, Q_m], [P_1, \ldots, P_n], [C_1, \ldots, C_m]],$$

let $CT = \{C_1, \ldots, C_m\}$ be the constraint set, $CS = \emptyset$ the construction sequence, $QS = \{Q_1, \ldots, Q_m\}$ the points with given construction order, and $PS = \{P_1, \ldots, P_n\}$ the remaining points. We assume that the problem is not over-constrained, i.e., we have $|CT| \leq 2 * |PS| - 3$.

2. Build the GIB as described in [14]. Then repeat the following steps first for QS and then for PS until both QS and PS become empty.

3. Take a point P from QS or PS. For each constraint $T_i \in CT$ involving P, decide the locus Lc of the points satisfying T_i, assuming that all points constructed in CS are known. We then obtain a set of triplets

$$\{(P, T_1, Lc_1), \ldots, (P, T_s, Lc_s)\}.$$

Three cases are considered. (1) $s = 0$. Point P might be an arbitrarily chosen (free) point. We add a new construction C = (POINT, P) to CS. (2) There exist $i \neq j$ such that $T_i \neq T_j$ and Lc_i and Lc_j are not parallel lines or concentric circles.[1] We add a new construction C = (INTERSECTION, P,

---

[1] We can check these conditions by using automated theorem proving methods such as Wu's method [9, 68] or the area method [13] which are quite fast.

Lc_i, Lc_j) to CS and remove T_i and T_j from CT. (3) Otherwise, point P is a semi-free point: it can move freely on a line or a circle. We add a new construction C= (ON, P, Lc_1) to CS and remove T_1 from CT.

4. Now we check whether the remaining problem is over-constrained, that is, whether $|CT| > 2 * |PS|$ (note this inequality is different from the one in Step 1). (1) If it is over-constrained then the construction sequence is invalid in this case. If P is from QS, the order given by the user can not be constructed and the process terminates. Otherwise, restore the removed constraints and repeat the preceding step for a new point from PS. (2) If it is not over-constrained, point P is constructed. We need to repeat the preceding step for a new point.

2.2 Owen's Graph Decomposition Method

In the graph analysis approach, a constraint problem is first transformed into a graph whose vertices and edges represent the geometric objects and constraints respectively. A constraint problem without loops can be represented as trees [62]. Owen introduced a graph analysis method that can handle constraint problems with loops [53]. Basically speaking, Owen's algorithm has two steps:

1. *Decompose the graph into sub-graphs.* A graph is called *bi-connected* if it is connected and we cannot find a vertex such that by removing this vertex the graph becomes two graphs. A graph is called *tri-connected* if it is bi-connected and we cannot find two vertices such that by removing these two vertices and the edge (if there exists one) connecting them, this graph becomes two graphs. The decomposition procedure is first decomposing the graph into bi-connected graphs and then decomposing the bi-connected graph into tri-connected graphs by using algorithms from graph theory [30]. At the final stage, every sub-graph must has less than three vertices. Otherwise, the algorithm cannot handle the problem.

2. *Assemble the sub-graphs into larger ones.* This step is the reverse of the first step and has two sub-cases for a well-constrained problem.

   (a) If two vertices are removed during the decomposition and both vertices are connected with the same unknown vertex V, then we can construct the geometric object represented by V in the next step. For instance, if the two known vertices represent lines and V represents a point P, then we will take the intersection of the two lines to obtain point P in the next step.

   (b) If a graph is decomposed into three sub-graphs in such a way that each pair of sub-graphs have a common vertex, then we can assemble the three sub-graphs together. Since the problem is well-constrained, each sub-graph represents a rigid part. Let the common vertices represent, say, three points. Since each pair of the points is in a rigid part, we can compute the distances between them. Then the assembling problem becomes the construction of a triangle with three known sides.

Let us consider the problem in Fig. 1. Figure 4 (a) is the graph of this constraint problem, where $l_1, l_2, l_3, l_4, l_5, l_6$ are lines (not segments) AB, BC, CX, XD, DE, EA respectively. We will solve this problem in two steps.

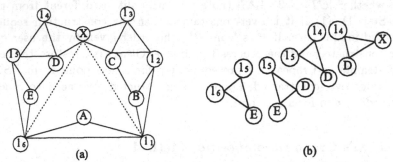

Fig. 4. Graph and graph decomposition for the diagram in Fig. 1

1. *Decompose the graph into sub-graphs.* We first decompose the graph into three sub-graphs by breaking the graph at vertices X, l_1, l_6. The sub-graph $l_1 l_6 A$ needs no further decomposition, since it contains three vertices only. The other two sub-graphs are similar. Figure 4 (b) gives the decomposition for one of them.

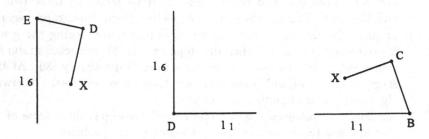

Fig. 5. Three sub-diagrams of the diagram in Fig. 1

2. *Assemble the sub-graphs into larger ones.* The three sub-graphs obtained in the first step of decomposition represent the three diagrams in Fig. 5, which are easy to construct with ruler and compass. The critical step is to assemble the three parts into one. Let us note that the respective common vertices are X, l_1, and l_6. Since each of the three parts is a rigid body, we know the distances from X to l_1 and l_6, and the angle formed by l_1 and l_6. Then we can easily draw a diagram containing X, l_1 and l_6. The remaining constructions are easy.

According to the above algorithm, we cannot solve the problem in Fig. 2.

3 Numerical Computation Approaches to AGDC

Numerical methods are usually used as the last resort in the constraint solving process, because they can be used to solve any kinds of equation systems. The most commonly-used method in the numerical approach is the Newton-Raphson method. It is fast, but has the instability problem: the method is sensitive to the initial values. A small deviation in the initial value may lead to an unexpected or unwanted solution, or to an iteration divergence. To overcome this difficulty, recently the homotopy method has been proposed and experimented with [45]. According to the report in [45], generally the homotopy method works much better in terms of stability. These two methods generally require the number of variables to be the same as the number of equations.

In [28], a method based on the optimization techniques for solving geometric constraint problems is proposed. Experiments with this method show that it is also quite stable. Furthermore, the method can *naturally* deal with under- and over-constrained problems.

Generally, a geometric constraint problem can be first translated into a system of equations:

$$f_1(x_1, \ldots, x_n) = 0,$$
$$f_2(x_1, \ldots, x_n) = 0,$$
$$\cdots$$
$$f_m(x_1, \ldots, x_n) = 0.$$

The optimization approach solves the equation system by converting it into finding the X at which the sum of squares

$$\sigma(X) = \sum_{i=1}^{m} f_i(X)^2$$

is minimal, where $X = (x_1, x_2, \ldots, x_n)$ is the set of variables. It is obvious that the equation system has a *real* solution X^* if and only if $\min \sigma(X)$ is 0. The problem of solving a system of equations is thus converted into the problem of finding the minimum of a real multi-variate function. The problem now can be solved by various well-developed numerical optimization methods [52, 17].

One obvious fact for this approach is that the number of equations m is not necessarily the same as the number of variables n. Thus for this approach, it is natural to deal with under- and over-constrained problems. Another advantage of this method is that it is not sensitive to the initial value. This feature is demonstrated in Fig. 6 where the three circles are specified to be mutually tangent and to be tangent to two neighboring sides of a triangle whose three vertices are specified to be fixed. Figures 6(a) and 6(c) are the initial diagrams sketched by the user. It is easy to see the differences between the initial guesses from the exact solutions of this problem. Also, this figure demonstrates how different initial values lead to different branches of the solutions.

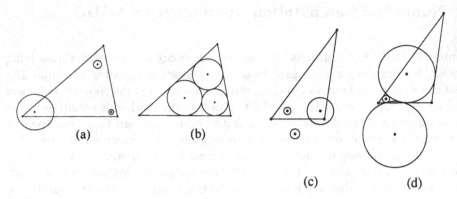

Fig. 6. The insensitivity to the inexact initial guesses of the solutions in the optimization method.

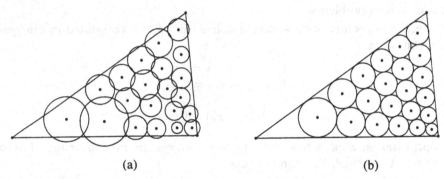

Fig. 7. A difficult problem. (a) is the initial diagram drawn by the user. (b) is the diagram generated after all tangent constraints are added

Figure 6 is the simplest case for the problem we call *the tangent packing problem*. The problem is to pack $n(n + 1)/2$ circles (n rows of circles) tangent to adjacent circles and/or the adjacent neighboring sides of a given triangle. Figure 7 is the case of $n = 6$, i.e., we need to pack 21 circles in the triangle. This difficult problem contains 174 variables (since we introduce auxiliary tangent points) and 6 linear equations and 168 quadratic equations which could not be block triangularized. Table 1 shows the running time for different n of this problem.

4 Symbolic Computation Approaches to AGDC

Theoretically, both the synthetic and numerical approaches are not complete. Synthetic approaches are not decision procedures even for the case of ruler and compass construction, i.e., for many problems that can be drawn with ruler and

Table 1. Running statistics for the tangent packing problems

# circles (# rows)	# equations	# variables	time
3 (2)	30	30	0.228
6 (3)	54	54	0.965
10 (4)	86	86	3.379
15 (5)	126	126	11.587
21 (6)	174	174	23.751

compass, they cannot give a drawing procedure. Using numerical approaches, if a solution is not found, we cannot assert that the system does not have a solution. Also, if we find one or two solutions of a system, we do not know whether these are all the solutions of the system. To solve these kinds of completeness problems, we may use the symbolic computation techniques such as Wu-Ritt's characteristic set method [68], the Gröbner basis method [7], the Dixon resultant method [39], and Collins' cylindrical algebraic decomposition method [15]. For instance, using Wu-Ritt's characteristic set method [68] we can give decision procedures ([22])

1. to construct a constrained diagram with ruler and compass (rc-constructibility);
2. to detect whether a constrained diagram is well-(under-, over-)constrained;
3. to detect whether a set parameters are conflicting, and if they are, find the relation among them; and
4. to detect whether a constraint is redundant.

Let us consider how to draw a regular pentagon $ABCDE$ with edge r. Let $A = (0,0), B = (r,0), C = (x_2, y_2), D = (x_3, y_3), E = (x_4, y_4)$. Since there are five points in the diagram, we need $2 * 5 - 3 = 7$ constraints. The constraint $|AB| = r$ is already satisfied automatically. We need six other constraints:

$$|AB| = |AE|: h_1 = y_4^2 + x_4^2 - r^2 = 0$$
$$|AB| = |BC|: h_2 = y_2^2 + x_2^2 - 2rx_2 = 0$$
$$|AB| = |CD|: h_3 = y_3^2 - 2y_2y_3 + x_3^2 - 2x_2x_3 + y_2^2 + x_2^2 - r^2 = 0$$
$$|AB| = |DE|: h_4 = y_4^2 - 2y_3y_4 + x_4^2 - 2x_3x_4 + y_3^2 + x_3^2 - r^2 = 0$$
$$|AD| = |DB|: h_5 = 2rx_3 - r^2 = 0$$
$$|AC| = |CE|: h_6 = y_4^2 - 2y_2y_4 + x_4^2 - 2x_2x_4 = 0.$$

Wu-Ritt's characteristic set method [68] can be used to represent the zero set of any polynomial equation system as the union of the zero sets of polynomial sets in *triangular form*. There are methods to find all solutions of a set of polynomial equations in triangular form. Let $Zero(PS)$ be the set of all zeros or solutions of the polynomial equation system PS and $Zero(PS/D) = Zero(PS) - Zero(D)$ for a polynomial D. Using Wu-Ritt's characteristic set method to the pentagon problem, we have

$$Zero(\{h_1, \ldots, h_6\}) = \cup_{i=1}^4 Zero(TS_i/I_i)$$

where TS_i and I_i are given below.

$TS_1 =$	$TS_2 =$	$TS_3 =$	$TS_4 =$
$y_4 - y_2$	$ry_4 + (2x_2 - 2r)y_2$	$y_4 - y_2$	y_4
$x_4 + x_2 - r$	$2x_4 - 2x_2 + r$	$x_4 + x_2 - r$	$x_4 + r$
$ry_3 + (-2x_2 + r)y_2$	$ry_3 + (2x_2 - 3r)y_2$	$ry_3 + (2x_2 - 3r)y_2$	$y_3 - y_2$
$2x_3 - r$	$2x_3 - r$	$2x_3 - r$	$2x_3 - r$
$4y_2^2 - 2rx_2 - r^2$	$4y_2^2 - 2rx_2 - r^2$	$4y_2^2 - 2rx_2 - r^2$	$4y_2^2 + 5r^2$
$4x_2^2 - 6rx_2 + r^2.$	$4x_2^2 - 6rx_2 + r^2.$	$4x_2^2 - 6rx_2 + r^2.$	$2x_2 + r.$

$I_1 = I_2 = I_3 = r$ and $I_4 = 1$. TS_4 does not have non-zero solutions, since it contains $4y_2^2 + 5r^2$. Each of TS_1, TS_2, and TS_3 contains six polynomials and has total degree four, i.e., for a non-zero value of r, each of TS_1, TS_2, and TS_3 generally gives four solutions to the problem (see Fig. 8).

4.1 Well-, Under-, and Over-Constrained Systems

A constraint system is called

- *well-constrained* if the shape of the corresponding diagram has only a finite number of cases;
- *under-constrained* if the shape of the corresponding diagram has infinite solutions;
- *over-constrained* if the corresponding diagram has no solution.

These properties are quite difficult to detect, so people usually use another kind of definition. A constraint system is called

- *structurally well-constrained* if each of its sub-diagram with n points and lines has $2 * n - 3$ constraints;
- *structurally under-constrained* if some of its sub-diagrams with n points and lines have less than $2 * n - 3$ constraints;
- *structurally over-constrained* if some of its sub-diagrams with n points and lines have more than $2 * n - 3$ constraints.

It is an interesting problem to study what kinds of structurally well-constrained systems are well constrained. In [54], some results are given. But in my viewpoint, these kinds of results are more theoretical-oriented than practical-oriented. Actually we can find many examples to show that structurally well- and over-constrained diagram could be under-constrained and structurally well-, under-, and over- constrained diagram could be over-constrained.

We use the pentagon example to show that sometimes we need more than $2n - 3$ constraints to define a constraint system with n points. The following table gives the result of checking three more geometric conditions on the three diagrams defined by TS_1, TS_2, and TS_3 by using the automated theorem proving methods given in [68].

	TS_1	TS_2	TS_3
$|AD| = |AC|$	true	false	false
$|BD| = |BE|$	true	true	false
$|EB| = |EC|$	true	false	true

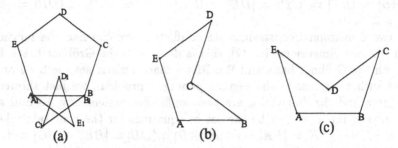

Fig. 8. Three components for the regular pentagon problem

It is clear that the pentagon problem is well-defined. But the seven independent constraints do not properly define a regular pentagon. To define a regular pentagon, we need to add at least another constraint

$$|AD| = |AC|: \quad h_7 = y_3^2 + x_3^2 - y_2^2 - x_2^2.$$

Now there is only one useful triangular set TS_1 in the decomposition:

$$Zero(\{h_1, \ldots, h_7\}) = Zero(TS_1/I_1) \cup Zero(TS_4/I_4).$$

But this constraint system is structurally over-constrained. Generally, we have [22]

Theorem 1. *Using Wu-Ritt's characteristic set method, we may decide whether a constrained diagram is well-, over-, or under- constrained over the field of complex numbers.*

4.2 Independent and Conflicting Constraints

A set of constraints is said to be *independent* if no one of them can be removed and the new constraint system still defines the same diagram. A set of constraints is said to be *conflicting* if no diagram in the Euclidean plane verifies this constraint system.

Let us first comment that in the general case there is no connection between the concepts of independency and the critical number $2n - 3$. Actually, there are conflicting and/or non-independent systems with less than $2n - 3$ constraints, as well as non-conflicting and/or independent systems with more than $2n - 3$ constraints.

For a constraint system CS, let PS be the corresponding algebraic equations of the constraints. The conflictness of CS means that PS does not have solutions. In the case of complex number field, this can be decided with Wu-Ritt's characteristic set method [68, 22]. For the case of real numbers, this can be done with the quantifier elimination algorithm [15]. For instance, if we want to draw a regular pentagon with edge 1 and diagonal 2, we will get a conflicting constraint system. In other words, $Zero(PS) = \emptyset$ where PS consists of equations for $|AB| = |BC| = |CD| = |DE| = |EA| = 1$ and $|AD| = |DB| = |BE| = |EC| = 2$.

If two dimensional constraints are conflicting, we can find the relation between the two dimensions. In [42], this is done with the Gröbner basis. In [11, 22], both the Gröbner basis and Wu-Ritt's characteristic set methods are used to find such relations. In the regular pentagon problem, we just showed that the edge r and the diagonal d are two conflicting dimensions. To find a relation between them, let PS be the set of equations for the constraints $|AB| = |BC| = |CD| = |DE| = |EA| = r$ and $|AD| = |DB| = |BE| = |EC| = d$. Using Wu-Ritt's characteristic set method, two such relations are found:

$$d^2 + rd - r^2 = 0 \text{ and } d^2 - rd - r^2 = 0.$$

Since d and r are positive, the two reasonable solutions are $d = \frac{1+\sqrt{5}}{2}r$ and $d = \frac{-1+\sqrt{5}}{2}r$. The first one represents the case where r is the edge and d is the diagonal. The second one represents the case where r is the diagonal and d is the edge.

Let PS be the equation set of a constraint system. This system is not independent iff there is a polynomial P in PS such that

$$Zero(PS - \{P\}) \subset Zero(P) \text{ or } Zero(PS - \{P\}) = Zero(PS). \qquad (4.1)$$

Note that to check (4.1) is actually to prove a geometry theorem. Let C_p be the geometric constraint of $P = 0$ and $C_1, \ldots, C_n$ the geometric constraints for other equations in PS. Then (4.1) is equivalent to the following geometry theorem problem

$$\forall \text{ points}[(C_1 \wedge C_2 \cdots \wedge C_n) \Rightarrow C_p].$$

This kind of *automated geometry theorem proving problems* has been studies extensively [68, 12, 38]. We may use these efficient algorithms to prove whether (4.1) is true.

4.3 A Decision Procedure for Rc-Constructibility

Let $\mathbf{Q}$ be the field of rational numbers. We allow ourselves to abuse terms by saying that the numerical solutions of a diagram can be constructed with ruler and compass if the corresponding diagram can be constructed with ruler and compass. It is well known [37] that a number η can be constructed with ruler and compass iff there exist n real numbers $\eta_1, \eta_2, \ldots, \eta_n = \eta$ and n quadratic equations

$$Q_1(x_1) = x_1^2 + b_1 x_1 + c_1 = 0$$
$$Q_2(x_1 x_2) = x_2^2 + b_2(x_1)x_2 + c_2(x_1) = 0$$
$$\cdots$$
$$Q_n(x_1, x_2, \ldots, x_n) = x_n^2 + b_n(x_1, \ldots, x_{n-1})x_n + c_n(x_1, \ldots, x_{n-1}) = 0$$

such that $Q_i(\eta_1, \eta_2, \ldots, \eta_i) = 0, (i = 1, \ldots, n)$. Since Q_i is of degree two, we can further assume that the b_i and c_i are linear in the variables $x_1, \ldots, x_{i-1}$. From the above equations, we have the well-known result: *An rc-constructible number must be the root of an equation of degree 2^k for some $k > 0$.* Thus *the root of a cubic polynomial equation is rc-constructible iff the polynomial is reducible.* Since we can efficiently factor polynomials with computer algebraic systems such as Maple [8], the only remaining problem is to decide whether the roots of polynomial equations of degree 2^k are rc-constructible.

Let us consider the simplest case: Decide whether the roots of a quartic equation

$$x^4 + h_3 x^3 + h_2 x^2 + h_1 x + h_0 = 0 \tag{4.2}$$

are rc-constructible. Let us assume that the quartic polynomial in (4.2) is irreducible. If a root y of (4.2) is rc-constructible, y can be written as the solution of the following two equations:

$$x_1^2 + bx_1 + c = 0$$
$$y^2 + (-mx_1 + f)y - nx_1 + g = 0.$$

Without loss of generality, the first equation can be reduced to $x_1^2 - 1 = 0$. From the second equation, we have $x_1 = \frac{y^2 + fy + g}{my + n}$. Substituting this into $x_1^2 - 1 = 0$, the numerator is $(y^2 + fy + g)^2 - (my + n)^2 = 0$. Comparing the coefficients of y in this equation with that of (4.2), we obtain a set of equations

$$2f - h_3 = 0$$
$$m^2 - 2g - f^2 + h_2 = 0$$
$$2mn - 2fg + h_1 = 0$$
$$n^2 - g^2 + h_0 = 0.$$

Eliminating f, m, and n, we have

$$8g^3 - 4h_2 g^2 + (2h_1 h_3 - 8h_0)g - h_0 h_3^2 + 4h_0 h_2 - h_1^2 = 0. \tag{4.3}$$

Let g_1 be a root of equation (4.3), f_1, m_1 and n_1 solutions for f, m and n with g being substituted by g_1. Then all the roots of equation (4.2) can be obtained from the quadratic equations

$$y^2 + f_1 y + g_1 \pm (m_1 y + n_1) = 0.$$

Therefore, the roots of equation (4.2) can be obtained by successively solving several quadratic and a cubic equations. Roots of a cubic equation are rc-constructible iff the cubic polynomial is reducible, i.e., it has a linear factor. Then we have a method of deciding whether the roots of a quartic equation can be constructed with ruler and compass. For the solution to the general case, please consult [22].

Let us consider Pappus' Problem: Given a fixed point P, a fixed angle AOB, and a fixed length d, draw a line passing through point P and cutting a segment of length d between lines OA and OB. Let the segment be XY. Use the following coordinates: $O = (0,0), A = (w,0), P = (u,v), B = (w,s), X = (x_1, y_1), Y = (x_2, 0)$. If we consider w, u, v, and s as non-zero constants, with Wu-Ritt's method, we can find a triangular set which contains the following quartic equation $f = s^2(s^2 + w^2)x_1^4 - 2ws^2(vs + wu)x_1^3 + (v^2 + u^2 - d^2)w^2s^2x_1^2 + 2d^2w^3vsx_1 - d^2w^4v^2 = 0$. With the method mentioned above, it is easy to check that this diagram generally cannot be drawn with ruler and compass. With the above method, we can also prove the following interesting fact: if point P is on the bisector of angle AOB then the diagram is rc-constructible.

5 Dynamic Geometry

Dynamic geometry is originated in the field of CAI (computer aided instruction) [20, 36], although it can be used in other areas such as linkage design [26] and computer vision [27]. Many geometry theorem provers also have diagram generation as an assistant part, e.g., [9, 66]. But these graphic editors are for constructive geometry statements. In this paper, we use dynamic geometry as an effective tool of generating geometric diagrams in GCS problems such as linkage design, computer vision, and CAI.

By *dynamic geometry*, we mean models built by computer software that can be changed dynamically [41]. The basic properties of dynamic models are: *dynamic transformation, dynamic measurement, free dragging, animation, and locus generation* [25, 26]. By doing dynamic transformation and free dragging, we can obtain various forms of diagrams easily and see the changing process vividly. Through animation, the user may observe the generation process of curves or the figures of functions and get an intuitive idea of the properties of the functions and curves. Furthermore, we can combine free dragging with locus generation to give a powerful tool of showing the properties of curves or functions.

Geometry Expert (GEX) [24] is a program for dynamic diagram drawing and automated geometry theorem proving and discovering. As a dynamic geometry system, GEX can be used to build *dynamic visual models* to assist teaching and learning of various mathematical concepts. As an automated reasoning system, we can build *dynamic logic models* which can do reasoning themselves. Logic models can be used for more intelligent educational tasks, such as automated generation of test problems, automated evaluation of students' answers, intelligent tutoring, etc.

With Geometry Expert, we can build the following classes of dynamic visual models.

1. Loci generated by diagrams constructed by using ruler and compass. This class includes conics, trigonometric functions and the following curves built with *Horner's construction* [25]:

   (a) Rational functions $y = \frac{f(x)}{g(x)}$.

   (b) Functions of the form $y = \sqrt[n]{f(x)}$.

(c) Diagrams for any *rational plane curves* defined by their parametric equations

$$x = \frac{P(t)}{R(t)}, y = \frac{Q(t)}{R(t)}$$

where P, Q, R are polynomials in t.

(d) **Interpolation polynomials.** For n distinct points $(x_i, y_i), i = 1, \ldots, n$, draw the diagram of a polynomial $y = f(x)$ such that $y_i = f(x_i)$.

2. Diagrams of functions using numerical computation. This class includes functions of the form: $y = f(x)$ where $f(x)$ could be any "elementary functions" – a^x, x^a, $log(x)$, trigonometric functions – and their arithmetic expressions and compositions. This part is quite similar to a "Graphic Calculator," but is more flexible and powerful.

3. Loci generated by linkages with rotating joints alone. As proved by Kempe [40], this class includes any algebraic curve $f(x, y) = 0$ where $f(x, y)$ is a polynomial in x and y.

To demonstrate how to use free dragging and locus generation, we consider the following problem. Let H be the orthocenter of triangle ABC. We fix points A and B and let point C move on a circle c. We want to know the shape of the locus of point H. Let (x_0, y_0) and r be the center and radius of circle c, and $A = (0,0)$, and $B = (d, 0)$. With the method in Section 6.1, we can derive the equation of this locus

$$((x - x_0)^2 + y_0^2 - r^2)y^2 + 2y_0(x - d)xy + (x - d)^2 x^2 = 0.$$

But from this equation, it is still difficult to know the shape of this curve. With dynamic geometry software like *Geometry Expert*, we can continuously change the radius of circle c by free dragging and observe the shape of the curve changing continuously (Fig. 10).

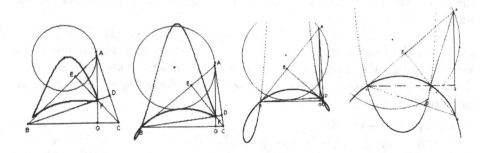

Fig. 9. Shapes of the locus of the orthocenter when the radius of circle c changes.

6 Applications: Engineering Geometry

We mentioned before that AGDC is first considered in the community of mechanical CAD. Here we want to show that by considering AGDC in a more general sense, it can be used to more engineering problems such as linkage design, computer vision, robotics [71, 35, 73] and geometric modeling [72, 23]. AGDC methods can also be used to solve problems from mathematics, physics, and education. We call the methods of AGDC and these applications *engineering geometry*.

6.1 Mechanical Formula Derivation

In geometry, the techniques for AGDC have close connection with the so-called *mechanical formula derivation*. There are two kinds of formula derivation problems. One is deriving geometry formulas [69, 10, 11, 65]; the other is finding locus equations [67, 65]. The problem of finding the relation between the edge and the diagonal of a regular pentagon in Section 4.2 is an example of finding geometry formulas. In general, all these problems can be considered as finding the manifold solutions of equation systems of positive dimension [70].

For a geometric configuration given by a set of polynomial equations

$$h_1(u_1, \ldots, u_q, x_1, \ldots, x_p) = 0, \ldots, h_r(u_1, \ldots, u_q, x_1, \ldots, x_p) = 0$$

and a set of inequations

$$\{D = d_1(u, x) \neq 0, \ldots, d_s(u, x) \neq 0\},$$

we want to find a relation (formula) between arbitrarily chosen variables $u_1, \ldots, u_q$ (parameters) and a dependent variable, say, x_1. Methods based on the Gröbner basis [11] and the characteristic set [69, 11, 65] are given.

An interesting example is Peaucellier's Linkage (Fig. 9). Links AD, AB, DC and BC have equal length, as do links EA and EC. The length of FD equals the distance from E to F. The locations of joints E and F are fixed points on the plane, but the linkage is allowed to rotate about these points. As it does, what is the locus of the joint B?

Let $F = (0, 0)$, $E = (r, 0)$, $C = (x_2, y_2)$, $D = (x_1, y_1)$, and $B = (x, y)$, n and m be the lengths of the projections of AD on the direction of line EDB and on the direction perpendicular to it when E, F and D are collinear. Then the geometry conditions can be expressed by the following set of polynomial equations H

$$
\begin{aligned}
&h_1 = y_1^2 + x_1^2 - r^2 = 0 && r = FD \\
&h_2 = y_2^2 - 2y_1y_2 + x_2^2 - 2x_1x_2 + y_1^2 + x_1^2 - n^2 - m^2 = 0 && CD^2 = n^2 + m^2 \\
&h_3 = y_2^2 - 2yy_2 + x_2^2 - 2xx_2 + x^2 + y^2 - n^2 - m^2 = 0 && CB^2 = n^2 + m^2 \\
&h_4 = y_2^2 + x_2^2 - 2rx_2 - n^2 - 4rn - m^2 - 3r^2 = 0 && EC^2 = (n + 2r)^2 + m^2 \\
&h_5 = (x - r)y_1 - yx_1 + ry = 0 && E \text{ is on } DB,
\end{aligned}
$$

together with the following set of polynomial inequations D:

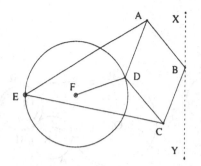

Fig. 10. A linkage generating a straight line

$$d_1 = x_1 - x \neq 0 \qquad\qquad\qquad B \neq D.$$

Selecting m, n, r, and y as parameters of the problem, we want to find the relation among m, n, r, y and x. Using the characteristic set method, we have found that $Zero(H/D)$ has only one non-degenerate component with the corresponding ascending chain ASC_1^*. The first polynomial in ASC_1^* is the relation $x + 2n + r = 0$, which tells us that the locus is a line parallel to the y-axis.

6.2 Parametric Mechanical CAD

A direct application of AGDC is to mechanical CAD systems. Here the advantage over the traditional CAD systems is not limited to a mere user friendly interface. In the constraint based system, the user may specify a design figure without knowing its shape previously which cannot be done in traditional CAD system. The constraint based system will find all the possible shapes of the specified system automatically. This is especially important in the phase of conceptual design.

For instance, for the diagram in Fig. 11, if the radius of circle B is known then this diagram is easy to draw. If we do not know the radius of this circle but instead we know that this circle passes through a known point A, then this diagram is not so easy to draw with ruler and compass. This problem is actually one of the ten Appolonius' construction problems [21]. A solution to this problem generated by the global propagation method is given below.

Suppose that O_1 and O_2 are two known circles and A is a known point. We need to construct a circle B which is tangent to circles O_1 and O_2 and passes through point A. Figure 11 (b) shows the solution given by the global propagation method. By a construction rule [21], line UV passes through the similarity center of circles O_1 and O_2, i.e., the intersection W of line O_1O_2 and a common tangent line T_1T_2 of the two circles (which is easy to construct). Let S be the intersection of the circle passing through T_1, T_2, A and the circle with O_2 as center and passing through T_2, H the intersection of lines AW and T_2S. Then O_2V is perpendicular to HV. Now, we can construct point V. Point U can be constructed easily.

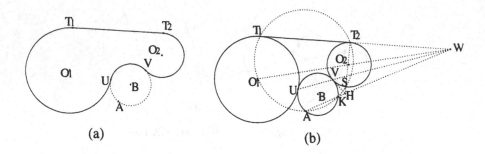

Fig. 11. Appolonius' construction problem and CAD

6.3 Linkage Design: A Realization of Kempe's Linkage

In Section 6.1, we introduce a linkage to generate a straight line. It is also able to design linkages to generate conics [26]. So it is natural to ask: What kinds of curves can be drawn with linkages? This problem has been answered theoretically by Kempe [40] who proved that any algebraic plane curve can be generated with a linkage. With the AGDC techniques reported in Sections 2, 3, and 4, we may design a linkage automatically that can be used to draw a plane algebraic curve. First, we need three kinds of tools.

1. The *Multiplicator* shown in Fig. 12 (a) consists of similar crossed parallelograms. Using a multiplicator, we may obtain integral multiples of any angle, e.g. $a\varphi$ or $b\theta$.

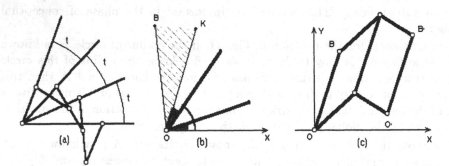

Fig. 12. Three tools for linkage construction

2. The *Additor*. Joining one multiplicator to another will produce the combination $a\varphi \pm b\theta$. This is the mechanism shown in Fig. 12 (b) where the plate BOK with angle β is connected rigidly to the bar. Thus we build up a linkage to produce $\angle BOK = a\varphi \pm b\theta \pm \beta$. If OB is taken equal to A, the x-coordinate of the point B is $A \times cos(a\varphi \pm b\theta \pm \beta)$.

3. The *Translator* shown in Fig. 12 (c) consists of parallelograms with OB pivoted at O. In this linkage, the bar $O'B'$ is always parallel to OB and can be moved freely within its limits.

Let the algebraic plane curve be defined by the following equation

$$f(x, y) = 0. \tag{5.1}$$

We construct a parallelogram (Fig. 13 (a)) such that OA and OB have lengths m and n and form angles θ and φ with the X-axis respectively. P is a point on the curve. Its coordinates are then given below:

$$x = m \times \cos\theta + n \times \cos\varphi; \tag{5.2}$$
$$y = m \times \sin\theta + n \times \sin\varphi = m \times \cos(\pi/2 - \theta) + n \times \cos(\pi/2 - \varphi) \tag{5.3}$$

Note that the products and powers of cosines can be expressed as the sum of cosines. When substituting equations (5.2) and (5.3) into (5.1), we shall have a sum of terms of the following form

$$f(x, y) = \sum_{i=1}^{k} A_i \times \cos(a_i\varphi \pm b_i\theta \pm \beta_i) + C = 0 \tag{5.4}$$

where A_i and C are constants; a_i and b_i are positive integers; and β_i equals 90, or 0.

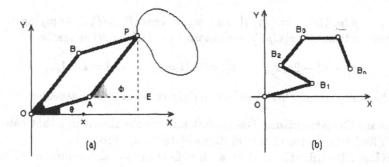

Fig. 13. Realization of Kempe's linkage

For each term $A_i \times \cos(a_i\varphi \pm b_i\theta \pm \beta_i)$, we can use the multiplicator and additor to construct a link OC_i such that OC_i is of length A_i and forms an angle $a_i\varphi \pm b_i\theta \pm \beta_i$ with the X-axis. Using the translators, we can construct a chain of links $OB_1, B_1B_2, B_2B_3, \ldots$, as shown in Fig. 13 (b), such that $B_1 = C_1$ and $OC_iB_iB_{i-1}$ is a parallelogram. Therefore, point B_n has x-coordinate:

$$X = \sum_{i=1}^{n} A_i \times \cos(a_i\varphi \pm b_i\theta \pm \beta_i) = f(x, y) - C \text{ (by equation (5.4))}.$$

If P is moved along the given curve, its coordinates x, y satisfy: $f(x, y) = 0$. Accordingly, the locus of the end point B_n of the chain is

$$X + C = 0,$$

a straight line parallel to the Y-axis. Conversely, if B_n is moved along this line (with the help of a Peaucellier cell, for instance) point P will generate the curve $f(x, y) = 0$. We thus finish the construction.

After the construction of the chain of points $B_1, \ldots, B_n$, we need to let point B_n move on line $X + C = 0$, and for a given position of B_n we find the positions of other points by using a numerical method, such as the optimization method introduced in Section 3. For examples of Kempe linkages, see [26].

6.4 Pose Determination: The PnP Problem

The Perspective-n-Point Problem (PnP), also known as the Location Determination Problem (LDP) [18] or the Exterior Camera Calibration Problem [31], originates from camera calibration [18]. It is to determine the exterior parameters: the position and orientation of the camera with respect to a scene object from n points. It concerns many important fields, such as computer animation, computer vision, image analysis and automated cartography, robotics, etc. Fischler and Bolles [18] summarize the problem as follows:

" Given the relative spatial locations of n control points P_i, and given the angle to every pair of control points from an additional point called the Center of Perspective (C_P), find the lengths of the line segments joining C_P to each of the control points."

Let $x_i = |C_P P_i|$, $a_{ij} = |P_i P_j|$, and $p_{ij} = \cos(\angle P_i C_P P_j)$. Applying the cosine theorem to triangles $P_i C_P P_j$, we obtain the PnP equation system:

$$x_i^2 + x_j^2 - 2x_i x_j p_{ij} - a_{ij}^2 = 0, (i = 1, \ldots, n; j = i + 1, \ldots, n).$$

The study of the PnP problem mainly consists of two aspects:

Diagram Construction. Design fast and stable algorithms that can be used to find all or some of the solutions of the PnP problem.

Solution Classification. Give a classification for the solutions of the PnP equation system, i.e., give explicit conditions under which the system has none, one, two, ... number of physical solutions.

There are many results for the first problem and the second problem is not solved completely.

All three kinds of AGDC techniques studied in this paper can be used to attack this problem. In [27], the synthetic approach and the symbolic computational approach are used to study the $P3P$ problem. In [75], symbolic computation methods are used to the solution classification problem.

In the symbolic computation approach, Wu-Ritt's zero decomposition algorithm [68] is used to find a complete solution decomposition for the $P3P$ equation system. The decomposition has the following implications. First, it provides a complete set of analytical solutions to the $P3P$ problem. Previous work usually consider the main solutions and omits many special cases. This might cause trouble when the given data is from the special case. Second, by expressing all solutions in *triangular form* it provides a fast and stable way for numerical solution. Third, it provides a clear solution space analysis of the $P3P$ problem and thus provides a good starting point for multiple solution analysis.

In the geometric approach, the three perspective angles are considered separately. Then the locus of the control point in each case is a torus and the control point is the intersection of the three tori. In this way, we give some pure geometric criteria for the number of solutions of the $P3P$ problem. One interesting result is

Theorem. *The P3P problem can have only one solution if all the three angles formed by the three control points are obtuse.*

For the special case $a_{12} = a_{13} = a_{23} = 1$ and $p_{12} = p_{13}$, the following solution classification for the P3P problem is given in [27]. For the sake of simplicity, we use two new parameters $p_{12} = r$ and $p_{23} = p$.

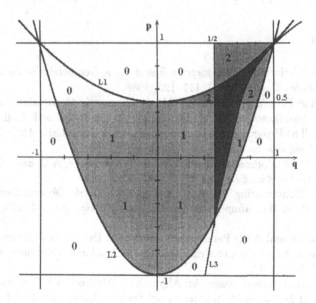

Fig. 14. Solution classification for $a_{12} = a_{13} = a_{23} = 1$ and $p_{12} = p_{13}$ L_1 is $p = \frac{1+q^2}{2}$, L_2 is $p = 2q^2 - 1$, and L_3 is $p = \frac{3}{2} - \frac{1}{2q^2}$. The number of solutions on the curves are omitted in this diagram.

1. Point P has four solutions, if and only if
$$1 > q > \tfrac{1}{2}, \quad \tfrac{1+q^2}{2} > p > \tfrac{1}{2}, \quad \text{and } p > \tfrac{3}{2} - \tfrac{1}{2q^2}.$$

2. Point P has three solutions, if and only if
$$\begin{cases} \tfrac{1}{2} < q < \tfrac{\sqrt{2}}{2} \\ \tfrac{3}{2} - \tfrac{1}{2q^2} < p \le \tfrac{1}{2}. \end{cases} \quad \text{or} \quad \begin{cases} \tfrac{1}{2} < q < 1 \\ p = \tfrac{1+q^2}{2}. \end{cases}$$

3. Point P has two solutions, if and only if

$$\begin{cases} 0 < q \le \frac{1}{2} \\ \frac{1}{2} < p < \frac{1+q^2}{2}. \end{cases} \quad \text{or} \quad \begin{cases} \frac{1}{2} < q < 1 \\ \frac{1+q^2}{2} < p < 1 \end{cases}$$

$$\text{or} \quad \begin{cases} \frac{\sqrt{2}}{2} < q < 1 \\ \frac{1}{2} < p \le \frac{3}{2} - \frac{1}{2q^2} \text{ and } p > 2q^2 - 1. \end{cases}$$

4. Point P has one solution, if and only if

$$\begin{cases} -\frac{\sqrt{3}}{2} < q < \frac{1}{2} \text{ or } \frac{\sqrt{2}}{2} \le q < \frac{\sqrt{3}}{2} \\ 2q^2 - 1 < p < \frac{1}{2} \end{cases} \quad \text{or} \quad \begin{cases} \frac{1}{2} < q < \frac{\sqrt{2}}{2} \\ 2q^2 - 1 < p \le \frac{3}{2} - \frac{1}{2q^2} \end{cases}$$

$$\text{or} \quad \begin{cases} 0 < q \le \frac{1}{2} \text{ or } \frac{\sqrt{2}}{2} \le q < \frac{\sqrt{3}}{2} \\ p = \frac{1}{2} \end{cases} \quad \text{or} \quad \begin{cases} 0 < q \le \frac{1}{2} \\ p = \frac{1+q^2}{2}. \end{cases}$$

References

1. B. Aldefeld, Variation of Geometries Based on a Geometric-Reasoning Method, *Computer Aided Design*, **20**(3), 117–126, 1988.
2. F. Arbab and B. Wang, Reasoning About Geometric Constraints, in *Intelligent CAD II*, H. Yoshikawa and T. Holden (eds.), pp. 93–107, North-Holland, 1990.
3. A. Borning, The Programming Language Aspect of ThingLab, *ACM Tras. on Programming Language and Systems*, **3**(4), 353–387, 1981.
4. W. Bouma, C. M. Hoffmann, I. Fudos, J. Cai and R. Paige, A Geometric Constraint Solver, *Computer Aided Design*, **27**(6), 487–501, 1995.
5. B. Brudelin, Constructing Three-Dimensional Geometric Objects Defined by Constraints, in *Proc. Workshop on Interactive 3D Graphics*, pp. 111–129, ACM Press, 1986.
6. S. A. Buchanan and A. de Pennington, Constraint Definition System: A Computer Algebra Based Approach to Solving Geometric Problems, *Computer Aided Design*, **25**(12), 740–750, 1993.
7. B. Buchberger, Gröbner Bases: An Algorithmic Method in Polynomial Ideal Theory, in *Recent Trends in Multidimensional Systems Theory*, D. Reidel Publ. Comp., 1985.
8. B. Char et al., *Maple V*, Springer-Verlag, Berlin, 1992.
9. S. C. Chou, *Mechanical Geometry Theorem Proving*, D. Reidel Publishing Company, Dordrecht, Netherlands, 1988.
10. S. C. Chou, A Method for Mechanical Deriving of Formulas in Elementary Geometry, *J. of Automated Reasoning*, **3**, 291–299, 1987.
11. S. C. Chou and X. S. Gao, Mechanical Formula Derivation in Elementary Geometries, in *Proc. ISSAC-90*, pp. 265–270, ACM Press, New York, 1990.
12. S. C. Chou and X. S. Gao, Ritt-Wu's Decomposition Algorithm and Geometry Theorem Proving, in *Porc. CADE-10*, M. E. Stickel (ed.), pp. 207–220, *LNCS*, Vol. 449, Springer-Verlag, Berlin, 1990.
13. S. C. Chou, X. S. Gao and J. Z. Zhang, *Machine Proofs in Geometry*, World Scientific, Singapore, 1994.

14. S. C. Chou, X. S. Gao and J. Z. Zhang, A Fixpoint Approach To Automated Geometry Theorem Proving, WSUCS-95-2, CS Dept, Wichita State University, 1995, To appear in J. of Automated Reasoning.

15. G. E. Collins, Quantifier Elimination for Real Closed Fields by Cylindrical Algebraic Decomposition, in *LNCS* vol. 33, pp. 134–183, Springer-Verlag, Berlin, 1975.

16. J. Chuan, Geometric Constructions with the Computer, in *Proc. ATCM'95*, pp. 329–338, Springer-Verlag, 1995.

17. K. H. Elster (ed.), *Modern Mathematical Methods of Optimization*, Akademie Verlag, 1993.

18. M. A. Fishler, and R. C. Bolles, Random Sample Consensus: A Paradigm for Model Fitting with Applications to Image Analysis and Automated Cartomated Cartography, *Communications of the ACM*, 24(6), 381–395, 1981.

19. I. Fudos and C. M. Hoffmann, A Graph-Constructive Approach to Solving Systems of Geometric Constraints, *ACM Transactions on Graphics*, 16(2), 179–216, 1997.

20. Gabri Geometry II, Texas Instruments, Dallas, Texas, 1994.

21. X. S. Gao and S. C. Chou, Solving Geometric Constraint Systems, I. A Global Propagation Approach, *Computer Aideded Design*, 30(1), 47–54, 1998.

22. X. S. Gao and S. C. Chou, Solving Geometric Constraint Systems, II. A Symbolic Computational Approach, *Computer Aided Design*, 30(2), 115–122, 1998.

23. X. S. Gao and S. C. Chou, Implicitization of Rational Parametric Equations, *Journal of Symbolic Computation*, 14, 459–470, 1992.

24. X. S. Gao, J. Z. Zhang and S. C. Chou, *Geometry Expert*, Nine Chapters Pub., 1998, Taiwan (in Chinese).

25. X. S. Gao, C. C. Zhu and Y. Huang, Building Dynamic Mathematical Models with Geometry Expert, I. Geometric Transformations, Functions and Plane Curves, in *Proc. of ATCM'98*, W. C. Yang (ed.), pp. 216–224, Springer-Verlag, 1998.

26. X. S. Gao, C. C. Zhu and Y. Huang, Building Dynamic Mathematical Models with Geometry Expert, II. Linkages, in *Proc. of ASCM'98*, Z. B. Li (ed.), pp. 15–22, LanZhou Univ. Press, 1998.

27. X. S. Gao and H. F. Cheng, On the Solution Classification of the "P3P" Problem, in *Proc. of ASCM'98*, Z. B. Li (ed.), pp. 185–200, LanZhou Univ. Press, 1998.

28. J. X. Ge, S. C. Chou and X. S. Gao, Geometric Constraint Satisfaction Using Optimization Methods, WSUCS-98-1, CS Dept, Wichita State University, 1998, *submitted to CAD*.

29. H. Gelernter, Realization of a Geometry-Theorem Proving Machine, in *Computers and Thought*, E. A. Feigenbaum and J. Feldman (eds.), pp. 134–152, Mcgraw Hill.

30. J. Hopcroft and R. Tarjan, Dividing A Graph into Triconnected Components, *SIAM J. Computing*, 2(3), 135–157, 1973.

31. R. Horaud, B. Conio and O. Leboulleux, An Analytic Solution for the Perspective 4-Point Problem, CVGIP, 47, 33–44, 1989.

32. A. Heydon and G. Nelson, The Juno-2 Constraint-Based Drawing Editor, *SRC Research Report 131a*, 1994.

33. C. Hoffmann, Geometric Constraint Solving in R^2 and R^3, in *Computing in Euclidean Geometry*, D. Z. Du and F. Huang (eds.), pp. 266–298, World Scientific, Singapore, 1995.

34. C. Hoffmann and I. Fudos, Constraint-based Parametric Conics for CAD, *Geometric Aided Design*, 28(2), 91–100, 1996.

35. Y. Huang and W. D. Wu, Kinematic Solution of a Steawrt Platform, in *Proc. IWMM'92*, (W. T. Wu and M. D. Cheng Eds.), pp. 181–188, Inter. Academic Publishers, Beijing, 1992.

36. N. Jakiw, *Geometer's Sketchpad*, User Guide and Reference Manual, Key Curriculum Press, Berkeley, USA, 1994.
37. N. Jacobson, *Basic Algebra*, Vol. 1, Freeman, San Francisco, 1985.
38. D. Kapur, Geometry Theorem Proving Using Hilbert's Nullstellensatz, in *Proc. SYMSAC'86*, Waterloo, pp. 202–208, ACM Press, 1986.
39. D. Kapur, T. Saxena and L. Yang, Algebraic and Geometric Reasoning with Dixon Resultants, in *Proc. ISSAC'94*, Oxford, ACM Press, 1994.
40. A. B. Kempe, On a General Method of Describing Plane Curves of the n-th Degree by Linkwork, *Proc. of L.M.S.*, 213–216, 1876; see also, Messenger of Math., T. VI., 143–144.
41. J. King abd D. Schattschneider, *Geometry Turned On*, The Mathematical Association of America, 1997.
42. K. Kondo, Algebraic Method for Manipulation of Dimensional Relationships in Geometric Models, *Geometric Aided Design*, **24**(3), 141–147, 1992.
43. G. Kramer, *Solving Geometric Constraint Systems*, MIT Press, 1992.
44. G. Kramer, A Geometric Constraint Engine, *Artificial Intelligence*, **58**, 327–360, 1992.
45. H. Lamure and D. Michelucci, Solving Geometric Constraints by Homotopy, *IEEE Trans on Visualization and Computer Graphics*, **2**(1), 28–34, 1996.
46. R. S. Latheam and A. E. Middleditch, Connectivity Analysis: A Tool for Processing Geometric Constraints, *Computer Aided Design*, **28**(11), 917–928, 1994.
47. J. Y. Lee and K. Kim, Geometric Reasoning for Knowledge-Based Parametric Design Using Graph Representation, *Computer Aided Design*, **28**(10), 831–841, 1996.
48. K. Lee and G. Andrews, Inference of the Positions of Components in an Assembly: Part 2, *Computer Aided Design*, **17**(1), 20–24, 1985.
49. W. Leler, *Constraint Programming Languages*, Addison Wesley, 1988.
50. R. Light and D. Gossard, Modification of Geometric Models through Variational Geometry, *Geometric Aided Design*, **14**, 208–214, 1982.
51. V. C. Lin, D. C. Gossard and R. A. Light, Variational Geometry in Computer-Aided Design, *Computer Graphics*, **15**(3), 171–177, 1981.
52. G. L. Nemhauser, A. H. G. Rinnooy Kan and M. J. Todd (eds.), *Optimization*, Elsevier Science Publishers B.V., 1989.
53. J. Owen, Algebraic Solution for Geometry from Dimensional Constraints, in *Proc. ACM Symp. Found. of Solid Modeling*, ACM Press, pp. 397–407, Austin, TX, 1991.
54. J. Owen, Constraints of Simple Geometry in Two and Three Dimensions, *Inter. J. of Comp. Geometry and Its Applications*, **6**, 421–434, 1996.
55. D. N. Rocheleau and K. Lee, System for Interactive Assembly Modeling, *Computer Aided Design*, **19** (1), 65–72, 1987.
56. D. Serrano and D. Gossard, Constraint Management in MCAE, in *Artificial Intelligence in Engineering: Design*, J. Gero (ed.), pp. 93–110, Elsevier, Amsterdam, 1988.
57. G. L. Steele and G. L. Sussman, CONSTRAINTS – A Language for Expressing Almost-Hierarchical Descriptions, *Artificial Intelligence*, **14**, 1–39, 1980.
58. H. Suzuki, H. Ando and F. Kimura, Geometric Constraints and Reasoning for Geometrical CAD Systems, *Computer and Graphics*, **14**(2), 211–224, 1990
59. G. Sunde, Specification of Shape by Dimensions and Other Geometric Constraints, in *Geometric Modeling for CAD Applications*, M. J. Wozny et al. (eds.), pp. 199–213, North Holland, 1988.
60. I. Sutherland, Sketchpad, A Man-Machine Graphical Communication System, in *Proc. of the Spring Joint Comp. Conference*, North-Holland, pp. 329–345, 1963.

61. A. Tarski, *A Decision Method for Elementary Algebra and Geometry*, Univ. of California Press, Berkeley, Calif., 1951.

62. P. Todd, A k-tree Generalization that Characterizes Consistency of Dimensioned Engineering Drawings, *SIAM J. of Disc. Math.*, **2**, 255–261, 1989.

63. R. C. Veltkamp, Geometric Constraint Management with Quanta, in *Intelligent Computer Aided Design*, D. C. Brown et al. (eds.), pp. 409–426, North-Holland, 1992.

64. A. Verroust, F. Schonek and D. Roller, Rule-oriented Method for Parameterized Computer-aided Design. *Geometric Aided Design*, **24**(3), 531–540, 1992.

65. D. M. Wang, Reasoning about Geometric Problems Using an Elimination Method, in *Automated Practical Reasoning: Algebraic Approaches*, J. Pfalzgraf and D. Wang (eds.), Springer-Verlag, Wien New York, pp. 147–185, 1995.

66. D. M. Wang, GEOTHER: A Geometry Theorem Prover, in *Proc. CADE-13*, New Brunswick, 1996, pp. 213–239, LNAI, Vol. 1104, Springer-Verlag, Berlin, 1996.

67. D. M. Wang and X. S. Gao, Geometry Theorems Proved Mechanically Using Wu's Method, Part on Elementary Geometries, *MM Research Preprints*, No 2, pp. 75–106, 1987, Institute of Systems Science.

68. W. T. Wu, *Mechanical Theorem Proving in Geometries: Basic Principles*, Springer-Verlag, Wien New York, 1994.

69. W. T. Wu, A Mechanizations Method of Geometry and its Applications I. Distances, Areas and Volumes, *J. Sys. Sci. and Math. Scis.*, **6**, 204–216, 1986.

70. W. T. Wu, *Mathematics Mechanization*, Science Press, Beijing, 1999.

71. W. T. Wu, A Mechanization Method of Geometry and Its Applications VI. Solving Inverse Kinematics Equations of PUMA-Type Robotics, pp. 49–53, MM Research Preprints, No 4, 1989, Institute of Systems Science.

72. W. T. Wu and D. K. Wang, On the Surface Fitting Problems in CAGD (in Chinese), *Mathematics in Practice and Theory*, 3, 1994.

73. L. Yang, H. Fu and Z. Zeng, A Practical Symbolic Algorithm for Inverse Kinematics of 6R Manipulators with Simple Geometry, in *Proc. CADE-14*, pp. 73–86, Springer-Verlag, Berlin, 1997.

74. L. Yang, J. Z. Zhang and X. R. Hou, Non-Linear Algebraic Equations and Theorem Machine Proof, ShangHai Science and Education Press, ShangHai, 1997 (in Chinese).

75. L. Yang, A Simplified Algorithm for Solution Classification of the P3P Problem, preprint, 1998.

A 2D Geometric Constraint Solver for Parametric Design Using Graph Analysis and Reduction

Jae Yeol Lee

EC/CALS Department
Computer.Software Technology Lab.
Electronics and Telecommunications Research Institute
Taejon, 305-350, South Korea
jaelee@etri.re.kr

Abstract. This paper proposes a DOF-based graph reduction approach to geometric constraint solving. The proposed approach incrementally solves a geometric constraint problem that is not ruler-and-compass constructible by incrementally identifying a set of constrained geometric entities with 3 DOF (degree of freedom) as a rigid body and determining the geometric entities in the rigid body using one of the two solving procedures: algebraic method and numerical method, instead of solving it simultaneously using a numerical method. However, the use of the numerical method is restricted to solve only those parts that must be solved numerically. By combining the advantages of algebraic solving with the universality of numerical solving, the proposed method can maximize the efficiency, robustness, and extensibility of a geometric constraint solver.

Keywords: Parametric design, variational design, graph reduction, constructive constraint solving.

1 Introduction

Parametric design is an approach to product modeling, which associates engineering knowledge with geometry and topology in a product design by means of geometric constraints [3]. It allows users to make modifications to existing designs by changing parameter values. For this reason, parametric design has been considered an indispensable tool in many applications such as mechanical part design, tolerance analysis, simulations, kinematics, and knowledge-based design automation: [11, 15, 21, 25, 26].

Many research efforts have been made toward improving parametric design functionality. One of the main efforts is to develop a geometric constraint solver that can solve a geometric constraint problem efficiently and robustly. There are three major approaches to solving a geometric constraint problem: 1) numerical approach, 2) constructive approach, and 3) symbolic approach.

In the *numerical approach*, geometric constraints are converted into a system of numerical equations: [12, 22]. Then, the system of equations is solved simultaneously by an iterative numerical method. This approach can solve a variety of geometric configurations including ruler-and-compass constructible and ruler-and-compass non-constructible configurations since any well-constrained problem that can be represented as a set of equations can be, in theory, solved by numerical techniques. However, along with this advantage come some significant shortcomings [17]:

- Numerical techniques have a number of problems related to numerical stability and solution consistency.
- The number of iterations required to solve a set of constraint equations can vary substantially, depending on initial conditions given to the solver.
- Numerical techniques are relatively inefficient.
- Numerical techniques cannot distinguish between different roots in the solution space.

Due to the limitations of the numerical approach mentioned above, most parametric design systems adopt the constructive approach as a fundamental scheme for solving geometric constraints.

In the *constructive approach*, geometric constraints are represented by a set of knowledge such as graphs or predicate symbols: [2, 4, 7, 9, 13, 14, 20, 23, 27–30]. In this approach, a constraint solver satisfies the constraints by incrementally processing the set of knowledge. Usually, the solver takes two phases of geometric constraint solving: a planning phase and an execution phase. During the first phase, a sequence of construction steps is derived using a graph-based technique or a rule-based technique. During the second phase, the sequence of construction steps is carried out to determine geometric entities. The main advantage of the constructive approach is that it separates the symbolic aspects from the numerical aspects so that those usual problems such as numerical instabilities associated with the numerical approach can be minimized.

Many research efforts have been made toward constructive geometric constraint solving. Owen [23] and Bouma *et al.* [4] presented graph reduction approaches to geometric constraint solving. Owen proposed a graph-based constructive solver in which a constraint graph is analyzed for triconnected components. DCM is a commercial constraint solver based on the Owen's method. Bouma *et al.* proposed a similar approach and, further, extended it to deal with more complex configurations. Moreover, Fudos *et al.* [7] extended Bouma *et al.*'s approach to deal with under-and over-constrained geometric constraint problems. Ait-Aoudia *et al.* [1] proposed a graph reduction method based on the bipartite graph analysis that decomposes a well-constrained system into irreducible ones. Latham *et al.* [19] proposed an algorithm for processing geometric constraints based on the connectivity analysis between constraints and geometric entities. These researchers have done excellent work on this approach. However, there is still much work to be done. For instance, they might have some strong assumptions; 1) the graph of constraints is biconnected [23], 2) it is hierarchically reducible into triangles or 3) all geometric entities have exactly 2 degrees of freedom (i.e. the radius of all circles must be known): [4, 7]. Moreover, there may be no effective way of solving each set of irreducible configurations that has been decomposed by the graph reduction analysis: [1, 19].

Lee and Kim [20] proposed a graph-based rule inferencing method, which can overcome an inefficient geometric reasoning process of rule-based inferencing methods. Nevertheless, it can only deal with ruler-and-compass constructible configurations. Gao *et al.* [9] proposed a rule-based approach to solving a large set of complex geometric configurations but could not overcome inefficient geometric reasoning.

In the *symbolic approach*, the constraints are translated into a system of algebraic equations. The system is then solved with symbolic methods such as Wu's method or Gröbner basis computation: [6, 10, 16]. The method can solve general nonlinear systems of algebraic equations but may require exponential running time.

Figure 1 shows two models that require sophisticated solving techniques. The triangle in Fig. 1(a) is well constrained, apart from rigid body translation and rotation. Though this configuration is seemingly very simple, it is difficult for some constructive approaches to solve the constraints since it requires reasonably sophisticated ordering of construction steps. The model in Fig. 1(b) cannot be solved by current constructive methods since it partially requires a numerical solving technique or another technique to determine the geometric entities. These examples show that the constructive method alone cannot solve a variety of geometric configurations.

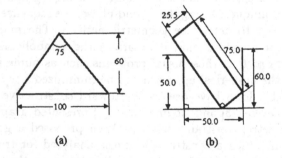

(a)   (b)

Fig. 1. Configurations that require sophisticated solving techniques

In this paper, we propose a DOF-based graph reduction approach to geometric constraint solving. The proposed approach employs a graph-based constructive approach globally and a numerical approach locally. The constraint solving process consists of two phases: (1) planning phase and (2) execution phase. In the *planning phase*, we *incrementally* identify a set of constrained geometric entities with 3 DOF (degree of freedom) as a rigid body that may be either *ruler-and-compass constructible* or *ruler-and-compass non-constructible*. A sequence of construction steps is generated as the result of this graph-based reduction process. In the *execution phase*, each construction step is evaluated sequentially to determine the geometric entities in the associated rigid body using one of three solving procedures: (1) algebraic procedure, (2) symbolic procedure, and

(3) numerical procedure. The *algebraic procedure* determines the geometric entities in the rigid body by calculating their positions and orientations if they are ruler-and-compass constructible and the *symbolic* or *numerical procedure* if they are ruler-and-compass non-constructible. Using the proposed approach, thus, we can incrementally solve a geometric constraint problem that is not ruler-and-compass constructible, instead of solving it simultaneously using a numerical method. However, the use of the numerical method is minimized. By combining the advantages of algebraic solving with the universality of numerical solving, the proposed method can maximize the efficiency, robustness, and extensibility of a geometric constraint solver.

The remainder of this paper is organized as follows. Section 2 describes an overview of the proposed geometric constraint solver. Section 3 presents the construction plan generation phase. Section 4 describes the plan evaluation phase. Section 5 shows implementation results. Section 6 presents a conclusion with some remarks.

2 Overview

A geometric constraint problem is defined by a geometric model consisting of a set of geometric entities and a set of geometric relations, called *constraints*. In this paper, we consider only well-constrained problems and exclude relations on dimension variables and inequality constraints. Geometric entities used in the paper include points, lines, circles, line segments, and circular arcs. Constraints include incidence, distance, angle, parallelism, concentricity, tangency, and perpendicularity. A geometric entity has its own degrees of freedom, which allow it to vary in shape, position, size, and orientation. For example, a point and a line have two degrees of freedom (DOF), and a circle has three DOF. On the other hand, a geometric constraint reduces the DOF of the geometric model by a certain number, called the *valency* of the constraint, depending on the constraint type as shown in Table 1 [2]. In order for a set of geometric entities to be constrained fully, all their degrees of freedom must be consumed or taken up by geometric constraints. The geometric model can be represented by a constraint graph in which nodes are geometric entities, and edges are geometric constraints.

The proposed constraint solving process consists of two phases: (1) planning and (2) execution. In the *planning phase*, a sequence of construction steps is generated by incrementally forming rigid bodies with three DOF (two translational, one rotational), called *clusters*. A rigid body is a set of geometric elements whose position and orientation relative to each other are known. At each clustering step, a rigid body with three degrees of freedom, consisting of geometric entities and/or clusters and geometric constraints, is identified and combined into a single merged cluster, R_i. This cluster formation process continues until the reduced constraint graph becomes a single cluster. In the execution phase, each construction step is evaluated to derive positions and orientations of the geometric entities in the cluster by selecting an appropriate procedure among

Table 1. Geometric constraints and their valency

Constraint Type	Associated Geometric Entities	Valency
Distance	Point, Point	1
	Point, Line	1
	Point, Circle	1
	Line, Circle	1
	Line, Line	2
Incidence	Point, Line	1
	Point, Circle	1
Coincidence	Point, Point	2
	Line, Line	2
Tangency	Line, Circle	1
	Circle, Circle	1
Angle	Line, Line	1
Parallelism	Line, Line	1
Concentricity	Point, Circle	2

the three constraint solving procedures described in Section 4, considering the clustering type.

Notations being used throughout the paper are summarized below:

- L_i, C_i, and P_i represent a line, a circle, and a point, respectively.
- G_i represents a geometric entity (or a cluster) with two DOF.
- R_i represents a cluster (or a geometric entity) with three DOF.

3 Plan Generation

If a geometric constraint model is well-constrained as shown in Fig. 2, a sequence of construction plan, as shown in Fig. 3, is generated by two phases; 1) preprocessing the pairs of adjacent geometric entity nodes constrained by the geometric constraints with two DOF as shown in Fig. 4, and 2) forming rigid bodies that have one of the clustering types shown in Fig. 5. Each set of nodes/edges in Fig. 5 is reduced into a cluster with 3 DOF. The preprocessing and clustering procedures are described in more detail below.

3.1 Processing the constraint graph

Most of geometric constraints take up one DOF, but there are some constraints that consume two DOF, as shown in Table 1 and Fig. 4. A distance dimension between two lines specifies both parallelism and distance so that it takes up two degrees of freedom. A coincidence constraint between two points also takes up two degrees of freedom, as does a concentricity constraint. These geometric

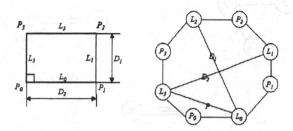

Fig. 2. A simple design and its constraint graph

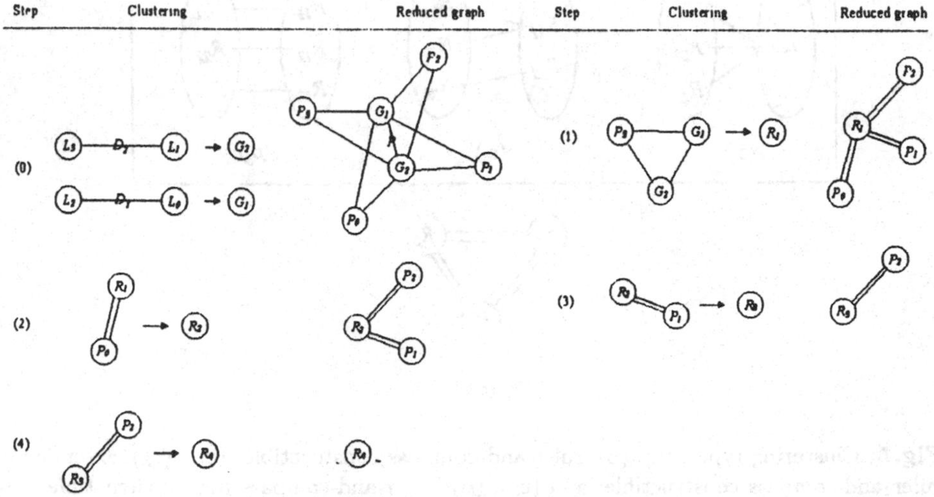

Step	Clustering	Reduced graph	Step	Clustering	Reduced graph

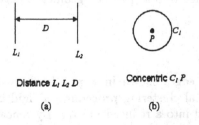

Fig. 3. Clustering steps for the design shown in Fig. 2

Fig. 4. Constraints that reduce two degrees of freedom

264

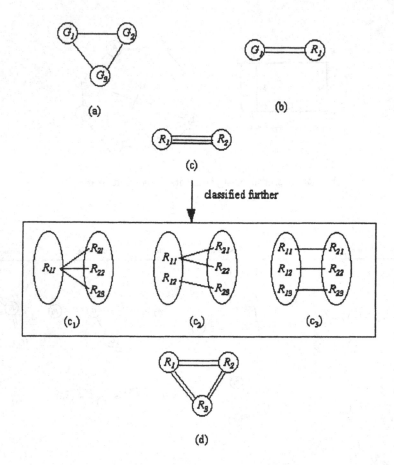

(a)

(b)

(c)

classified further

(c₁)   (c₂)   (c₃)

(d)

Fig. 5. Clustering types: (a), (b): ruler-and-compass constructible; (c₁), (c₂): extended ruler-and-compass constructible; and (c₃), (d): ruler-and-compass non-constructible

constraints and their associated geometric entities are combined into a special type of clusters with 2 DOF. In the proposed approach, a cluster with 2 DOF is treated as a *pseudo* geometric entity. During the preprocessing, thus, the set of a geometric constraint with 2 valency and its two associated geometric entities is identified and combined into a (pseudo) geometric entity as shown at step 0 in Fig. 3.

3.2 Cluster forming

Each set of nodes and edges shown in Fig. 5 forms a rigid body with three DOF. In this incremental clustering procedure, a rigid body with three DOF is identified and merged into a reduced cluster. By repeating these identifying and merging procedures, the constraint graph may be reduced to a single cluster node as shown in Fig. 3. The clustering process is classified into three types: (1) ruler-and-compass constructible or RCC, (2) extended ruler-and-compass constructible or ERCC, and (3) ruler-and-compass non-constructible or RCNC. An appropriate solving procedure is developed for each of the clustering types as shown in Table 2 [1].

Table 2. Solving procedures according to clustering types

Clustering Types	Type Descriptions	Related Graphs in Fig. 5	Solving Procedures
One connecting edge	Three G nodes	a	**A**
Two connecting edges	One G node and one R node	b	**A**
	Three R nodes	d	**C**
Three connecting edges	One geometric entity constrained by three constraints	c_1	**B**
	One geometric entity constrained by two constraints	c_2	**B**
	Each geometric entity constrained by one constraint	c_3	**C**

The geometric configurations shown in Fig. 5(a) and 5(b) are ruler-and-compass constructible. Thus, they can be efficiently determined by an algebraic solving procedure [20]. The geometric configuration shown in Fig. 5(c) is not ruler-and-compass constructible. To solve this configuration effectively, the cluster formation process is further classified into three types, according to the relations between geometric entities in two clusters: (1) one-to-three, (2) one-to-two, and (3) one-to-one, as shown in Fig. 5(c_1), 5(c_2), and 5(c_3), respectively[2].

One-to-three and one-to-two type configurations are effectively solved by an extended ruler-and-compass method, whereas one-to-one type clusters are not. For example, the geometric configurations shown in Fig. 6(a) and 6(b) are extended ruler-and-compass constructible. The one-to-one type configuration shown in Fig. 7 is solved effectively by a numerical procedure.

4 Plan Execution

Each construction step is evaluated to derive the relative positions and orientations of geometric entities in the cluster (or rigid body) by executing an appropriate solving procedure described below. A ruler-and-compass constructible

---

[1] **A**, **B**, and **C** represent procedures for solving ruler-and-compass constructible (RCC), extended ruler-and-compass constructible (ERCC), and ruler-and-compass non-constructible (RCNC) configurations, respectively.

[2] In Fig. 5, G_i represents a geometric entity or cluster with two degrees of freedom. R_i represents a cluster or geometric entity with three degrees of freedom. R_{ij} represents a geometric entity in the cluster R_i.

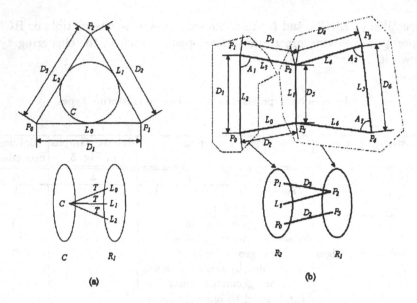

Fig. 6. Extended ruler-and-compass constructible: (a) one-to-three, (b) one-to-two

configuration is determined by an algebraic solving procedure: [20, 21]. This procedure sequentially calculates the position and orientation of each undetermined geometric entity within the cluster by solving the algebraic equations associated with the constraints. An extended ruler-and-compass constructible is solved by an extended algebraic solving procedure. This procedure determines the geometric entities by finding a sequence of rotations and translations to satisfy the geometric constraints. A ruler-and-compass non-constructible cluster is solved by a symbolic or a numerical method according to its topological structure. These solving methods are explained below.

4.1   Solving the RCC clusters

For the configuration shown in Fig. 5(a), we calculate the position and/or orientation of each geometric entity G_i, relative to the other entities within the rigid body, by solving algebraically the equations associated with the constraints (that is, evaluating inferred rules, see [20]). For the configuration shown in Fig. 5(b), we compute the position and orientation of the geometric entity G_1, relative to the cluster R_1, by solving the algebraic equations associated with the constraints between geometric entities in G_1 and R_1.

4.2   Solving the ERCC clusters

Considering the geometric entities and their relations in the clusters, an appropriate procedure is developed for each type of the extended ruler-and-compass constructible clusters. Each procedure specifies a sequence of rotation & translation operations that transforms one cluster R_1 relative to the other cluster R_2 to

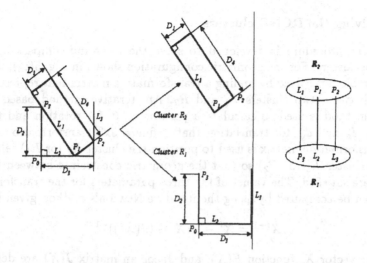

Fig. 7. Ruler-and-compass non-constructible

satisfy the geometric constraints. As an example, the procedure for the extended ruler-and-compass constructible cluster shown in Fig. 6(b) is summarized as follows. In the figure, (i) P_2 and P_5 in R_1, (ii) L_3, P_1, and P_0 in R_2, and (iii) P_2 is connected to P_1 and L_3.

PROCEDURE ONE_TO_TWO$(R_1(P_2, P_5), R_2(L_3, P_1, P_0))$
INPUT: two clusters R_1 and R_2
OUTPUT: a merged cluster R consisting of R_1 and R_2
$L = \text{line}(P_2, \text{normal}(\text{direction}(L_3)))^3$;
$IP_1 = \text{intersect}(L_3, L)$;
$\text{translate}(R_2, \text{vector-differ}(P_2, IP))$;
$C_1 = \text{circle}(P_2, D_3)$;
$IP_2 = \text{intersect}(L_3, C_1)$;
$\text{translate}(R_2, \text{vector-differ}(IP_2, P_1))$;
$D = \text{distance}(P_2, P_0)$;
$C_2 = \text{circle}(P_2, D)$;
$C_3 = \text{circle}(P_5, D_2)$;
$IP_3 = \text{intersect}(C_2, C_3)$;
$A = \text{angle}(\text{vector-differ}(P_0 - P_2), \text{vector-differ}(IP_3, P_2))$;
$\text{translate}(R_2, -P_2)$;
$\text{rotate}(R_2, -A)$;
$\text{translate}(R_2, P_2)$;
END_PROCEDURE

---

[3] A line is assumed to be defined by its direction vector (normal(direction(L_3))) and a point on the line (P_2). normal(direction(L_3)) represents a vector perpendicular to direction(L_3).

4.3 Solving the RCNC clusters

An efficient procedure is developed to solve the ruler-and-compass non constructible clusters. For the geometric configuration shown in Fig. 5(c_3), the constraint problem is solved by finding a transformation matrix that represents the relation between two clusters R_1 and R_2. An iterative method based on the Newton method is used to calculate a parameter, θ, for rotation and two parameters, d_x and d_y, for translation that define a 3x3 transformation matrix. This transformation matrix is used to position the cluster R_2 (or R_1) relative to the base cluster R_1 (or R_2) so that the geometric constraints between the two clusters are satisfied. The values of the three parameters for the transformation matrix can be computed by using the iterative Newton's method given by

$$X^{i+1} = X^i - F(X^i) \cdot (J(X^i))^{-1}$$

where the vector X, function $F(X)$, and Jacobian matrix $J(X)$ are defined as follows

$$X = \begin{pmatrix} \theta \\ d_x \\ d_y \end{pmatrix} \quad F(X) = \begin{pmatrix} f_1(X) \\ f_2(X) \\ f_3(X) \end{pmatrix} \quad J(X) = \begin{pmatrix} \frac{\partial f_1(X)}{\partial \theta} & \frac{\partial f_1(X)}{\partial d_x} & \frac{\partial f_1(X)}{\partial d_y} \\ \frac{\partial f_2(X)}{\partial \theta} & \frac{\partial f_2(X)}{\partial d_x} & \frac{\partial f_2(X)}{\partial d_y} \\ \frac{\partial f_3(X)}{\partial \theta} & \frac{\partial f_3(X)}{\partial d_x} & \frac{\partial f_3(X)}{\partial d_y} \end{pmatrix}.$$

A numerical solving procedure can also handle the 3-cluster configuration shown in Fig. 5(d). Since three clusters are involved, six functions for the Newton method are necessary to be defined to compute two transformation matrices for R_2 and R_3, relative to R_1. The proposed procedure improves the reliability of the numerical approach by reducing the number of equations to be solved. For the 2-cluster merging shown in Fig. 5 (c_3), for example, we solve three equations to determine the geometric entities by computing a transformation matrix in this procedure, instead of 12 equations in conventional numerical methods where 6 geometric entities or 12 DOF are involved. Thus, we can minimize the numerical instability of a numerical solver. As explained above, the one-to-one type cluster can be solved effectively by a numerical method. Among these clusters, however, the clusters with the same configuration as that shown in Fig. 8 can be effectively solved by a symbolic approach such as Gröbner basis computation [4]. Eliminating some variables with Gröbner's base computation generates a 4th order polynomial equation. This equation can be easily solved algebraically. The difference between the two clusters shown in Fig. 7 and Fig. 8 lies in the constraint relation in each cluster. The configuration shown in Fig. 8 has a cyclic relation among geometric entities in each cluster. On the other hand, the configuration in Fig. 7 has no such a relation.

4.4 Extending clustering types

The clusters shown in Fig. 5 can solve a large set of geometric configurations including RCNC. However, it is still needed to extend the solver since it cannot

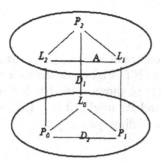

Fig. 8. One-to-one type cluster with a cyclic relation

deal with some complex configurations if they require more than three clusters to satisfy their geometric constraints. Figure 9 shows such an example [9]. The solver cannot continue the reduction operations because there is no rigid single loop in the reduced graph. However, the whole graph is well-constrained so that it can be solved. In this case, the solver takes another type of the rigid loop identification procedure called *multi-circuited loop identification*. This procedure takes several loop identifications and reductions until it finds a rigid subgraph. The identified rigid subgraph in this procedure is called a multi-circuited cluster. For the example in Fig. 9, the solver first finds a loop consisting of R_1, R_2, and L_6, and then reduces it as a cluster C_1. At this time, C_1 is under-constrained. After then, it finds another cluster C_2 consisting of C_1 and R_3. Note that C_2 is over-constrained. This compensates for the free DOF remaining in C_1 such that the 2-circuited cluster is well-constrained. Finally, the 2-circuited cluster can be solved algebraically or numerically according to the geometric structure.

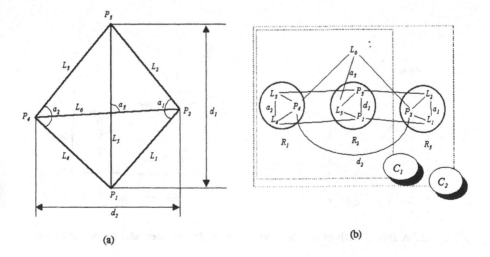

(a) (b)

Fig. 9. Solver extension: 2-circuited cluster identification

5 Implementation

The proposed geometric constraint solving procedures have been implemented in Visual C++ on Windows NT. Figure 10 shows a snapshot of the developed constraint solver. Figure 11–14 shows some implementation results. Figure 14 shows a mechanical part and its modified one that are modeled by using the feature-based parametric modeling system [20].

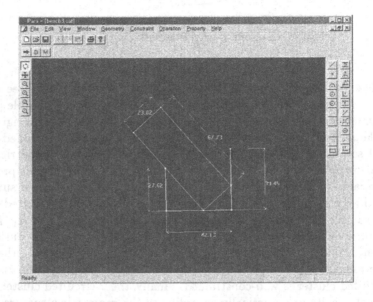

Fig. 10. A snapshot of the developed constraint solver

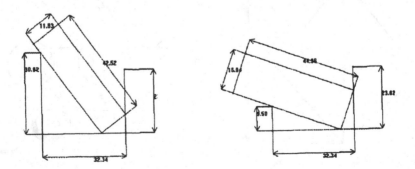

Fig. 11. A model with one-to-one type of a RCNC cluster and its modification

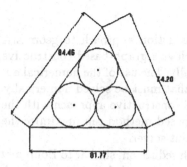
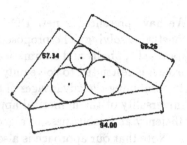

Fig. 12. A model with a multi-circuited cluster and its modification

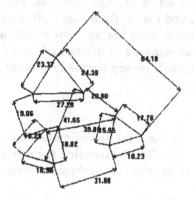
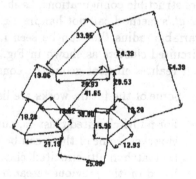

Fig. 13. A model of the type shown in Fig. 5(d) and its modification

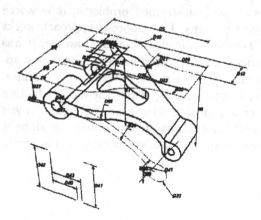
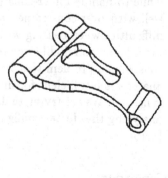

Fig. 14. 3D feature-based parametric modeling: a hinge design and its modification

6 Discussions

We have presented a new DOF-based graph reduction approach to geometric constraint solving. The proposed approach employs a graph-based constructive approach globally and a numerical approach locally. The use of the numerical approach is restricted to solve only those clusters that must be solved numerically. By combining the advantages of the graph-based constructive approach with the universality of the numerical approach, the proposed method can maximize the efficiency and robustness of a geometric constraint solver.

Note that our approach is also based on graph reduction, similar to Bouma *et al.*'s approach [4]. However, our method is different from their method in graph reduction (or cluster formation) process. While their graph reduction method is based on analyzing *node-sharing among clusters*, our reduction method is based on analyzing *edge-sharing among clusters* and *their degrees of freedom*. This allows us to deal consistently with a larger set of ruler-and-compass non-constructible configurations, as shown in Fig. 5 and Fig. 9. For example, Bouma *et al.*'s method can not handle a geometric configuration having a circle with a variable radius that can be seen in Appollonius' drawing problems and multi-circuited clusters as shown in Fig. 12. Our approach is very simple in concept, but realizes efficient geometric constraint solving.

Some of the future works are listed below:

- For multi-circuited clusters, currently, most of them are solved numerically. However, some of them can be algebraically [9]. We are extending the solver to match multi-circuited clusters with rule graphs for algebraic solution based on the previous research work [17].
- Although we have used a modified Newton-Raphson method with global convergence property to improve the reliability of the numerical constraint solving procedure [24], further work is still needed. It may be considered to adopt the homotopy method [18] for numerical solving.
- Although we have only dealt with well-constrained problems, it is worth while to handle under-constrained problems. The proposed approach works well with over-constrained and well-constrained problems. However, it has difficulties when dealing with under-constrained problems because there appears to be no reliable way to locally add constraints to transform an under-constrained problem to a well-constrained one [8]. Thus, the local reduction analysis is insufficient for under-constrained problems, and a global analysis is needed. We are trying to deal with under-constrained problems by globally analyzing the cluster configurations derived by the graph reduction analysis: [1, 19].

References

1. Ait-Aoudia, S., Jegou, R., and Michelucci, D., Reduction of constraint systems, In *Compugraphics*, Alvor Portugal, (1993) 83–92

2. Aldefeld, B., Variation of geometries based on a geometric reasoning method, *Computer Aided Design*, (1988) **20**(3) 117–126

3. Anderl, R. and Mendgen, R., Parametric design and its impack on solid modeling applications, *Proc. 3rd Symp. Solid Modeling Foundations & CAD/CAM Applications*, ACM Press, (1995) 1–12

4. Bouma, W., Fudos, I., Hoffmann, C. M., Cai, J. and Paige, R., A geometric constraint solver, *Computer Aided Design*, (1995) **27**(6) 487–501

5. Bruderlin, B., Constructing three-dimensional geometric objects defined by constraints, *Proc. Workshop on Interactive 3D Graphics*, ACM, (1986) 111–129

6. Buchberger, B., Collins, G., and Kutzler, B., Algebraic methods for geometric reasoning, *Ann. Rev. Comput. Sci.*, (1988) **3** 85–120

7. Fudos, I. and Hoffmann, C. M., Correctness proof of a geometric constraint solver, *International Journal of Computational Geometry & Applications*, (1996) **6**(4) 405–420

8. Fudos, I. and Hoffmann, C. M., A graph-constructive approach to solving systems of geometric constraints, *ACM Transactions on Graphics*, (1997) **16**(2) 179–216

9. Gao, X. S. and Chou, S. C., Solving geometric constraint systems, Part I: A global propagation approach, *Computer Aided Design*, (1998) **30**(1) 47–54

10. Gao, X. S. and Chou, S. C., Solving geometric constraint systems, Part II: A symbolic approach and decision of Rc-constructiblity, *Computer Aided Design*, (1998) **30**(2) 115–122

11. Gossard, D. C., Zuffante, R. P. and Sakurai, H., Representing dimensions, tolerances, and features in MCAE systems, *IEEE Comput. Graph. & Applic.*, (1988) **5**(3) 51–59

12. Hillyard, R. and Braid, I., Analysis of dimensions and tolerances in computer-aided mechanical design, *Computer Aided Design*, (1978) **10**(3) 161–166

13. Hoffmann, C. M. and Joan-Arinyo, R., Symbolic constraints in constructive geometric constraint solving, *J. Symbolic Computation*, (1997) **23** 287–299

14. Hsu, C. and Bruderlin, B., A hybrid constraint solver using exact and iterative geometric constraints, *CAD Systems Development: Tools and Methods*, Roller and Brunet (eds.), Springer, (1997) 265–279

15. Kondo, K., PIGMOD: Parametric and interactive geometric modeller for mechanical design, *Computer Aided Design*, (1990) **22**(10) 633–644

16. Kondo, K., Algebraic method for manipulation of dimensional relationships in geometric models, *Computer Aided Design*, (1992) **24**(3) 141–147

17. Kramer, G. A., *Solving Geometric Constraint Systems: A Case Study in Kinematics*, MIT Press, Cambridge, Massachusetts, (1992)

18. Lamure, H. and Michelucci, D., Solving geometric constraints by homotopy, *Proc. 3rd Symp. Solid Modeling Foundations & CAD/CAM Applications*, ACM Press, (1995) 263–269

19. Latham, R. S. and Middleditch, A. E., Connectivity analysis: A tool for processing geometric constraints, *Computer Aided Design*, (1996) **28**(11) 917–928

20. Lee, J. Y. and Kim, K., Geometric reasoning for knowledge-based parametric design using graph representation, *Computer Aided Design*, (1996) **28**(10) 831–841

21. Lee, J. Y., A knowledge-based approach to parametric feature-based modeling, *Ph.D. Thesis*, POSTECH, South Korea, (1998)

22. Light, R. A. and Gossard, D. C., Modification of geometric models through variational geometry, *Computer Aided Design*, (1982) **14**(4) 209–214

23. Owen, J. C., Algebraic solution for geometry from dimensional constraints, *Proc. 1st Symp. Solid Modeling Foundations & CAD/CAM Applications*, ACM Press, (1991) 379–407

24. Press, W. H., Teukolsky, S. A., Vetterling, W. T., and Flannery, B. P., *Numerical Recipes in C*, Cambridge Univ. Press, (1992)
25. Roller, D., An approach to computer-aided parametric design, *Computer Aided Design*, (1991) **23**(5) 385–391
26. Solano, L. and Brunet, P., Constructive constraint-based model for parametric CAD systems, *Computer Aided Design*, (1994) **26**(8) 614–622
27. Sunde, G., Specification of shape by dimensions and other geometric constraints, *Geometric Modeling for CAD Applications*, North-Holland, (1990) 199–213
28. Suzuki, H., Ando, H., and Kimura, F., Geometric constraints and reasoning for geometric CAD systems, *Computers & Graphics*, (1990) **14**(2) 211–224
29. Todd, P., A k-tree generalization that characterizes consistency of dimensioned engineering drawings, *SIAM J. Discrete Math.*, **2** (1989) 255–261
30. Verroust, A., Schonek, F., and Roller, D., Rule-oriented method for parametrized computer-aided design, *Computer Aided Design*, (1993) **25**(10) 531–540

Variant Geometry Analysis and Synthesis in Mechanical CAD

Zongying Ou and Jun Liu

CAD and CG Lab
Dalian University of Technology
116024 Dalian City, China
ouzyg@gingko.dlut.edu.cn

Abstract. This paper deals with modeling and problem solving of variant geometry in mechanical computer aided design. Enhanced undirected graph is used as a schema for modeling a constraint geometry system. Two important techniques called equivalent line segment method and separable entity group approach are introduced in this paper. These techniques developed by the authors are used to unify and simplify variant geometry problem solving.

1 Variant Geometry Problems in MCAD

Variant geometry deals with flexible geometry systems in which the geometry entities are subjected to variant geometric constraints. A variant geometry problem can be formulated as

[geometry entities, constraints].

Variant geometry has varieties of applications in mechanical computer aided design (MCAD) [4, 1, 11, 2, 3, 9, 10].

A. Parametric and feature CAD

More than seventy percent of practical designs are adaptive or innovative designs including series product design, adapted to new working condition design, extending and adding new function design, etc. All of these designs usually begin with or imitate to a prototype design; they share some common features (topology and/or shape) but formed with some variant differences. To effectively deal with these variant designs, a new generation of design approach called parametric modeling and/or feature-based modeling has been developed. A new design can be generated by changing specific parameters and/or by combining appropriate predefined features in an intelligent CAD system. The typical application cases include:

2D parametric drafting. System entities are adjustable and changeable drawing primitives: line, arc, circle, etc. The subjected constraints are either explicit or implicit. Explicit constraints are annotated in drawing like dimensional annotations. Implicit constraints are defined by default drawing conventions like some lines keeping horizontal orientation, some lines keeping vertical orientation, some related entities tangent to each other.

Parametric (also feature-based) solid modeling. System entities are adjustable and changeable geometric feature objects such as faces, edges, vertices, etc. The constraints are distance dimension constraints, relative position constraints and other relations.

B. Evaluation and Synthesis of Kinematics of Mechanisms

A mechanism is also a system of geometry objects subjected to some constraints related to each other. Geometry objects are components and component-like subassemblies. The constraints are rotary connection, sliding connection, specified kinematics feature limitation including fixed to frame or specified kinematics control, etc.

Keeping a mechanism conforming to its given kinematics constraints and sequentially changing the position of its driving component, the variant geometry processing system will generate history positions of the whole mechanism; this is the simulation procedure for the mechanism. Designers can easily try different variances of a mechanism and evaluate the different kinematical behaviors instantly. This is very useful for optimization of mechanism design.

All the geometry applications are intelligent processes based on automated deduction in geometry. The processes include the following steps: automatically establishing the variant model; transforming the constraint conditions into processable (calculating and reasoning) relations; generating and evaluating the variant system by solving the relations. The main issues of a variant geometry processing are the modeling method and solving approach. Techniques presented below are developed by the authors for variant processing of parametric drafting and variant processing of planar linkage mechanism simulation.

2 Model Graph

A variant geometry system is a system composed of a number of geometry entities related to each other with variant constraint relations. The system structure can be modeled with an undirected graph [9, 10, 8]. Nodes of a model graph are geometric entities of a variant geometry system, and edges connecting the nodes in the graph represent the constraint relations between the corresponding entities. A model graph relates the data within a system and is the structure skeleton in this system database.

For a planar linkage mechanism, the model graph can be express as

$$mechanism = Graph(B, L, C).$$

In the above expression, B is the collection of links, $B = B_1, B_2, ..., B_m$, where m is the number of links; L is the collection of constraints, $L = L_1, L_2, ..., L_n$, where n is the number of constraints; C is the constraint relations,

$$C = ..., L_i(R, B_j(P_j), B_k(P_k)), ..., L_t(S, B_u(P_u), B_v(P_v, P_w)), ...,$$

where R is a rotary connection, P_j, P_k are pivot points, S is a sliding connection, and P_u, P_v and P_w are points on a sliding surface. Figure 1 shows a planar linkage mechanism and its model graph, the model graph can also be expressed as Graph

$\{B_1, B_2, B_3, B_4, B_5, B_6\}$, $\{L_1, L_2, L_3, L_4, L_5, L_6, L_7\}$, $L_1(R, B_1(P_1), B_2(P_1))$,
$L_2(R, B_2(P_2), B_3(P_2))$, $L_3(R, B_3(P_3), B_4(P_3))$, $L_4(R, B_4(P_4), B_5(P_4))$,
$L_5(R, B_5(P_5), B_1(p_5))$, $L_6(R, B_4(P_6), B_6(P_6))$, $L_7(S, B_6(P_7), B_1(P_8, P_9))$.

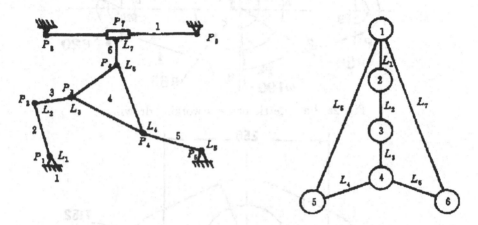

Fig. 1. Linkage mechanism and its model graph

The number of geometry entities in an engineering drawing is much more than in a mechanism. The constraint relations in a parametric drawing are also more complicated than in a mechanism. We design an enhanced graph structure for modeling the system of a variant drawing. Figure 2 shows an example of parametric drawing, which is composed of 12 geometry entities subjected to dimensional and other constraints. Fig. 3 shows the graph data structure for the drawing. The relation and constraint conditions (tangency, symmetry, co-cyclic, concentric) are organized as special link list structures for convenient search and extract in processing, the data of key points for featuring entities are also included in the data structure.

Users can define and modify a constraint system interactively; the system will update the model graph automatically.

3 Equivalent Line Segment Entity Method

The basic types of geometry entities in engineering drawing are line, arc and circle. A line segment usually is defined with two end points. Every point has two coordinate parameters. So, the number of degrees of freedom for a line segment is 4. A circle can be defined with one center point and one radius parameter, so the number of degrees of freedom for a circle is 3. An arc can be defined with one center point, one radius parameter, and two independent parameters

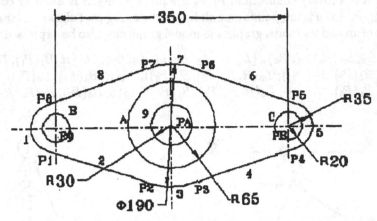

Fig. 2a. Parametric process working drawing

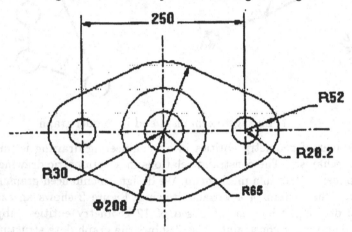

Fig. 2b. Drawing after parametric process

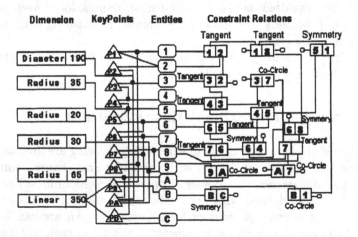

Fig. 3. Graph data structure for the drawing in Fig. 2

to locate the start and end points of this arc, so the number of degrees of freedom for an arc is 5. To effectively deal with different types of entities, a new method called equivalent line segment entity method has been developed [9, 5, 6, 8]. With this method, all line, arc and circle entities can be transformed into equivalent entities that are line segments or combination of line segments (Fig. 4).

The equivalent entity of a line is the line segment itself.

The equivalent entity of a circle is a line segment with length equal to the radius of the circle and one end point coinciding with the center point of the circle; the other end point is movable and the orientation of this line is not constrained.

The equivalent entity of an arc is a combination of two equivalent line segments. These two equivalent line segments are drawn from the center of the arc to two end points of the arc respectively.

All the above equivalent entities have the same number of degrees of freedom and the same feature points as their original entities. During the variant processing, an entity is treated as an object in object-oriented programming. Every entity has two forms: original form and equivalent form. These two forms can be inter-transformed from one form into another form based on the common feature points. The feature point data and the inter-transformation operation procedures are encapsulated within entity objects.

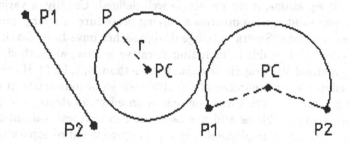

Fig. 4. Line, circle, arc and their equivalent entities

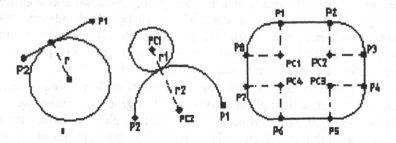

Fig. 5. Drawing examples of equivalent entity transformation

Figure 5 shows some drawing examples of equivalent entity transformation. A circle and a tangent line were transformed into two equivalent lines contacted at tangent point. Two tangential circles were transformed into two equivalent lines meeting at tangent point. A contour composed of lines and arcs was transformed into a contour composed of lines only. The standard form equation for a circle or an arc is a quadratic polynomial equation; and the equation for a line is a linear equation. When a circle or an arc maintains tangent or intersection relations with other entities, it is quite cumbersome to process these relations based on quadratic equations. In the above cases, replacing connected circles or arcs with equivalent lines drawn from the center point to related feature points (tangent points, intersection points, etc.) will lead to simplification of establishing relation equation procedures. Furthermore, the equations are all linear equations. Transforming arbitrary drawings composed of different types of geometry entities into drawings composed of unified equivalent line segments will unify the data structure, regulate the relation equations and significantly simplify the variant processing.

4 Separable Entity Group Approach

All the geometric parameters of a geometric constraint system should be conformed to the constraint relations. The whole number of unknown geometric parameters for a solvable constraint system should be equal to the number of total constraint equations, if the system is well defined. Usually, a variant processing for a constraint system involves a solving procedure of a large number of simultaneous conditions. Several intelligent approaches have been developed for solving this constraint problem, including iterative numerical method, geometry reasoning method and graphic constructive method [4, 1, 11, 2]. However, the difficulty of solving a large number of simultaneous equations exists in all variant processing systems. Divide and conquer is an effective strategy in practice for solving large size problems and has been used in different variant processing systems with different implementations. An approach called separable entity group approach is described in the following, which is also developed based on the divide and conquer strategy. A large variant system will be decomposed to several separable entity groups step by step [9, 10, 7, 8]. A separable entity group is a subsystem of a constraint system, which can be solved independently. The separability criterion of an entity group is that the total number of unknown geometric parameters of this group is equal to the number of equations defined by the related constraint conditions inside this group.

All the geometry entities and constraint relations have been schematically represented in a model graph. A separable entity group is a linked node cluster satisfying the separability criterion in a model graph. The basic principle for extracting separable entity group is searching the possible linked node cluster in the model graph and checking the cluster with separability criterion. The searching procedure is iterative during the variant processing. There might be multiple-choice of separable entity groups in processing, we prefer that the separable entity groups with least number of entities (least number of nodes) should be extracted and processed first.

The procedure of extracting separable entity groups is incorporated in the variant processing procedure; the main points of the procedure are as follows

1. Establish two entity lists: the defined entity list and undefined entity list. An entity is defined if all the necessary parameters for defining this entity are known.
2. Interpret and transform system constraint conditions into equations. Establish system constraint equation list.
3. Sequentially check each equation in system constraint equation list. If there is an equation including only one unknown variable; then solve it, update entity lists and eliminate this equation from system equation list.
4. Set the limited number of entities in separable entity group $NE = 1$.
5. Establish new working (candidate) entity group. Select an entity from undefined entity list which is adjacent (related) to a defined entity. Set this entity as first entity in working entity group.
6. Check the working entity group with separability criterion. If the criterion is satisfied, go to step 15.
7. Add a new undefined entity to the working entity group, the new undefined entity should be adjacent (related) to the entities of working entity group. It is better if the new entity is also adjacent to a defined entity.
8. Set NW equal to the number of entities in working entity group.
9. If NW is not greater than NE, then go to step 6.
10. If NW is greater than NE and the first entity in working entity group is not the last entity in undefined entity list, then go to step 5 but try to set other undefined entity as first entity in working entity group.
11. If NW is greater than NE and all the entities of undefined entity list had been tried as first entity in working entity group, then set $NE = NE + 1$.
12. If NE is not greater than the number of undefined entities, then go to step 5.
13. If NE is greater than the number of undefined entities, then stop and output message: "the variant system is under-constrained."
14. If there is no new entity added to defined entity list after a whole iterative cycle of variant processing procedure, then stop.
15. Extract the working entity group as a separable entity group; solve the constraint equations; and update defined entity list and undefined entity list.
16. If undefined entity list is empty, then stop.
17. Go to step 3.

The running result of linkage mechanism showed in Fig. 1 will be as follows

1. Assume specifying link 1 as the frame (fixed) link and link 5 as the driving link. The system will update system database and add link 1 and link 5 into defined entity list.
2. Extract link cluster 46 in the model graph as the first separable entity group.
3. Extract link cluster 32 in the model graph as another separable entity group.

The running result of drawing showed in Fig. 2 might be as follows

1. Assume specifying point PA as fixed point and specifying the coordinates of point PA.
2. Solve linear dimension equations and get the coordinates of point PB and point $P9$.
3. Extract entity circle No. 9 as separable entity group.
4. Extract entity circle No. A as separable entity group.
5. Extract entity circle No. B as separable entity group.
6. Extract entity circle No. C as separable entity group.
7. Extract entity cluster (arc No.1-line No.2-arc No.3-line No.4-arc No.5-line No.6-arc No.7-line No.8) as separable entity group.

Usually, a candidate for separable entity group is an entity cluster linked between defined entity nodes in a model graph. The proposed approach is efficient since the extracting processing is performed by directly searching and checking the separability criterion along the relation path in a system model graph. For further increasing the efficiency, the following content-oriented features can be used as rules of thumb for guiding the extracting processing.

- If entities in a node chain form a contour in a drawing, this node chain is probably a separable entity group.
- If entities in a node cluster form an independent geometric part in a geometry object, this node cluster is also probably a separable entity group.
- The entities linked between driving components (there might be multiple driving sources) and fixed component is a good beginning choice as a basic cluster for further searching in kinematics analysis.

5   Implementation

The CAD and CG Research Institute of Dalian University of Technology already developed a prototype expert system running on MS-Windows environment

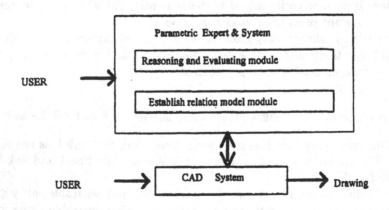

Fig. 6. System architecture

for parametric drafting and simulation of linkage mechanism based on the above approach. The module for simulation of linkage mechanism was implemented by using PROLOG language, and the parametric drafting module was implemented by using C++ language. The prototype system co-works with a conventional CAD system. The system architecture is showed in Fig. 6. Figure 7 shows a working display of linkage mechanism simulation and Fig. 8 shows a working display of parametric drafting. The implementation verified the correctness and effectiveness of the above method and approach.

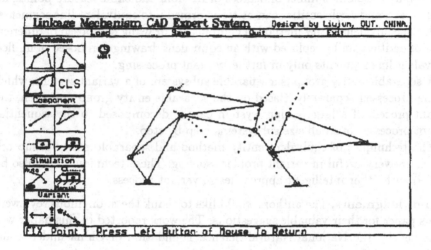

Fig. 7. Working display of linkage mechanism simulation

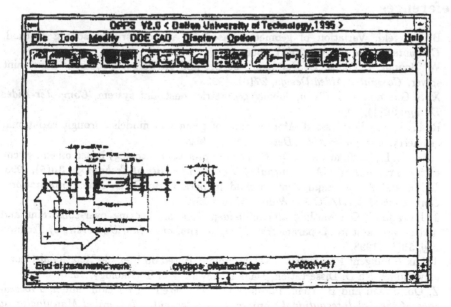

Fig. 8. Working display of parametric drafting

6 Conclusion

Variant geometry has a very important variety of applications in MCAD, ranging from geometry design to kinematics in 2D and 3D. Parametric, feature-based modeling design and variant kinematic analysis are applications of automated deduction in geometry.

An enhanced graph structure has been designed as structure skeleton for modeling variant geometry system.

Keeping the same number of degrees of freedom and same feature points, all the line, arc, and circle entities can be transformed into equivalent line segments. With the equivalent entity method, a complicated drawing composed of different types of entities can be replaced with an equivalent drawing composed of unified equivalent line segments only in further variant processing.

A separable entity group is a separable subsystem of a variant system, which can be processed separately. Based on the separable entity group approach, the variant process of a large variant system can be decomposed to the sequential variant processes of small size subsystems step by step.

The techniques of equivalent entity method and separable entity group approach are very useful in variant problem solving. These techniques can also be applied with other intelligent approaches in variant process.

Acknowledgments. The authors would like to thank the anonymous reviewers of this paper for their valuable suggestions. The work reported in this paper was supported by the National Natural Science Foundation of China under grant number 69174034.

References

1. B. Aldefeld, Variation of geometry based on a geometric reasoning method, *Computer-Aided Design*, **20**(3), 1988.
2. W. Bouma, I. Fudos, C. Hoffmann, J. Cai and R. Paige, Geometric constraint solver, *Computer-Aided Design*, **27**(6), 1995.
3. X. S. Gao and S. C. Chou, Solving geometric constraint system, *Computer-Aided Design*, **30**(1), 1998.
4. R. Light and D. Gossard, Modification of geometric models through variational geometry, *Computer-Aided Design*, **14**(4), 1982.
5. J. Liu, L. Li, B. Yuan and Z. Ou, Constraint processing strategy based on equivalent entity in parametric CAD, *Journal of Dalian University of Technology*, **35**(5), 1995.
6. J. Liu and Z. Ou, Equivalent method in parametric CAD, *Proc. International conference of CAD/CG 95*, Wuhan, China 1995.
7. J. Liu and Z. Ou, Method and implementation of checking over-constraint and under-constraint in 2D parametric CAD, *Journal of Dalian University of Technology*, **38**(5), 1998.
8. J. Liu, Intelligent Parametric CAD Technique Based on Flexible Relational Geometry Model, Ph.D. Dissertation, Dalian University of Technology, 1996.
9. Z. Ou, J. Liu and B. Yuan, Parametric CAD based on relation model, *Proceedings of the 3rd International Conference on Computer Integrated Manufacturing (ICCIM'95)*, World Scientific Publishing Co., Singapore, 1995.

10. Z. Ou, J. Liu and J. Liu, Analysis and simulation expert system of planar linkage mechanism, *China Journal of Computer-Aided Design and Computer Graphics*, **7**(1), 1995.
11. A.Verroust, F. Schonek and D. Roller, Rule-oriented method for parameterized computer aided design, *Computer-Aided Design*, **24**(10), 1992.

Author Index

Lecture Notes in Artificial Intelligence (LNAI)

Lecture Notes in Computer Science